THE SEARCH FOR LIFE ON MARS

THE SEARCH
FOR LIFE ON
MARS

THE GREATEST SCIENTIFIC
DETECTIVE STORY OF ALL TIME

ELIZABETH HOWELL
AND
NICHOLAS BOOTH

ARCADE PUBLISHING • NEW YORK

First Edition

Arcade Publishing books may be purchased in bulk at special discounts for sales promotion, corporate gifts, fund-raising, or educational purposes. Special editions can also be created to specifications. For details, contact the Special Sales Department, Arcade Publishing, 307 West 36th Street, 11th Floor, New York, NY 10018 or arcade@skyhorsepublishing.com.

Arcade Publishing® is a registered trademark of Skyhorse Publishing, Inc.®, a Delaware corporation.

Visit our website at www.arcadepub.com.
Visit Elizabeth Howell's site at elizabethhowell.ca.

10 9 8 7 6 5 4 3 2 1

Library of Congress Cataloging-in-Publication Data is available on file.
Library of Congress Control Number: 2020934878

Cover design by Erin Seaward-Hiatt
Cover photograph courtesy NASA/JPL/MSSS

ISBN: 978-1-950691-39-5
Ebook ISBN: 978-1-950691-66-1

Printed in the United States of America

"If we are interested in Mars at all, it is only because we wonder over our past and worry about our possible future."

—Ray Bradbury

CONTENTS

AUTHORS' NOTE

This book has its origins in the 1990s, when Nicholas Booth was working in British newspapers writing stories on science and technology. The most exciting evening in his decade as a reporter occurred when the discovery of suspected fossilized life in the Allan Hills meteorite was announced in August 1996. It was every newspaper writer's dream story. Originally intending to do a quick tie-in book, Nick began squirreling away notes. Prior to this, he had spoken with many of the first generation of researchers who were able to send instruments to Mars. Alas, a number of interviewees have since passed away. We hope that their words here reflect their significant contribution to the story. From 1996 onward, Nick continued more formally. By the time he left newspapers, he had started to talk to most of the leading researchers and many who have since become foremost experts in their field.

Elizabeth Howell has been reporting on space since 2004, freelancing for various publications since 2006. Since 2012, she has freelanced full-time for clients in Canada, the United States, and Europe. Her reference clients include Space.com, *Forbes*, and the Canadian Broadcasting Corp. Awarded a PhD in the summer of 2019, she also teaches communications at Algonquin College in Ottawa, Canada. While she hasn't been to space (yet), she has been to the Red Planet—or, at least, the Mars Desert Research Station in Utah, where she pretended to be an astronaut for two weeks in 2014.

This book, then, is a unique collaboration. From the outset, we decided not to write a dry, dull chronology, but rather to attempt to tell the story in a different way. Our overarching aim has been readability. It has been a great privilege to talk to many people who are involved in cutting-edge research about the Red Planet. That has meant, for us at least, a greater agony familiar to all authors in having to discard whole sections to make the length of the book manageable. As a result, the Martian moons, Phobos and Deimos, get scant mention, as does the astonishing story of the first Soviet Mars missions; nor could we cover the individual triumphs and failures of the (roughly) two hundred other instruments that have been sent to Mars and the people who have built them over the last six decades. We apologize profusely for not being able to cover everything.

It goes without saying that at times we have made glib and often oversimplified statements. No doubt others would write a different type of book, but here, we have tried to give a flavor of research for people who have no technical background. Any errors of interpretation and fact are ours, and we bear responsibility for them. We will endeavor to correct them in subsequent editions of this book.

PREFACE

Mars has always been full of surprises.

Over the last fifty or more years of direct exploration by spacecraft, the Red Planet has had a remarkable capacity to spring surprises. When the first successful missions shot past Mars in the 1960s, it provided only fleeting glimpses of a battered, cratered terrain. Mars appeared much like the Moon, a surprise at the time. For most of 1971, the whole of the Martian surface was blanketed in a thick layer of dust. It took weeks for the planet below to gradually reveal itself. When it did, at the start of 1972, the first successful orbiter, *Mariner 9*, was ready and waiting with its television cameras.

What it saw was another complete surprise. Gigantic volcanoes rose from the equator, revealing that the Red Planet was far from dead. Dried-up outflows of water scarred the surface, giving hope that life might have evolved in formerly aqueous environments. A massive canyon, later named Valles Marineris after its discoverer, stretches across a fifth of the planet's circumference. This "Grand Canyon" on Mars dwarfs Earth's equivalent, at roughly five times the size of its namesake in Arizona.

Mars had clearly been geologically active. Something had driven extensive volcanism, which could also have been a source of energy for life, perhaps surviving or hibernating as microbes close to the surface. A pair of Viking spacecraft landed on the surface in the summer of 1976 to find out. Their instruments, which were revolutionary for the time, sifted the rusty soils, only to discover a peculiar, completely alien chemistry that baffled its investigators, as it still does today.

More importantly, there were no signs of life.

The results of the Viking biology experiments remain controversial, though. One scientist involved in the mission has claimed to have found evidence for life in samples that were deemed sterile. Belatedly, there has been another great surprise. There is the very surreal possibility that some of the material examined by heating it within a sensitive chemical oven on the Viking landers actually *caught fire.* The answer might have turned to ashes before the experimenters' eyes.

The next steps in exploration are about to be taken by a new generation of rovers that are being sent to the surface of the Red Planet in 2020. Appropriately, NASA has named the largest and most sophisticated Perseverance, which describes what many scientists have had to exhibit over the years. Their experiences, their triumphs and failures, are told in their own words in the following pages.

* * *

At various points in history, a number of people have reported that they have discovered life on Mars, only to have their findings shot down publicly. This, too, is part of the greater narrative covered in this book. The likelihood of there being life on Mars has been akin to a swinging pendulum of popular perception.

Today, that pendulum is pointing unequivocally toward life.

For the first time since the Viking missions in the seventies, scientists are seriously discussing the possibility that microbial life may be found on Mars. At its heart, then, this book is a detective story. It has needed investigators' careful gathering of evidence to produce definitive answers. At times, the narrative is as convoluted as any other police procedural, with false starts, unreliable witnesses, red herrings, forlorn hopes, bizarre occurrences, and improbable happenings. And many more will assuredly follow.

If there is life on Mars, it might be buried far below the surface, in a polar ice cap, or even in a pool of briny water that can survive the subzero temperatures of the Martian nights and winters. Nobody can say where exactly it is, or what form it will take, but the hunt is on.

This is the story of human ingenuity and exploration at its greatest, one that promises to spring ever more amazing surprises in the years to come. If there is life on Mars, it will be found quickly. That final surprise—the greatest scientific discovery of our age—is within our grasp.

THE SEARCH FOR
LIFE ON MARS

1

FROZEN IN TIME

"To understand how life would evolve on Mars, you have to go to Antarctica," says Dr. Christopher McKay. "There is no other place like it."

Widely seen as the most eloquent spokesman for the tantalizing possibilities of life on Mars, Chris McKay has spent decades studying life in Antarctica to gain a greater understanding of how microbes could exist on the Red Planet. Working out of NASA's Ames Research Center, south of San Francisco in the suburb of Mountain View, McKay is a leading astrobiologist. Tall, genial, and with a voice so *basso profundo* it seems to emanate from somewhere below the floor and boom in empty spaces, McKay made his first journey down to "the ice" in 1980. "Antarctica is like a second home to me," he says today.

McKay is acknowledged as a voice of reason in a field of research that has sometimes split into contentious factions, especially where life on Mars is concerned. "That question would be easier to answer if we could understand the evolution of life on Earth," he says, "or even if there was a consensus on the origins of life. Mars is going to help us with this riddle."

Though he has also investigated life in Siberia and in the Atacama Desert in Chile, McKay believes that the high, dry valleys of Antarctica are the only place on Earth where conditions are sufficiently extreme to mimic the Red Planet. The key ingredient is water. "People often say how amazingly robust life is," McKay says. "My reaction is the opposite. It always needs water. If we had the trick of learning to live without water, life would be hardier."

If we can't look for life directly, then searching out water is the next best thing. That has informed the scientific rationale behind the most recent and the next missions to the Red Planet. Without liquid water, life would have been unthinkable on Earth as it would have been on Mars. Given the fact that most water on Mars is concentrated in the form of ice at its poles, McKay believes that is where life is most likely to be found.

The polar ice caps of Mars have beguiled and enticed astronomers since their discovery in the eighteenth century. Their waxing and waning showed that, like Earth, the Red Planet undergoes seasons as it alternately tilts away from and toward the Sun. The seasonal ice caps grow and retract with the passage of the seasons. It was later found that the Martian tilt is 25°, similar to the 23.5° value for Earth.

However, it is very difficult to reach the Martian poles. Tricky maneuvers would be required to touch down far from the easier-to-reach equator of the planet, requiring greater amounts of fuel at the expense of scientific instruments. Any attempt would be severely constrained by weight limitations and the extreme temperatures. Actually landing a probe amid the ice there is even more hazardous than exploring the poles of our own world.

NASA's first attempt to do so, in 1999, failed. The Mars Polar Lander crashed somewhere in the southern polar regions, likely as a consequence of a software error that affected its landing system. Nine years later, the *Phoenix* lander, named for the ancient bird that rises from the ashes, successfully made it all the way down at 67°N ("which is like Iceland on Earth," says one observer).

Over the northern winter of 2008–2009, *Phoenix* found evidence that snow accumulates on the surface and detected what are known as perchlorates (a possible "food" for some microbes) in the soils. It also showed beyond doubt that there is a solid ice layer immediately underneath where *Phoenix* landed. "Nobody really knew how much ice was lurking just below the surface," says Professor Jack Holt, a glaciologist at the University of Arizona. "So that was kind of a surprise to a number of people."

That discovery has, he says, opened the door to discovering that there are much more extensive icy deposits below the surface, which have been remotely detected across whole swaths of the Red Planet. Ice on Mars, the result of changes to the planet's climate over geological time, is no longer confined to the poles. Have these icy deposits always been so prevalent, or are they, as some believe, more of a recent phenomenon?

* * *

Antarctica is as alien as it gets on Earth.

Conditions are scored for extremes. During the summer, the average temperature in coastal regions hovers around freezing point, while it varies between −15°C to −30°C (−5°F to −22°F) inland. In the central plateau, temperatures range from −40°C to −70°C (−40°F to −95°F). The lowest temperature ever recorded on Earth was −89.6°C (−130°F) in the winter of 1983 at the Soviet Vostok research center there. Small wonder it has been referred to as "the Gulag of the South" by those who have willingly stayed at the center in the name of scientific duty.

Though it only covers about 10 percent of the total landmass on Earth, the South Pole contains about 90 percent of the world's supply of ice. The Antarctic continent is shaped like a squat, lopsided letter *Q*, with the lower squiggle forming the Antarctic Peninsula. It points like a crooked finger toward South America, five hundred miles (some thousand kilometers) distant. Antarctica's ice lies on a foundation of rock, most of which is hidden from view.

Ice flows in strange ways on this southern continent. Around the edge of Antarctica is a ring of mountains through which continental ice is forced to pass. Eventually, it falls into the sea, but first the frozen hulk tends to form ice shelves that are glued to the landmass by the freezing cold. Some ice chunks are as large as small countries. At times these massive sheets break off to form large icebergs; this process has accelerated with recent climate change, which is warming Earth's poles rapidly.

The climate of Antarctica is unique: its air is trapped for most of the year under a giant anticyclone. As a result, winds descend around its outer extremities and flatten as the air flows outward. The winds, immortalized by mariners as the Screaming Sixties, whip up sudden storms and squalls. Thankfully, on the shelf-like coast of the continent around the main ice sheet, a thousand miles from New Zealand, the weather is distinctly better.

It was both the clement conditions and the sheltered inlet of this area that commended it as a stopping-off point for one of the most important voyages of discovery of the nineteenth century. James Clark Ross, a dashing officer in the British Royal Navy, had already discovered the magnetic north pole when he set off to find its southern equivalent in the late 1830s. In the large sailing ships *Erebus* and *Terror*, his expedition ventured farther south than anyone had ever done before.

They happened upon the Antarctic coastline after "a magical journey of towering mountains and shining glaciers," in the memorable phrase of one chronicler of their travels. Ross's own diary records that on January 28, 1841, the sea that now bears his name appeared like a sheet of frozen silver in the uncharacteristically good weather and blue skies. His crew were openmouthed upon finding "a perpendicular cliff of ice between 150 and 200 feet above the level of the sea, perfectly flat and level at the top and without any fissures or promontories on its seaward face."

Because the ice cleared faster here than anywhere else they had happened upon in Antarctica, it became an obvious point of contact for future explorers. Scientists today head for this sheltered sound, which was named after the senior lieutenant of the *Terror*, Archibald McMurdo. Now visitors have the luxury of flying in on modified Hercules transport aircraft on flights from New Zealand. Once they're deposited, the plane often doesn't stick around in case its delicate engine parts freeze over.

Only in the direst of circumstances will the authorities ever attempt a Win-Fly, as the flights in during the dead cold of winter are known. One such situation took place in the austral winter of 2017 to rescue an eighty-seven-year-old man who was experiencing breathing difficulties at the geographical South Pole. Later shown recuperating in a hospital in Christchurch, he smiled for the cameras after his ordeal. That he wore a T-shirt with the phrase GET YOUR ASS TO MARS, made famous in the original *Total Recall* movie, and that he was the second human being to walk on another world had a lot to do with the impression of sangfroid he gave.

But then, Buzz Aldrin has always dreamed of an encore: walking on Mars. "I think we can all say with confidence that we are closer to Mars today than we have ever been," Aldrin had said earlier that same year.

* * *

Antarctica is tough to explore, but it has nothing on Mars.

The Red Planet is roughly 1.5 times farther from the Sun than we are. Mars orbits the Sun at an average distance of 142 million miles (228 million kilometers), compared to 93 million miles (150 million kilometers) for Earth. Its orbit is much more elliptical than Earth's. The Red Planet takes twenty-three of our months—nearly two years—to complete one orbit. It also receives much less heat than Earth. Conditions on Mars would make Vostok station look positively balmy. The average temperature on Mars is about –60°C (–76°F), and though there are places where it can fleetingly hover around the freezing point of water, temperatures can plunge down to –150°C (–240°F) during the polar night.

Atmospheric pressure distinguishes Mars from Antarctica, even though our own southern continent is one of the highest regions on Earth. (The thinner atmosphere there caused the problems with Buzz Aldrin's breathing.) Because of Antarctica's altitude, one Soviet researcher at the same Vostok station where the coldest temperature was measured was astounded to find that potatoes took three hours to cook through. They boiled at 88°C (190°F). On the Red Planet, the average atmospheric pressure is less than a hundredth of that on Earth. There is so little atmosphere on Mars that water molecules would rush out in a mass exodus. If you took a pan of water outside, it would burst outward in a freezing explosion.

Mars has a lower atmospheric pressure because it is roughly half the size of Earth and a tenth of its mass, so its gravitational influence is smaller, roughly 40 percent of ours. Throughout its history, Mars was not able to hold on to its primordial reserves of water, which evaporated or were lost to space. Today, this also means that the Red Planet cannot hold on to as thick an atmosphere as Earth. The average atmospheric pressure on Mars is 6.1 millibars, compared to 1013 millibars at sea level on Earth. The range of pressures on the Red Planet varies, running from nearly 9 millibars at the bottom of the largest basin to 2 millibars at the top of the highest volcanoes.

The atmosphere of Mars is composed almost entirely—95 percent—of carbon dioxide. The gas traps sunlight on the planet's surface, lifting the average temperature there some 5°C (41°F), compared to 35°C

(95°F) on Earth. Mars is also almost completely dry. Even more so than Antarctica, it lives up to the nickname of "freezing desert."

* * *

The most Mars-like places in Antarctica are the remarkable Dry Valleys, close by McMurdo Sound. Here the temperature averages –20°C (–4°F) and rarely rises above the freezing point of water. The Dry Valleys receive less annual precipitation than the Gobi Desert. They were discovered by Captain Robert Falcon Scott on one of his first journeys to the South Pole, in a region known as Victoria Land in honor of the monarch of the time. The Dry Valleys are separated from the remorseless encroachment of Antarctic glaciers by the Transantarctic Mountains.

When Scott and his team happened upon them, they were astounded. "The hillsides were covered with a coarse granitic sand strewn with numerous boulders," he recorded in his diaries. "It was curious to observe that these boulders, from being rounded and sub-angular below, gradually grew to be sharper in outline as they rose in level."

Scott later investigated this area during his more famous, ultimately tragic expedition in 1912. Two of the valleys are named after scientists attached to his expedition, Thomas Griffith Taylor and Charles Wright. The valleys receive at most four inches (ten centimeters) of snow per year, precipitation that is blown away by the harsh winds whistling through the region. They are the coldest and driest places on Earth.

Until the 2000s, scientists had found no trace of life in these harsh valleys. In the early 1970s, when NASA was preparing its first missions to land on Mars, the *Viking* spacecraft, the Dry Valleys were chosen as a test site for some of the life-detecting instruments. If they could find microbes in the Dry Valleys, they would be able to find them on Mars. However, their findings in Antarctica were ambiguous. What resulted was an almighty row between factions within the Viking biology teams, with some claiming that the valleys were entirely sterile and others that they weren't.

Today, cooler heads have prevailed. The original argument was based on biologists' ability to culture any living material from samples of the soil. The greater truth is that nothing could be cultured *from the soils* in the Dry Valleys, hardly surprising given that 90 percent of organisms in any soil cannot be grown in this way. With a sensitive enough probe, though, biologists have subsequently found plenty of evidence for microorganisms throughout the Dry Valleys. Whether that life resulted from material blown in from elsewhere or was indigenous and actually growing there remained a matter of debate until more recent times.

Closer examination reveals thriving microbial ecosystems in the Dry Valleys. Rocks act like little greenhouses and often trap water. Just below the surface of Sun-facing sandstone rocks are layers of lichen and algae that can survive because the dark surface of the rocks is warmed above air temperature. Pores within the rock trap whatever liquid water is available from the occasional snow flurries. The organisms are cocooned from the cold and receive enough sunlight to allow photosynthesis to take place.

At the bottom of the valleys are lakes and ponds, which were also discovered in Scott's time. Some are replete with thick, salty waters that are fed by the annual buildup of snow. Uniquely, they do not drain away. Rather, their liquid content evaporates due to the fearsome winds that constantly blow through the valleys. Around the shorelines of the lakes may be found microbial life in the form of algae, upon which populations of yeast and molds may feed. These microbes support microscopic protozoa, rotifers, and tardigrades, all tiny organisms that congregate at the very base of the food chain.

Whatever their origin, the microorganisms in the Dry Valleys provide clues to ones we may ultimately find on Mars. If we can better understand how such life originally formed here, biologists will be able to get a much better handle on what may have happened on the planet next door. In 2011, when the most powerful camera ever sent to the Red Planet detected what looked like fresh flows of water from orbit, one scientist's comment to the press was especially pertinent: "Mars looks more like the Dry Valleys of Antarctica every day."

* * *

Chris McKay was in graduate school when the Viking missions landed on Mars in 1976. Though they found no evidence for microbial life, he was more intrigued about the absence of organic molecules in the

Martian soil. These complex chains of carbon are crucial in the evolution of life as we know it. The singular fact of their complete absence led to an absolute change in scientific opinion about the possibilities for life on Mars. Taken at face value, the lack of organics implied it would be pointless looking for life there. In very simple terms, there was no biochemical backbone on which life could have formed.

"After Viking, there was a general lack of interest in the scientific community," he says. "Viking immediately suggested to many people that there isn't life on Mars, nor could there have ever been. I don't think there was a really objective scientific assessment of the results. The initial disappointment was too much."

Now, in the twenty-first century, the pendulum is swinging back the other way. Over the last eight years, NASA's Curiosity rover has uncovered organics on several occasions on the Martian surface; the latest, in the summer of 2018, were tough organic molecules—"tough" in the sense they had survived for so long—buried in three-billion-year-old rocks that likely had originally accumulated from sediments in a lake.

More recent missions have raised the stakes further, revealing that water once flowed and conditions were probably right for life to have formed in the ancient past. Hematite, a form of iron that is oxidized in the presence of water, has been discovered in various places across the Martian surface. Many rovers, most recently Curiosity, have discerned telltale signs of flooding by dramatic flows of water. From orbit, other spacecraft have observed deeply cut canyons that look like ravines and outflows carved by water. There is also some evidence for an extensive shallow sea in the ancient past that may have covered sizable swaths of the planet's northern hemisphere.

There may even be fresh flows of water on Mars today. Seasonal changes have been observed on the slopes of some Martian craters, although what is causing these dark streaks to appear and disappear is a matter of much debate. Hydrated minerals have been observed in these streaks. Some scientists argue it is because of water from the atmosphere, while others say this might be the result of flowing, briny water. We won't know for sure until some future mission ventures up close to these features, which are called recurring slope lineae.

Scientists are by their very nature conservative, cautious, and slow to adapt. There are times when the shock of the new can cause a radical change in opinion, known as paradigm shifts, but change is rarely sudden. When Alfred Wegener first proposed the notion of plate tectonics on Earth in the 1920s, he was largely ignored. It took nearly four decades for his work to be unequivocally accepted.

Those who study life on Mars cite a similar paradigm shift after the general gloom of Viking's apparent inability to detect life. The shift resulted from a paper that Chris McKay presented with colleagues in 1984 when he was a postgraduate student at the University of Colorado, Boulder. It catalogued the competing theories for the origins of life on Earth and explored whether they would also work on Mars, where many of the same early conditions may have existed. The answer was a resounding yes.

"That's a bit of a weaselly argument," McKay says with a knowing smile today, "because we can't really say what happened on Mars without knowing how life evolved on Earth. But all the theories didn't contradict anything we already know about Mars."

Ever the optimist, McKay says this implicit confirmation increases the likelihood that life could have existed on the Red Planet. Not only do all the theories about the origin of life on Earth apply to Mars, but McKay also points out that life must have evolved pretty quickly. Once life has come about, it is hard to get rid of—no matter what the original conditions were like or how hostile the environment then becomes.

McKay believes that in the ancient past, survival of the fittest would dictate that Martian microbes would have migrated underground. Even if life died pretty quickly after it formed, say within the first billion years of Martian geological history, then its biochemical signature might still be around in the rocks or close below the surface. Certainly, the missions to Mars being launched in 2020 will seek out these markers or "biosignatures" that may still be present. "I think we'll only find organic preserved remains," McKay cautions. "There will be morphological structures, but the organisms themselves will long since have been and gone."

His reasoning may seem obscure, but it comes from the benefit of his own experiences. Chris McKay was involved in an intriguing series of experiments in the permafrost of Siberia, carried out in 1991 by a team of Russian and American astrobiologists. In northeastern Siberia, subzero conditions have persisted for over three million years, a stitch in time compared to Martian geological history. Nevertheless, what they found reveals just how astoundingly hardy life can be in even its lowliest microbial manifestations. Large numbers of bacteria have been effectively freeze-dried. When thawed out, they resume their life functions. The Siberian results show that there are up to 100 million bacteria per gram of frozen soil. Even more remarkably, after being frozen for three million years at a temperature of −10°C to −12°C (10°F to 14°F), they do not seem particularly harmed by the experience.

"They are viable," McKay says. "What keeps them ticking is the natural radioactivity of the rocks, which over ten million years or less is not a problem."

But Mars is a far harsher environment: could microbes have persisted there for such a long period tied up in the permafrost? In principle, yes. Half of the Red Planet has been "dead," geologically speaking, since the intense cratering that immediately followed the birth of the solar system. The southern hemisphere of Mars bears testimony to ubiquitous bombardment, with surface features that have not been removed by subsequent geological activity. If life ever started, its presence may still be there.

"Those are almost ideal preservation conditions," McKay says. "Dry, frozen, and at low atmospheric pressure. You couldn't ask a curator to do a better job of preserving any samples from that time." He cautions, however, that any microorganisms may have been assaulted by accumulated radiation within the rocks and from space that has bathed the Martian surface, since the atmosphere does a poor job of protection.

Biologists don't have a particularly good understanding of what happens when organisms are frozen for millennia, and the effect of cumulative exposure of microbes to low levels of radiation is largely unknown. The situation is further complicated because the dosage received—on Earth or Mars—would fluctuate over time, making it difficult to extrapolate. Most experiments to date in this field have been with bursts of very high levels of radiation over short periods of time. Today, more precise measurements of radiation are coming from NASA's Curiosity rover in and around Gale Crater. McKay is thus working to better understand how cumulative exposure to natural radioactivity would affect possible Martian microbes.

* * *

If life did begin on Mars three and a half billion years ago, any water would have been locked up as permafrost as its climate subsequently evolved and became cooler. Organic material would have become incorporated into these frozen sediments and remained frozen in place, both physically and chemically.

There is also another factor favoring preservation. Mars does not have plate tectonics, so the upper layers of its surface did not recycle the original material that made up its primordial crust. It has been estimated that erosion and burial rates at the Martian surface are approximately one meter per billion years. Perhaps two-thirds of the Martian surface is older than, and has remained unchanged for, three and a half billion years. This means that even the most ancient of permafrost should be fairly easily accessible to robotic or human explorers. Material a few meters below the surface would also have been protected from dangerous cosmic radiation and ultraviolet rays from the Sun in the epochs since.

Some researchers have even gone so far as to say that underneath the Martian surface, there may well be extant life-forms today. Others are more cautious. Nevertheless, the intriguing possibility remains that there will be compelling evidence for ancient biological activity preserved in the Martian permafrost.

Frozen water is the key. Some Martian craters seem to have formed when they were accompanied by what looks like muddy slurries. Larger impacts were needed to create telltale torrents of mud. If the ground contained significant amounts of water, the craters from the impacts would have been frozen in place thereafter. Closer to the poles, even the smallest of craters seem to be surrounded by telltale signs of ancient muddiness. This suggests that, in these regions at least, the ice is nearer to the surface.

Nearer the equator, however, the relative warmth of the Sun means that the ice migrated farther below the surface. The presence of ground ice becomes apparent from the midlatitudes up toward the poles. "We know there's ice from observations by our radar on spacecraft in orbit," says veteran Mars researcher Richard Zurek of the Jet Propulsion Laboratory (JPL). "There is ice beneath the surface as well as at the polar caps."

Under present conditions, it has been calculated that the surface of Mars is frozen to depths of about a mile (about two kilometers) at the equator and about three miles (six kilometers) at the poles. In 2015, a team led by Ali Bramson, then at the University of Arizona, mapped the location of nearly two hundred craters across an otherwise featureless plain north of the equator, Arcadia Planitia. They were chosen because radar observations from NASA's Mars Reconnaissance Orbiter had measured reflections from under the surface that are characteristic of ice. Many exhibited "terraces" that could only have formed if there was ice under the surface.

To fit the observed "splats" and the radar measurements, the Bramson team estimated that there is a thirteen-story-deep ice sheet buried underneath Arcadia Planitia. A year later another team, led by Cassie Stuurman at the University of Texas at Austin, used a similar radar technique to examine another northern plain called Utopia, where the second *Viking* lander had landed in September 1976. Terrain that has cracked into telltale polygonal shapes and scalloped depressions suggests they were formed by extensive water ice that had sublimated or thawed out. The Stuurman team estimated there were 145,000 square miles (375,000 square kilometers) of ice below the surface.

Taken with the earlier work, these studies show that recent climate change has led to the accumulation of ice at the midlatitudes. Such deep reserves exist because surface ice in the equatorial regions is unstable and will easily sublimate away. At the poles, however, it won't. Higher than 40° latitude in either hemisphere, ground ice will be found closer to the surface. The nearer to the Martian poles you look, the more likely you are to find it. That, in essence, is why the poles could be such a mecca for microbes.

Another tantalizing glimpse of the subsurface water surrounding the south pole of Mars has been inferred from orbit. An unnamed crater in a region called Noachis Terra, in the vicinity of the Martian south pole, reveals the strongest evidence yet for seepage of underground water. Just thirty miles (fifty kilometers) wide, this crater contains dark, tiger-toothed features within its rim that look similar to glacial water seepage seen in Iceland and around Mount St. Helens. This strongly suggests that the water was exposed when the crater was excavated by the impacting body that gouged it out of the surrounding surface.

Not only that, the dark floor of the crater is smooth, which suggests it could well have been covered with a pool of water. The edge of the inside wall of the crater reveals that there are "islands" of material poking up through the floor. The most likely explanation is that water did indeed burst out of the crater rims, and then formed a lake that either evaporated or froze in place.

Certainly, life as we know it depends on water of a certain chemistry—not too salty, for example—so finding life-friendly water could lead to life itself. On Mars, a great deal would have certainly percolated through the soil to form the layer of permafrost or else remained frozen on the surface and thereafter become covered by accumulations of dust.

As such, ice—not least in the form of glaciers—is an important bridge to the past, buttressing theories of what might have happened when the Red Planet was warmer. Could conditions have been clement enough for a large ocean to have covered sizable portions of the planet, as inferred from the geologic evidence? How wet was Mars in its earliest days? Did the planet's ancient climate suddenly change forever, or did it oscillate between warmth and cold?

As ever with the Red Planet, there may well be other factors that are not yet understood. One important insight comes from perhaps the most remarkable discovery ever made about the poles of Mars: the laminated terrain.

* * *

They stretch as far the eye can see in every direction: regular, relentless, and undulating along the perimeters of both polar ice caps on Mars. They look like strange icy layers in a cake created by a cosmic chef with an enhanced sense of aesthetics. Nothing quite like them has been seen elsewhere in the solar system, and that includes the even colder, icier moons of the outer planets.

This is the mysterious layered terrain surrounding the Martian poles that even today defies detailed understanding. The alternating layers of dust and ice are seen at latitudes greater than 80° in both hemispheres and stretch uninterrupted for many hundreds of miles. They appear as fine bands along slopes where the Sun's rays and wind have removed seasonal ice, and they occur in relatively smooth, undulating landforms into which the polar ice has cut steep slopes and scarps. When they were discovered in the 1970s, they caused no end of amazement. Their regularity hints at how the climate has changed in the past. The relative "thickness" of deposited dust and ice has faithfully mimicked climatic conditions throughout Martian history.

Nobody knows how long it would take to deposit an icy layer. Theoretical calculations have shown it could be anywhere from a year to many hundreds of years. Nevertheless, the broader outlines of how the layers formed have come from some remarkable astronomical detective work. By divining the motions of the Red Planet as it has moved through space over recent millennia, climatologists have determined why dust was more likely to be deposited than ice and vice versa at different times in its history.

Planets are not the perfectly defined spheres of childhood drawings, nor are they uniformly dense. Earth, for example, is distinctly pear-shaped. The equator is out of true with the poles by many tens of miles. That irregularity means planets don't spin as perfect spheres would. Mars, too, is oddly shaped, like a slightly squashed egg. It, too, wobbles as it spins, and this has had a pronounced effect on its climate.

When the Martian orbit is at its most elliptical, more sunlight will fall on the hemisphere undergoing summertime, as it is so much closer to the Sun. At present, that is the southern hemisphere. During the southern summer, global dust storms tend to kick up and deposit dust on the northern pole, due to the quirks of atmospheric circulation on today's Red Planet.

This tilt, which is more formally known as obliquity, also has an important effect on climate. As noted earlier, the current Martian tilt (25°) is similar to Earth's (23.5°). Jupiter's immense gravitational influence may have tugged the axis to as high as 46° within the first half billion years of the planet's history. By comparison, Earth's axial tilt has changed only by about one degree over geological time.

Even after the emergence of four enormous volcanoes on the Martian equator, which dampened down the tilt of Mars, the obliquity was reduced to only 35°, still large enough to have had a distinct climactic effect over the succession of seasons. Perhaps today's tilt of 25° is the lowest ever amount by which Mars can roll on its axis. At present, it seems that changes can vary by up to 10° over a period of 100,000 or a million years.

As a result, some have argued that the current deep freeze seen on Mars may just be a temporary phenomenon, so far as geological time is concerned. When the axial tilt is higher, the summer pole would be far hotter. Much less atmospheric carbon dioxide would freeze out on the winter pole in the annual yin and yang of the seasons. More dust would be deposited because there would be less ice around. The opposite would be the case when the planet tilted by much less.

Such changes would obviously contribute to the layering seen in and around the Martian poles. Stronger summer heating at one pole may have released greater amounts of greenhouse gases from the polar caps. Permafrost that is rich in frozen carbon dioxide would have sublimated to allow a temporary greenhouse effect to result. What that also means is that the deep freeze would be temporary in any given region on Mars. The changes to the tilt would cause extensive glaciations, which would then also move down to the midlatitudes or back up to the poles.

If liquid water were available, it could have flowed across the surface before it froze or evaporated. Chemical reactions with the atmosphere would have formed carbonates or salts, reducing the overall atmospheric pressure.

When the planet's tilt was reduced and its orbit became more elliptical, Mars would have been distinctly cooler. At the poles, there would have been a greater chance for snow and glaciers to be more likely. Based on his extensive experience studying glaciers on Earth, Professor Jack Holt, a glaciologist at the University of Arizona, is amazed at the estimated range of ages of the ice which is still present on Mars—anywhere from a few hundred million to several hundred million years. "We don't have anything remotely close to that preserved anywhere on Earth," he says. "On Mars, I think that this ice actually contains a great deal of past climate information that is so far missing from our understanding."

Much more scientific detective work is needed to discern the exact details of the more recent climate history of the Red Planet. While the ice is very old in terms of human experience, it is geologically quite young. Nevertheless, scientists are akin to detectives looking at a crime scene that has long since been altered. They are hampered by not knowing what exactly has taken place. To date, they have inferred details from the equivalent of fleeting snapshots and unreliable witness statements. Worse still, much of the evidence for the evolution of climate on Mars is either circumstantial, contradictory, or plain confusing. One thing most experts agree upon is that the poles of Mars may unlock the mysteries of climate and, hence, the possibility for ancient life. They beckon to us across the interplanetary void, tantalizing and often ambiguous, in the vital information they contain. They will help unravel many of the enduring mysteries about the ancient history of the world next door.

* * *

Since the start of the twenty-first century, ever more detailed suites of instruments have been peering down at the poles of Mars from spacecraft as they orbit overhead. They have been scrutinizing the ice caps in visible light, in the infrared, using laser reflections, and—most revealingly—by radar. Taken altogether, our portrait of the polar ice has been fundamentally redrawn before our very eyes. "We've actually seen exposed cliffs of ice," says Rich Zurek of the Jet Propulsion Laboratory. "And you go, 'Wait a minute, those must be rapidly sublimating into the atmosphere today. So how do you keep this there for any length of time?'" There are many places where there is subsurface ice and a lot of debris surrounding it. "And that insulates the ice and keeps it there," he adds.

One of the persistent riddles is that the northern pole seems warmer than the south yet also seems to have more water ice on display. The explanation is that appearances may be deceptive. The south pole is covered by a frosting of carbon dioxide, which is believed to be quite thin compared to the vast bulk of water ice below.

The Mars Global Surveyor spent most of a decade in orbit around the Red Planet after arriving in September 1997. It accurately scanned the surface below using a laser that fired pulses to build up a detailed map of the terrain. By accurately timing how long the signal took to return, the laser signals revealed the extent of depressions or mountains beneath. If the surface was comparatively lower, the laser pulse took longer to return, and vice versa if it was higher. The Surveyor's orbit was better placed to observe the north pole, and thanks to that, a three-dimensional map of the Martian arctic has been possible.

"Without [the laser instrument], we would have had a lot of trouble with our interpretations because it's our baseline reference for all our current data," admits Jack Holt. His recent work, for example, has used radar that can penetrate below the surface. But making sense of the radar signals requires various corrections that depend on the state of the atmosphere through which they have traveled (such as electrical activity in its outermost layers, which have been found to create aurorae). With a laser instrument, however, what you saw was what you got. Without the laser ranging information, Holt says, "it would have been really impossible to do the level of interpretation that we have."

What laser altimetry has shown, particularly in its greater vertical resolution, is that the north polar cap on Mars is gouged by canyons and troughs that are as deep as 0.6 miles (1 kilometer) beneath the surface. They appear to be carved by the winds and the evaporation of ice.

The statistics stagger when comparing that part of Mars with Greenland on Earth. The shape of the arctic cap on Mars indicates that it is composed primarily of water ice, with a volume of 300,000 cubic miles (1.2 million cubic kilometers) and an average thickness of 0.6 miles (1 kilometer). "Combined together, the Martian polar caps contain about the same amount of water ice as there is in Greenland," notes Dr. Frances Butcher, a planetary glaciologist based at the University of Sheffield in Britain. On Earth, there is much more ice at the poles. Less than 1 percent of Earth's glacial ice is elsewhere; on Mars, 10 percent of its water ice is in the nonpolar regions. What that implies is that ice has played a very important role in the story of Martian climate. To understand that better, researchers need to peel back the outer layers of the ice and peer at what may be lurking below.

* * *

Like Earth, the Martian poles have permanent caps and seasonal ones. The seasonal ice comes from carbon dioxide freezing out onto the winter pole, reducing overall atmospheric pressure in the process (by up to 20 percent as measured on the surface). On the winter pole, the seasonal cap formed by accumulating carbon dioxide is about a meter thick. Depending on how dusty the atmosphere is, the seasonal caps at both poles consist of dirty ice, with dust mixed in. The extent of the mixing depends on how relatively warm or cold it is.

Over the last two decades, new radar instruments have revealed unsuspected details about the structure within the layers. They have also allowed glaciologists to make comprehensive three-dimensional maps of exposed layers of the scarps seen at the poles. In the summer of 2019, researchers in Texas and Arizona presented a new estimate for all the ice currently contained at the north pole of Mars. They examined a deposit of water ice and sand that lies beneath the current permanent cap. These layers formed during many past ice ages on Mars. Hitherto, it had been thought any evidence for this kind of ancient ice would have been lost.

"There is quite a bit more ice [in that basal unit] than people thought," says Professor Jack Holt, who led the study. "We don't have a good age constraint upon it. It probably spans a large range of ages from a billion years up to just five million years."

And it is where some of the missing water might still reside. To their amazement, the researchers estimated there was enough to fill a layer at least five feet (1.5 meters) deep right across the whole planet. Such a large amount of water is not easily explained by current understanding of how climate has evolved on Mars. It could also mean that much of the "unaccounted for" water resides below the surface in a massive frozen aquifer.

In particular, the "shallow radar" aboard the NASA Mars Reconnaissance Orbiter has been able to penetrate down to 1.5 miles (2.5 kilometers) below the surface. The total volume locked up in these buried deposits is about the same as that seen in glaciers and buried ice across the rest of the Red Planet. "There is clearly a process of exchange going back and forth," Holt says. "Or maybe that's just coincidence."

The bottom line shows "there is some process by which significant amounts of ice can get trapped at the pole during these exchanges." It is a one-trick pony: because the water is frozen, it is not available to flow freely ever again. With each subsequent freezing, less is available for the climate. And that, ultimately, might provide scant comfort in the search for extant life.

"You can have all the right conditions for life," says Stefano Nerozzi, Holt's PhD student who has mapped this ancient ice, "but if most of the water is locked up at the poles, then it becomes difficult to have sufficient amounts of liquid water near to the equator."

* * *

By comparison, the south pole on Mars is very different. One important factor is that it is on average more than three miles (six kilometers) higher than the north pole. That extra height alone might account for the very obvious differences between them. In Jack Holt's estimation, the ice at the south pole appears much older than that at the north. From repeated radar observations, it appears that the wind has had a greater role in shaping the polar caps than had previously been known. Many of the observed patterns of ice and dust at both poles show telltale signs of having been eroded by wind.

But there is a frustrating problem. The highest resolution camera in orbit around Mars today can see down to ten inches (twenty-five centimeters) at the surface (better than Google Earth, in fact). At best, the shallow radar instrument that flies along with it can only see down to thirty-three feet (ten meters). "We can't quite connect them," says Professor Holt. "We haven't been able to positively identify an exposed layer and outcrop that you can trace and connect with [the shallow radar]." Some of these layers can be tracked across the entire polar cap, "which is amazing, and we see all this structure," he says. "The holy grail is still to have a definite correlation between the [changes to the orbit] and the layers."

The way in which the ice has grown and retreated has been modeled at both poles. The ice at the north pole is probably less than five million years old. There has been growth, retreat, and growth. "You see a nice correlation of that in the radar data," Holt says. Some of the layers are cut off mostly around the edges, "where you clearly have had that ice growth and then you have had these periods when you start removing the ice."

In the high-resolution radar observations, "aeolian processes"—as the cumulative effect of wind is known—have played a significant role in shaping the cap. Further work is needed to refine these measurements, with researchers knowing only too well that the finding of finer, thinner bands within the layers may well cause a rethink of how the Martian climate has changed.

* * *

The European Mars Express Orbiter arrived in 2003, three years before NASA's Mars Reconnaissance Orbiter did in 2006. Both have been maneuvered into near-polar orbits around the Red Planet. As they move around Mars, the whole of the planet can be spotted below them (apart from a "collar" of 3° around the geographical poles themselves). Both have radar instruments that have observed distinct differences in the polar regions.

"Our shallow radar works really well for the north pole," says Professor Jack Holt. "It sees lots of internal layering and structure, and you can see all the way down to [the basal unit]."

This instrument has greater resolution at the surface but at the expense of looking far below: "we have shorter wavelengths and higher resolution." The European radar operates at a lower frequency, which means it can penetrate further. "For a reason we don't understand yet," Holt explains, "our signals get scattered by something within the south polar layered deposits. It appears to be a subsurface kind of shallow layer or layers that we just don't understand."

The term they use is "a fog," not in the literal sense but rather a diffuse signal that obscures everything else returned in the radar profile. In other words, there is no useful information that can be interpreted, in much the same way fog obscures a surface seen with the naked eye.

"The European radar does better in the south and not so well in the north," he says. This radar instrument on Mars Express hasn't yet clearly delineated much internal structure within the northern polar layer deposits. It sees the top and the bottom, and, depending on how the information is analyzed, some sort of structures within appear to move up and down.

How the two radars complement each other is interesting, Jack Holt says, although "it's a little frustrating at times." Certainly, that has been the case with one of the more sensational discoveries in recent years: that there is a vast layer of water underneath the south pole of Mars. A team of mostly Italian scientists used the Mars Express radar information to infer that there is a lake-sized reservoir of water with sediment below it. As a likely habitat for biological material, it became front page news in the summer of 2018 when their research was published in the journal *Science*.

"This discovery is changing again the view of the possible presence of liquid water," said one of the team, Elena Pettinelli, a researcher at Roma Tre University in Italy. During dedicated passes between May 2012 and December 2015, the European radar saw "anomalously bright subsurface reflections" beneath the icy Planum Australe region.

Within an area twelve miles (twenty kilometers) wide, something lurks below. The Italian team have interpreted this as a possible sediment-infused body of liquid water buried under the surface, pooling roughly less than a mile (one and a half kilometers) under the frozen ice layers. Other scientists, especially those working with data from the other radar aboard NASA's Mars Reconnaissance Orbiter, remain unconvinced.

Theoretically, the American spacecraft's radar should also be able to see this feature, too. To date, it has not. Everything hinges on the way that the radar "echo" is returned and how it is interpreted. The Mars Express radar instrument can penetrate the top three hundred–plus feet (one hundred meters) below the surface, while the other instrument on the NASA spacecraft sees down to thirty feet (ten meters). There is something strongly attenuating the radar signals, not allowing them to see the bottom half of the ice sheet.

What happens next depends on how these observations are modeled. And, as in so many areas of science, the results depend on just how many assumptions are made at each step along the way. Philosophically, you might say, that depends on just how Earthlike the Martian poles are. "In Antarctica, for example," notes Jack Holt, "when you have a lake, for it to be sustained over any period of time, you have to have melting off the face of the ice sheet."

That will also tend to warp the ice above it, but warping has not been observed in any of the European radar images. If it's an ephemeral feature, it also seems hard to explain. Though salinity (the amount of salt needed to keep the water as liquid) has been inferred, several critics have pointed out that the observations don't seem to match anything known in nature. "There's no amount of salt that can cause an amount of liquid to exist underneath the ice sheet," Holt adds, "even if you made it purely salty."

The simplest way to keep water as liquid is with "a heat anomaly." On Earth, there are such features in the form of volcanoes lying below ice sheets. But on Mars, nobody has a detailed understanding of how much heat is available. Professor Holt points out that if it was some sort of magma chamber that could provide enough warmth, there is no supporting evidence. "The best-case scenario that if there is liquid there, it could be a kind of icy sludge," he says.

The Italian authors, nevertheless, remain confident in their findings as they have taken years to collect the data. As ever with cutting-edge research, it is also a matter of elimination. They looked at other possibilities that might explain the radar reflection before turning to water. They rejected the idea of the layer being due to carbon dioxide or water ice. It is unlikely both could form at the base of the ice cap under those conditions either.

At this stage, it is probably fair to say the jury is still out. A more accurate assessment might come from new information about the interior heat flow within the planet as a whole, which would tell scientists whether pools of liquid water could be possible. For that, measuring the heat flowing just below the surface is crucial. As we will see in the next chapter, attempts to do this are ongoing with the most recent visitor to the surface of Mars, the NASA InSight lander.

Fundamental disagreements are part and parcel of scientific inquiry. Clearly, a deluge of information is helping fine-tune some theories and leading to others being thrown out. One ugly fact, it has often been said, can destroy the beauty of what may appear to be an elegant theory. These are the sorts of riddles which will be addressed by the next missions to Mars. If past experience is anything to go by, the answer will be far stranger than anyone could have ever imagined.

* * *

Liquid water beneath either of the poles has important ramifications for life on Mars. For that reason, the polar ice might provide a new and fertile hunting ground. "The permanent polar caps of Mars are frozen water and would act as a splendid 'cold trap' where organic molecules could condense," wrote one prominent biologist in the early 1980s. "The scavenging oxidizing agents would largely be absent, so that there is an odds-on chance of finding life-forms." Professor Jack Holt adds, "If there's any life near to the surface of Mars, it's going to be in the icy deposits. But maybe the ice is going to be too young to preserve it."

Certainly, midlatitude glaciers formed where conditions were warmer and may well be easier to reach. And there is another very good reason to examine ice below the surface. Something similar may take place there as has been found on Earth. "Any time we've gone to deep ice on Earth," Jack Holt says, "there are always things living there."

Ultimately, it won't be a single mission that will tell the story of how the polar regions of the Red Planet have changed over time—and whether life has played any part in it. One obvious mission would be to have a small spacecraft land and attempt to take ice cores in the laminated terrain. "With precise dating, you could correlate the precise sequence of how they were lain down," says Professor Holt.

Chris McKay has been working for a number of years on the Ice-breaker Life mission concept, which would look for telltale signs of biological activity on the northern plains of Mars. It would be a rerun of the *Phoenix* mission with a new payload designed to look for any hints of biochemical processes. Though not chosen by NASA in 2015 nor again at the start of 2020, it may eventually be taken up as a project in the years ahead. But there would be little chance for any current technology to actually land farther north on the polar ice caps themselves and survive for any length of time.

The mystery of the Martian poles may not be solved any time soon. According to many researchers, trying to land at either Martian pole is, for now, science fiction. Yet, only a few decades ago, so was establishing a base in the Antarctic or sending a spacecraft to Mars. In the past, the icy poles on Earth were called terra incognita, as they were unknown and unknowable for so long. Thanks to pioneers like Ross, Scott, and, in more recent times, dedicated scientists who live on the ice for months at a time, the poles on Earth are no longer mysterious. The same can be said for Mars.

The Red Planet has been observed from a distance, in orbit, and on the surface by landers and rovers. Its geology and general characteristics are reasonably well known. But many more mysteries remain. The most recent US arrival on Mars has taken another leap into the future by attempting to unravel another important aspect of its past. "I'm looking forward to making the first map of the inside of the planet," says one involved with the Mars InSight mission. In the spring of 2019, it found something long suspected that finally showed that we can examine a planet—and a red one, at that—from the inside out.

2

INSIDE OUT

As discoveries go, it was more of a whimper than a bang, a low-frequency rumble akin to a sustained hum rather than a planet-shattering wave. On Saturday, April 6, 2019, the first-ever "marsquake" was recorded by the Red Planet's most recent robotic visitor. Five months after its landing, NASA's Mars InSight had found exactly what it had been sent to search for.

As the first American Mars mission launched from California, InSight was picking up truly good vibrations—in this case, ones far below the range of hearing in humans. The vibrations lasted for more than ten minutes and were caused by motions deep within the interior of the Red Planet. Up to this point, InSight had "been collecting background noise," in the words of one official, mainly caused by winds and vibrations of the body of the lander itself.

Then suddenly a clear signal emerged. That Saturday in April, they hit the jackpot. The seismic vibrations—immortalized as the "Sol 128 event" by the research team—measured magnitude 3 on the scale the scientists have developed, though in truth, it was a slight tremor that would barely have been felt by anyone standing nearby.

As had long been suspected, here was proof not only that the Red Planet was undergoing internal activity but, for the engineers who had struggled to build the challenging hardware, that such elusive signals could indeed be detected. InSight's seismometers are among the most sensitive ever built. Their findings are systematically "taking the outer wrappings off the planet," in the phrase of one of the researchers involved.

Subsequently, there has been on average at least one marsquake every other day. Most have not been quite as large as the first, but the total is far greater than had been expected, numbering 450 by the end of February 2020. Their observed patterns don't seem to make much sense as yet. They have revealed something very different than expected, with many new mysteries thrown up and one of the scientists involved admitting that "they're super puzzling."

* * *

Until Mars InSight landed in November 2018, what lay beneath the surface was utterly mysterious. The lander is situated in a quiet corner of the Red Planet. Its measurements are shedding light on how Mars has evolved over geological time and, equally importantly, what that means for conditions when life could have formed in the ancient past.

The Martian interior holds vital clues. As the Red Planet is smaller than Earth, it is less active, and its internal dynamo, the core, effectively switched off far sooner. If life did form and die out, its signature will not have been erased by any subsequent geological activity. Microbes might well have hibernated somewhere underground, but for now that is just a guess.

Known by the acronym standing for Interior Exploration using Seismic Investigations, Geodesy, and Heat Transport, InSight is less geological explorer, more medical examiner. The mission has been taking the pulse and reflexes and attempting to take the internal temperature of the Red Planet. Its instruments are like stethoscopes and sonograms to open up a unique window into the past.

In another sense, the InSight mission is more like a time machine, looking back to see how Mars has evolved by observing what principal investigator Bruce Banerdt calls "all these fingerprints of these early processes [that] are still retained in the deep interior." How large is the Martian core and how long was it active? When exactly did geological activity shut down? How thick is the crust? How much heat did it generate to sustain the kind of chemical reactions needed to create life? By looking at how current internal activity is manifesting itself as seismic waves, InSight is piecing together this unknown past of the Red Planet.

Though simpler than the more recent rovers, Mars InSight has been a "tough mission." Delayed once due to problems with one of the seismometers, its total cost was nudging $830 million by the time it was launched in May 2018. In the depth and range of information it is returning, however, it has already paid for itself in terms of the scientific bonanza.

For seismologists, the clock has been reset after they spent forty years in the wilderness, thanks to a curious afterthought to the otherwise spectacularly successful Viking missions. In the seventies, seismometers were added to the *Viking* landers only at the last minute. Consequently, they could only receive leftover power and were bolted on where there was available room. That happened to be the top of the lander, roughly four feet off the ground. Engineers thought that, even that high up, they would still be able to spot evidence for seismic activity. Alas, they didn't.

The *Viking 1* seismometer failed, as it did not uncage properly. *Viking 2* caught some vibrations, but only from the passing winds. The seismometers inadvertently became wind detectors. Great for atmospheric scientists, but not so good for those interested in the Martian interior. So this time around, InSight carried a package of seismometers specifically designed to be lowered onto the surface. There they have had to be meticulously isolated against not just changes in heat—even a drop in temperature of a few tenths of a degree would affect their sensitivity—but winds, the biggest problem of all.

As if to prove the point, within days of its landing, InSight's instruments picked up the rattling of the spacecraft's own solar panels. A small pressure sensor has also picked up the sign of hundreds of dust devils—the atmosphere was very dusty when it landed in November 2018—and has found greater wind activity at night. These and other external influences have all had to be eliminated from their readings before the scientists could begin their seismic inquiries.

After a certain amount of head-scratching, the seismologists now pronounce themselves delighted. Though they were expecting more Earthlike events on a Moon-like time frame, the activity seems different from the seismic vibrancy of the former and the dormancy of the latter.

"Mars is more moonlike," acknowledges one, "the crust is more broken up than a lot of people thought." Which means they are now exploring new territory every bit as uncharted as the Red Planet itself was when it was first spotted on an unmarked night by unknown observers in the ancient world.

* * *

Those first Martian explorers were Earth-bound, trying to divine meaning from the regular passage of stars that paraded before them in the skies. Our earliest ancestors first saw Mars as a bright object in the night sky. As it was the color of blood, it soon became known as a portent of war, whose baleful presence cast an unnerving malevolence throughout the heavenly firmament.

The invention of the telescope brought the Red Planet closer. When the first astronomers trained their instruments toward it in the seventeenth century, they saw ocher-colored deserts upon which darker, greener features could be made out. Some interpreted them as vegetation, supported by observations that they seemed to change with the seasons.

"The analogy between Mars and the Earth is, perhaps, by far the greatest in the whole solar system," wrote the Enlightenment astronomer William Herschel, who also discovered the Martian polar caps. Herschel could have had no inkling just how important the waxing and waning of their bright iciness would become in our current speculations about life on Mars.

A century later, the new science of spectroscopy had come into its own. Split light into its constituent rainbow spread of colors through the action of a prism, and in fine detail you will see thousands of dark lines, as distinct and unique as fingerprints. Because such lines appear in the spectra of gases burning in a laboratory on Earth or deep within a star, they allow scientists to compare and contrast their presence throughout the universe. Astronomers could thus learn about the chemical makeup of distant objects from the records of laboratory chemists.

But the spectra of Mars were elusive and illusory. Painfully slowly, it became apparent that the Red Planet was colder, drier, and less conducive to life than had been supposed initially. This was confirmed by the first space missions that scrutinized the Martian surface in the 1960s and 1970s. The Red Planet appeared dead, heavily peppered by craters, and with a poisonous carbon dioxide atmosphere about a hundredth the pressure of Earth.

In time, evidence was amassed for volcanism and water flows in the ancient past. But there seemed to be no magnetic field—the result, it implied, of a core in which the sloshing of molten material was largely absent or had ceased altogether. That lack of what may be termed an internal electric dynamo also meant that energetic particles from the Sun were unhindered by any magnetic field, as on Earth. The surface of Mars was thus bathed in violent radiation that flowed in unimpeded.

Today, new information is gleaned by very different means.

Planets reveal so much more than how they appear in the night sky. Where once astronomers had to attach spectrometers to telescope eyepieces, today those same instruments can be taken directly down to the Martian surface. Instead of remotely trying to read spectra, samples can be taken in situ with what are known as mass spectrometers, which allow for the analysis of any molecules present based on their constituent chemical composition. We no longer have to look up at the night sky; our robotic envoys can look down from orbit while still others can land on the surface itself.

Finally, instruments are now scrutinizing the *inside* of the Red Planet. "We want to look at the actual physical structure of the planet, what it's made of and what the depth and thickness of the various layers are," says Bruce Banerdt, based at the Jet Propulsion Laboratory, mission control for Mars InSight. What they are now doing is examining just what is happening to the Red Planet from the inside out.

* * *

Nerves were heightened on the last Monday in November 2018 when the InSight spacecraft reached Mars. It came in heavier and faster than its nearest equivalent mission a decade earlier. Its landing site was about 4,900 feet (1.5 kilometers) higher than the average height of the surface. That meant there was appreciably less atmosphere available for it to decelerate safely.

That atmosphere was extremely dusty. For most of 2018, the Red
Planet had been engulfed in the fiercest global dust storm in living memory. It had already consumed the NASA Opportunity rover, which hadn't
phoned home, like all good aliens, since May of that year. "We were
absolutely terrified what the dust might do," says one participant.

Publicly, however, officials were upbeat. For all the buildup and worries, the InSight landing, while tense, was anticlimactic. Shortly after
noon on that Monday afternoon, mission control in Pasadena, California, erupted into cheers. On Mars, work started straightaway.

Unlike NASA's more famous and mediagenic rovers, InSight was
designed to sit perfectly still. Within a few hours of landing, a precise sequence of events began. The first, and most important, was the
unfolding of the solar panels. Without the power they generate, InSight's
onboard batteries would die for want of further energy from the Sun.

Within minutes, cameras on the body of the craft and at the end
of its robotic arm revealed a flat, completely unremarkable landscape.
Elysium Planitia is an ancient lava plain that cooled long ago. Where
InSight had come to rest was soon christened "Homestead Hollow," as
there was a slight indentation left over from a crater that had been filled
in over time. Such a feature was fortuitous. "These eroded craters are
fairly common on the Martian plains," says Bruce Banerdt. "I think we're
pretty fortunate, actually, to have landed in one because it made our
deployment somewhat easy."

The surrounding terrain was exactly as advertised. *Boring.*

The very fact that there wasn't much going on there had commended
itself. InSight's instruments could take a background reading of the completely average conditions on Mars. "Boring" meant more representative
of the planet as a whole. InSight's sensitive instruments are thus located
as far as possible from any sources of surface activity: avalanches, meteorite impacts, landslides, and ice flows, all of which have been observed to
take place on the Red Planet today.

The first findings were of a low rumble, which came from the northwest over to the southeast, as wind blew across the lander at roughly ten
to fifteen miles per hour. These gusts rattled the solar panels, and the
seismometer package, firmly anchored to the "deck" of the spacecraft,
picked up the noise at the lowest reaches of human hearing. "It's like

InSight is cupping its ears and hearing the Mars wind beating on it," said one project official.

What followed in a slow, precisely choreographed sequence was the "instrument deploy phase." On Wednesday, December 19, 2018, engineers commanded the lander arm to reach out as far as it could and place the seismometer package exactly where the geologists wanted it. "Having the seismometer on the ground is like holding up a phone to your ear," said the elated French principal investigator.

For the dozens of scientists involved in the InSight mission from all over the world, a "marsquake" service has been developed to alert them to anything unusual. Once the first data came in, "we stayed up late to watch all the signals." While the initial thrill may have subsequently died down, it has allowed what principal investigator Bruce Banerdt terms "exquisitely precise measurements" to be made on the surface of another world.

*   *   *

In the Mars InSight seismometers, French and British scientists have built two instruments for the price of one. A traditional seismometer made by the French listens for low-frequency seismic waves. The British instrument picks up shorter, higher-frequency signals. "On Mars, both are needed to record the diverse signals we are seeing," says Dr. Anna Horleston of Bristol University, a member of the marsquake service.

The French instrument is heavier and covers a wider range of frequencies. As with all traditional seismometers, it is essentially a magnet and a coil. Any vibrations shake the coil at a very minuscule level that can be detected. Because it has a wider range of "hearing," it is the instrument that has to date picked up evidence for the first marsquakes. The British-built components are smaller and listen for shorter-period seismic activity. Essentially, they comprise small masses on a spring, each the size of a digital camera's SD card, which are equally highly sensitive. The springs are fashioned from slivers of silicon half the width of a human hair.

The seismometers are capable of picking up vibrations on the scale of a shaking atom and have to be protected from anything that might

affect them. The seismometer package has several ways of insulating itself
against the extreme changes to the local environment. "It is sort of like
a Russian doll," Bruce Banerdt has explained. Temperatures vary more
than a hundred degrees during each day, and winds can reach up to fifty
miles per hour; the low pressure of the mainly carbon dioxide atmosphere conducts heat slowly. Within the instrument, if some components
expand and contract, others do the opposite to counteract them. The
seismic sensors themselves are encased within a vacuum-sealed enclo-
sure within a titanium container. The container has honeycomb-shaped
walls that trap the surrounding atmosphere to insulate itself against any
changes in heat.

In Bruce Banerdt's phrase, temperature was one of their greater bug-
aboos. "Think of the shield as putting a cozy over your food on a table,"
he said in an interview. "It keeps [the seismometer device] from warming
up too much during the day or cooling off too much at night."

What looks like an upturned wok (with a hexagonal, copper-colored
covering) is in fact an aerodynamic dome that covers the instruments. It
was designed to reduce any further shaking from winds swirling around
the lander. Finally, a chain-mail "skirt" and thermal blankets around the
base of the instrument prevent any updrafts from below. Once this skirt
was put in place, the ambient noise was reduced. Despite the extreme
changes in temperature over the course of a Martian day, everything was
in "a nice, stable state," says Anna Horleston. Even better for the seismologists, there were "a number of days where we saw the quieter periods
get quieter."

* * *

It was a development that Inspector Morse might have savored. As one of
the university departments involved in building the British seismometer,
researchers at Oxford's Clarendon Laboratory on Parks Road carried out
a development test in the basement one Sunday. To their surprise, the
team observed a very peculiar signal: a small spread of seismic waves, definitely there in the data. Slowly it dawned on them what they were seeing.
Or rather hearing: bell ringers practicing in a nearby college chapel. The
instrument was so sensitive that it could pick up such activity in a deep

basement. To understand how it did so also neatly explains what the seismologists have subsequently been measuring on Mars.

The noise that bells make are vibrations. These sound waves not only
transmit through the air, they also pass through the ground. Known as
pressure waves, they travel outward and forward (think of a stone dropped
in a pond and the spread of ripples on the surface). Even the hand movements of the bell ringers create vibrations that also travel through the air
and ground. The trick is to measure everything and identify the contributions from each.

In the case of Mars, any activity within the planet will generate elas-
tic pressure waves. These are seismic vibrations that shake the ground
back and forth in the same direction in which the wave moves. These
compressive waves can travel through solids and liquids. How they do
so subtly alters the observed signal. As well as forward motion, there are
"shear waves" from any sideward motions and "surface waves" that result
from how the uppermost layers of the Martian crust are shaken.

Shear waves, however, cannot travel through liquids. The absence of
shear waves as measured by terrestrial seismometers during an earthquake
indicates that Earth's outer core is definitely liquid. Seismologists are now
attempting to make the same kind of determination for the interior of
Mars. Since they could pick up the vibrations from the Oxford college
chapel bells, they knew they could do as much on the Red Planet. Seismic activity is dictated by exactly the same kind of physical principles.

In the case of the seismic symphony from Mars, however, the researchers did not know from the outset what the score is, exactly what all the
instruments are, less still where they are located, how big the orchestra
is, and whether each instrument is coming in on time or in tune. As
occurs on Earth, there is a constant background hum, but on Mars it is
at a higher pitch (2.4 hertz). Underlying this is a confounding mixture
of pressure drops, winds, temperature variations, the buzz of electronics
within the lander body, and what are termed "unknown resonances." All
have had to be eliminated, and that has taken some time.

At first sight, the seismic score is a series of plots across time and
frequency. Each of the underlying signals has to be isolated and removed
one by one. So, for example, the InSight spacecraft has been shaking in
the cold. Even the groans and strains within the seismometer shield itself

have been observed. When it is windy, there is a lot more background noise. There are also clear "resonances"—in essence, spikes in the signal that have gradually been identified. Some come from the operating frequency of the lander's electronics (1 hertz).

Late in the evening, another "extended resonance" has been observed that is likely a peculiarity of Homestead Hollow. ("It's weird that we don't actually know what they are," says one researcher.) They also have to check the spacecraft's logs, detailed information on what the lander is doing and whether anything is moving or being moved around by its arm or other instruments.

Gradually, an underlying seismic soundtrack has emerged. "There are a lot of things we've been able to rule out because we know that's wind, that's pressure, that's temperature," explains Anna Horleston. "They tend to happen every day at the same time so we can eliminate them."

It should be quieter when the Sun is down. There is no direct heating nor any associated air currents. But, in fact, there is a clear daily cycle of underlying noise, especially when the teams thought things would quieten down. The raw signals show innumerable small vibrations within which are dips, caused by the passage of atmospheric dust.

Indeed, InSight is equipped with the most sophisticated automatic weather station ever built, which has already sensed any number of dust devils passing through. They change the surrounding pressure in a unique but very characteristic way, "like a little vacuum cleaner pulling up on the surface," explains Bruce Banerdt. "It sucks the dust from the surface up into the atmosphere and actually raises the surface slightly and causes it to tilt." That causes a perceptible dip in the pressure readings. Several hundred dust devils have now been observed by the weather station and also by the seismometers. Most of these have occurred at night, and their tracks appear the next morning. "We can see with our cameras where the ground has been sort of cleaned off in a line where the dust devil has gone by," says Banerdt.

At night, winds are blowing in from Elysium Mons, a volcano roughly eight hundred miles (1,500 kilometers) to the east from the flat plain where InSight landed. Downslope winds, as they are known, were not expected to reach this far away. "We thought, 'Oh, we're in Kansas. We're not going to have any of that,'" says Sue Smrekar, the deputy project scientist at the Jet Propulsion Laboratory. "But apparently from Elysium Mons, it's creating these nighttime winds, which is not good news for our seismometer."

* * *

How "noisy" it has been after sunset in Homestead Hollow has been a surprise. "The lander vibrates when it's windy," notes Dr. Horleston. "The signals decline a little bit as we go into the night, and then they'll rise again as you go into the noisier period."

These vibrations are not so large during the day but represent more noise than expected. It tends to be noisy in the wee small hours of the morning, then a lull, sometimes around sunrise, "but it's not consistent at the moment." After sunup, there is more noise for six to seven hours in the middle of the day. "That's when the wind is at its strongest," says Dr. Horleston, "and it's very hard to see anything." Around sunset, the noise decreases markedly ("We have four hours of beautiful quiet," she says with something like relief), but then the noise ramps up at midnight for reasons not understood.

All this activity is right across the board, measured by both the broadband French instrument and the British one. Once this background noise is eliminated, the team can start to understand how the Martian crust has been traumatized by repeated activity. "We're basically looking for signals which are not consistent with the normal pattern," Dr. Horleston says.

While it is still early days, the observed marsquakes show that the seismic waves have been scattered and smeared out in time as they travel outward from the interior. How they're altered reflects how broken up the outermost layers of the planet are. Though Homestead Hollow appears relatively boring, that's because the surface has been "smoothed" out by the flow of lava that resulted from volcanic activity in the past. The geologists believe the basaltic material they're seeing has covered up any underlying material that speaks of ancient tumult. Below the lava, Elysium Planitia has been bombarded and broken up, and the seismic waves are characteristic of how fragmentary the upper crust is in this region.

The Martian crust is more jumbled up than had been thought. "It doesn't look like anything we've ever seen," says Anna Horleston, "or been expecting." The seismic events—the marsquakes—suggest lots of broken-up material at the surface, which was expected; what is unusual is that there is also less solid rock below, which helps propagate the seismic signals better. "We're seeing a very different distribution of signals to what we see on Earth," Anna Horleston says. "We're confused by their diffusion: yes, it's more Moon-like, but there is more activity than we thought there would be."

* * *

One of the greater puzzles about Mars is just how geologically active it was in the ancient past. Certainly, the most spectacular expression of that geology came in the extensive volcanism that has affected much of the northern hemisphere of the planet. Lying across a region known as Tharsis on the Martian equator are four huge volcanoes, the largest of which has been called, appropriately enough, Olympus Mons. It is the most extensive volcanic edifice seen anywhere in the solar system.

All four volcanoes are considered large "plugs" of lava that emerged out of the crust in the ancient past. Did this Tharsis bulge, as it is known, rise gradually and slowly? Or were there "bursts" of activity separated in time? And just exactly how was the volcanism generated? One indication of how intense it may have been comes from faults in the surface, where it has literally cracked open. To date, mapping of geological faults and estimates of how the Red Planet subsequently cooled have just been inspired guesswork.

"The faulting is, we think, mostly driven by the contraction of the planet," explains Bruce Banerdt. "As the heat flows out of the planet, it contracts just like anything else as it cools off, and as it contracts, it crumples a little bit. That crumpling is manifested in quakes and seismic activity."

Though the crust of Mars has split open, it is still relatively thick and rigid when compared to Earth's. Whereas the terrestrial crust has broken up into distinct plates, on Mars it has not. So the vast Martian volcanoes grew and grew over the same hot spots where magma had built up more or less continuously from the interior. The crust was clearly thick enough to support their increasing weight over geological time.

There is a growing body of opinion that suggests there may have been short-term episodes of massive volcanic activity on Mars. Taken at face value, the InSight findings suggest there may be some sort of activity going on now. While the first-ever recorded marsquake was significant, it was also very surprising, and of limited usefulness. "It's probably from a relatively small quake," Banerdt explains, "and relatively close by. And relatively close by means a few hundred kilometers away."

The direction is easier to work out because the seismologists can trace it back. "The beginning of the wave trains are coherent, and they are polarized," Bruce Banerdt says. "That means we can actually figure out what the direction of motion of the wave is." Using the minuscule time difference between the arrival of the pressure and shear waves, they can estimate how far away they both could have come from.

"We can get a location," says Banerdt. "Both of them are located in the Cerberus Fossae region." This is a shallow depression named for the hound that stopped the dead from leaving Hades in Greek mythology. Cerberus Fossae consists of steep-sided sets of troughs associated with Elysium Mons. Its steep slopes are known to undergo landslides (which have been observed from orbit), though this is a red herring so far as marsquakes are concerned. "We don't think that landslides would have enough energy to create a signal that we would see," says Bruce Banerdt. "A landslide or rockfall just doesn't have enough vibrational energy to create a signal that we would see that far away."

What InSight has observed is more likely to have been quakes caused by shaking below the surface. "So our assumption is that these are marsquakes that occur on faults," says Banerdt. From studying how their energy is modified, seismologists should be able to determine whether this area is a smaller, Red Planet version of San Andreas. The question is how long and how active this region may be now—and may have been in the recent past. "We want to know what has been keeping it going in terms of geological activity," says Smrekar. "So we hope to learn more about the processes that are allowing some volcanism to continue until at least a million years ago."

The quakes may be the result of another process. The steep troughs are likely to have let lava and water be released across the surface in the past. Deep underground, some of those liquids may still be present, which could cause the tremors seen today. Others believe the seismic activity could result from a heating effect from the Sun as it rises over the troughs, cracking and shattering subsurface ice.

Cerberus Fossae was volcanically active roughly two million to ten million years ago, and possibly more recently than that. "It's very interesting to be recording these events from a tectonically active, volcanically active area," adds Sue Smrekar. "Even before this, it was difficult to really understand why was there such recent volcanism on the surface of Mars, when most of it is so old."

* * *

The first harvest of seismic data is intriguing. Most of the observed events—numbering nearly five hundred in total by the spring of 2020—are smaller and not so clearly defined. "Most of them are enigmatic," admits Bruce Banerdt, "very small, little events that may not even be seismic events. We have about a dozen that we think are almost certainly seismic."

These smaller marsquakes may be more localized. If there was any activity on the other side of the planet, their spread of frequencies would have a different shape than those observed. Signals from farther away would also tend to attenuate as they pass through the planet's interior. "They die off with distance," Bruce Banerdt explains. "And so, you're only going to see the very largest quakes at a distance."

A number of these events seem to last between five to ten minutes and are dominantly of one frequency, 2.4 hertz, the background "hum" noted earlier. "We think they are some type of process in the subsurface," says Sue Smrekar, "but we're still not sure if they are, in fact, a tectonic event or if it's some type of a resonance with the atmosphere."

They are tough to explain scientifically because they last longer than a typical earthquake and seem to scatter more effectively. They have no clear periodicity—that is, they aren't regular and don't last the same amount of time—nor do they seem to correlate with any particular time of day. And while these observations may be a peculiarity of the landing site, nobody has come up with an explanation that keeps everyone else happy. As Sue Smrekar notes, colleagues who originally studied the quakes seen on the Moon—observed by instruments left behind by the Apollo astronauts—were similarly puzzled. "So, I'm encouraged by the fact that it takes a while to fully understand this kind of data for a different planet."

Certainly, the seismologists didn't expect Mars to be like Earth. "It doesn't have plate tectonics," Bruce Banerdt says. "We can see that because there's no evidence of that kind of faulting on the surface." But then to just glibly state Mars is more Moon-like is not quite accurate either. Most lunar quakes are the result of tides induced by Earth, which flex the Moon's surface. The Red Planet isn't subject to the same level of external forcing. Any internal cooling, which has also been observed within the lunar interior, isn't thought to cause everything they are seeing from InSight.

"We had hoped Mars would give cleaner signals," Anna Horleston says. "Perhaps not as clean as on Earth, but certainly not as blurred as they are."

* * *

If exploring Mars has taught researchers one thing, it is surely patience. Another InSight investigation was designed to drill into the surface to measure the heat flowing from the interior, which, in Bruce Banerdt's phrase, "is really a measure of the energy that's driving the geology of the planet." For that, a "self-hammering mole," formally known as the Heat Flow and Physical Properties Probe, was included on the lander to measure the way residual heat is flowing upward. The plan was to drill down to sixteen feet (five meters) and determine the internal temperature at regular intervals.

At least, that was the plan, but unfortunately, the mole has been ratty. Once the seismometer was settled and stable, the mole was ready for work on the last day of February 2019. When it dug, there was a surprise. It got only as far as one foot (thirty centimeters) into the soil. The scientists had been expecting to reach more than two feet (seventy centimeters) on that first day. Every time it drilled, the mole bounced back due to a lack of friction available to pull it down in the loose soil, which was something they hadn't expected based on past missions.

It is hardly brute-force engineering. The mole weighs less than a pair of shoes and operates on less power than a standard Wi-Fi router. No sooner did they push the mole down so that it was flush with the surface than it kept popping back out again. What Troy Lee Hudson, the lead systems engineer for the mole at the Jet Propulsion Laboratory, terms "insanity"—"just hammering over and over again, expecting a different result"—caused them to stop. No frictional contact, no hammering.

By the fall of 2019, the InSight engineers had cleverly pinned the mole against the side of the shallow hole by using the spacecraft's robotic arm (which was, in simple terms, designed to pick things up and not bash things down) to keep it from popping out again. "We're very lucky we have a robotic arm to interact with the mole," says Hudson. "If we didn't have that, none of this would have been possible."

The engineers had spent months using test versions of the mole in Pasadena and another at the German space agency, DLR, which had actually built the device. So that their work could be synchronized, the German engineers have been working nights "and that's been tough on them," Hudson says. Just before November 2019—the arbitrary deadline they had given themselves—the pushing and shoving technique finally worked. The mole found enough friction to start burrowing again. If they pushed too hard, the tether, which would carry the temperature measurements, might snap. They kept on pushing, though—"increasing the friction between the mole and the soil on the other side"—and it worked. Troy Hudson termed it "an amazing and joyful moment."

"The big thing we have learned is the material is more cohesive to a deeper depth than we've seen anywhere else on Mars," says Hudson. Everywhere else on the Martian surface has been characterized by what he terms a "duracrust," lightly cemented soil that behaves like sand on a beach when it has been scraped by robotic arms or rover wheels. They expected to find similar soils a few millimeters thick, loose, and weakly bound.

It is akin to trying to dig a stable hole in dry sand on Earth, though the reason why remains elusive. Over geological time, the Elysium region could have been colder for longer, its underlying surface soils cemented together by salts that are more tightly packed, "kind of like dirt clod" in Bruce Banerdt's estimation. As the mole has gone down and vibrated, it has likely broken up some of the material surrounding its hole. "At a certain point, we thought the ground would take over," Hudson says. But it hasn't. This material has been compressed, falling farther down the hole, and doesn't provide enough friction for the mole. Even stranger, a small cavity surrounding the mole has also opened up, for reasons the engineers are still struggling to explain.

The point of the mole is to obtain a more accurate estimate of the energy that has driven the internal evolution of Mars over the last four and a half billion years or so. To date, the mole's tentative thermal measurements are still within the range of the diurnal change of temperature seen above. The cold of winter and warmth of summer will affect the background temperature measured within the first few feet of the surface, but a full measurement depth of sixteen feet (five meters) would ensure that readings aren't affected by the yin and yang of the Martian seasons.

"It's been an emotional roller coaster," admits Hudson. "When we had the success where we dug into the ground, I was so elated. When it backed out, I was crushed again. It's been exhausting emotionally to lead this effort. I've invested a lot in the mission. It led to a depression for a couple of months."

In a last-ditch effort, the engineers pinned and pushed the mole again in the first few weeks of 2020, using the scoop on the lander arm to bash it down. "Mars and the 'mole' continue to make our lives . . . how should I say . . . interesting," wrote the lead engineer in Germany on February 21. They are using the robotic arm for one last hurrah to see, finally, if they can burrow below the surface, in the hope that they will be able to take the measurements they need.

* * *

Modeling the various seismic waves seen by InSight will eventually allow the size of the Martian crust and the composition of the mantle to be estimated. From the outset, there is another puzzle. "We feel we ought to have seen surface waves, but we haven't seen a clear signal at the surface," Anna Horleston says. "The fact we don't see surface waves is being interpreted that these sorts of quakes might be deep."

Energy "rises" and vibrates the surface, which InSight should be able to pick up. In Bruce Banerdt's estimation, the reason it hasn't is that there hasn't been a large enough event yet. Shallow quakes tend to generate larger wave trains—a rapid succession of waves—than deep quakes, and these have not yet been observed. "So our sort of general assumption is that [the ones we have seen] are relatively deep," says Banerdt, "meaning maybe twenty-five miles or more deep. But depth is one of the hardest parameters to constrain."

Surface waves are more strongly excited within quakes that occur on relatively shallow faults—less than ten miles (twenty kilometers) or so in depth. "And if on Mars the events happen at greater depths, which some theories predict, then we would see less efficient generation of surface waves," Banerdt explains. "So we're really waiting for a larger event in order to see these surface wave trains, but time will tell."

The dream would be to match an event with the seismic signal it generates. That way the seismologists could calibrate their computer models for accuracy. Indeed, many on the InSight team pronounce themselves stumped as to what they are actually observing. As they record everything and watch how the signals play out, they want to untangle the various waves present within the data. "As they pass through the planet," Anna Horleston explains, "these vibrations pick up information—that is, depending on whether they pass through solids or liquids, the waves will be changed as they travel."

The waves are already presenting an interesting picture. Preliminary analysis of the shear waves observed so far haven't revealed that there is a liquid core but have started to hint what the outer layers of the planet may be like. "We are seeing a change in the waves when they pass between materials of different densities," says Dr. Horleston. That doesn't necessarily mean a change from liquid to solid, but "much more likely to be from a compositional change between different types of rocks."

More information will help them narrow down the size of the core at the heart of it all. "Knowing the thickness of the core is just a super important piece of information," says Sue Smrekar, "because it then allows us to really constrain the composition." Only certain materials will create any "phase transitions" seen in the way the signals are attenuated. "So it may sound like the size of the core isn't that much information, but it really is a huge piece of understanding the overall evolution of the planet."

The size of the core will reveal clues about how convection works within the Martian interior. Researchers can then calculate just what kind of material is actually present, and they can work backward in time to estimate the thermal evolution of the planet. They just have to be very patient. "We want to ensure that this geophysical observatory on Mars lasts as long as possible so that we can get as much data as we can to look into the planet," Bruce Banerdt says.

In the meantime, JPL used some venerable, ancient life-forms to publicize their mission. The Rolling Stones, now with a combined age of over three hundred years, gave a concert at the nearby Rose Bowl in late August 2019. Actor Robert Downey Jr. announced, on behalf of JPL, that a rock in Homestead Hollow had been named after the band. When InSight landed, its rocket jets blasted a small rock "about the size of a golf ball, and it had rolled along and had left a little trail," explains Banerdt. Downey "sort of officially presented this rock" to the band. Mick Jagger was delighted. "NASA has given us something we've always dreamed of," the singer said on stage, "our own rock on Mars. I can't believe it. I want to bring it back and put it on our mantelpiece."

* * *

InSight is carrying out another experiment that Bruce Banerdt says is as "close to magic" as possible while still remaining science. By plotting the radio signals sent from the lander on Mars, mission scientists can measure exactly how the Red Planet is rotating. The magic comes from just how accurate those measurements are. From a hundred million miles away, they can pinpoint the lander's position down to a few inches on the surface. Determining subtle details of how Mars wobbles from how the radio signal is changed over time will show how "the core is sloshing inside the planet." The wobble will reveal how big and dense the core is.

Previous radio measurements from the surface have helped pin down exactly how Mars rotates. For example, the axis of rotation (think of an arrow projecting outward from the poles) of the Red Planet shifts every 165,000 years along a cone shape, wobbling like a slowed-down spinning top. From watching exactly how those wobbles occur, it is possible to infer what sort of core may be found at the Martian center. Once that is known, the structure can be deduced. "Something where most of the mass is close to the center wobbles more frequently than if it is distributed over the whole body of the object," Bruce Banerdt explains. "You can think about a spinning skater: when they pull in their arms, the skater spins faster and faster. So with a planet, this tells us the size and the mass of the core. If it's a very large core, a lot more of the [planet's] mass is closer to the center."

As the Red Planet rotates, radio signals can be monitored with extreme accuracy by JPL's worldwide network of radio telescopes, the Deep Space Network. This tracking has been so precise that previous landers have been pinpointed down to less than a meter from where they were on the surface. Such accuracy allows the detection of minuscule changes to the rotation of the planet caused by additional CO_2 freezing out onto the winter poles of the planet.

The most recent landers have used a greater number of frequencies in order to make the measurements less susceptible to error. Since the 1970s when the *Viking*s landed, the direction in which Mars rotates has changed very slightly. This process is called precession, which is the change in the axis of rotation. The change in direction is an important indicator about how material is spread below its surface; the rate of precession is dictated by the distribution of material inside the planet. Known as the moment of inertia, it was a complete unknown before the *Pathfinder* mission in 1997.

By measuring the way that *Pathfinder*'s radio signal varied as Mars rotated and comparing them to the *Viking* results, geologists were able to infer the Red Planet had a crust, a mantle, and an iron core. It was also possible to make a rough estimate of how large a Martian core would have to be to have created the kind of rocks seen on the surface. Geologists have attempted to put "numbers" onto the interior conditions and temperatures within the core. But they are only educated guesses.

To date, the best guess is that the Martian crust is formed of iron, magnesium, aluminum, calcium, and potassium in a zone that stretches between six and thirty miles (ten to fifty kilometers) underneath the surface. Below that is a rocky mantle between 770 and 1,170 miles (1,240 to 1,880 kilometers) in thickness. Then in the center is the core, believed to be dense—a conglomerate of iron, nickel, and sulfur about 930 to 1,300 miles (1,500 to 2,100 kilometers) in radius.

InSight's radio science experiment is much more precise than previous missions, since the transmission technology has improved. So Bruce Banerdt expects the spacecraft will also be able to discern the subtle wobbles of the pole that happens in a little less than a Martian year. Assuming the core at the center of Mars is liquid, scientists will be able to measure its interactions with the solid, rocky mantle above it.

From InSight's initial seismic observations, there are hints of distinct layering immediately below the surface. The outermost layer is perhaps some five miles deep, which may be the extent of the broken-up zone of the crust. There may also be another layer, perhaps the base of the crust, at seventeen miles' (thirty kilometers') depth. Over the next year, further measurements will refine these early estimates from watching how the seismic signals have been attenuated.

InSight investigators also expect to learn more about the core's density. From that they will be able to determine just how much residual heat would have been left over to fire up plenty of geological—and possibly biological—activity over the eons. Knowing this value will also allow them to work out what chemical elements would be present. To some, the chemistry of the mantle, in particular oxidation, is important for releasing the kinds of gases and elements that are conducive to life. The state of the interior provides important clues about the early history of Mars. To gain a greater insight, investigators are tying in observations from other spacecraft that have also been scrutinizing the Red Planet from above.

* * *

The discovery of the magnetic field of Mars has been late in coming. Because the first successful mission, *Mariner 4*, did not detect one as it sped past in 1965, subsequent probes did not carry magnetometers. This remained a missing but crucial piece of the puzzle. In early 1989, a partially successful Soviet probe, *Phobos 2*, changed things from "maybe yes, maybe no" to "maybe yes."

A state-of-the-art magnetometer aboard NASA's Mars Global Surveyor, which arrived eight years later, was therefore expected to either see something—or nothing. What it found was both a revelation and the genesis of a new line of inquiry that has culminated with the measurements now being taken by the InSight lander at the Martian surface.

Within a month of its arrival in September 1997, the Surveyor happened upon a magnetic field, but one that was weaker than Earth's. The spacecraft passed through a substantial, permanent magnetic field that was global in its reach. In places, the observed magnetism contained some of the strongest localized fields ever seen in the solar system, including compared to Earth. "We were blown away by the results," said Dr. Daniel Winterhalter, a member of the magnetometer team at JPL. Indeed, they found that the Martian magnetic field was overall just 1/800th the strength of Earth's. It was not a complete field, rather the "memory" of one, concentrated into lumps of extreme magnetic field strength, forty times greater than the more uniform terrestrial field. Like giant magnets trapped within the crust, these magnetic "concentrations" were about a hundred miles across. They were distinctly odd.

"If you're a Boy Scout on Mars with a compass, you are lost," was how another NASA scientist declared the findings to the press that October.

These initial results were subtly encouraging with respect to the kind of environment that might have been capable of sustaining life. If Mars did have an active dynamo and a significant internal heat source in its earliest epochs, then it could have generated a magnetic field to stand off the Sun's influence. A field would mean a shield. That would have limited the damage caused by energetic particles from the Sun streaming down to the surface unimpeded.

In late 1998, the Mars Global Surveyor entered a much lower orbit than had been originally intended. Traveling only sixty miles above the surface, the Surveyor measured these magnetic anomalies much more accurately. As it came in low over the north pole, it detected nothing over the northern hemisphere. "All of a sudden, as we approached the equator we saw these incredible features," recalls Daniel Winterhalter. "They were about ten times larger than the Earth's field."

Along a band estimated to be roughly fifteen hundred miles (two thousand kilometers) in length, there appeared to be alternating bands in the direction of the magnetic field. It was a pattern that repeated itself up to six times. These local fields probably originated when the Martian core was still molten. That electromagnetic "dynamo" imprinted its magnetic field within these localized concentrations. As the molten material cooled, the surface rocks effectively acquired a "memory" of that primordial magnetization.

Clearly, not everywhere was magnetized to the same extent. Geologists have subsequently estimated that any surface features on Mars older than 4.1 billion years ago—very early in the planet's evolution—retained the field. Anywhere younger did not. The simplest explanation is the most obvious: the core must have frozen out around four billion years ago.

* * *

More recently, further details have been revealed from orbit by another dedicated NASA mission called MAVEN (Mars Atmosphere and Volatile EvolutioN). Since 2014, it has been examining how the Red Planet interacts with the Sun's influence, how that affects the Martian atmosphere, and how these localized regions of magnetism in the crust have a "knock-on" effect, too.

Mars has a weak field of influence. The Red Planet is bereft of the global protection that Earth enjoys today. Its weak magnetic field simply cannot prevent the Sun's energy from eroding the outermost layers of its atmosphere. Worse, the solar wind, a cocktail of lethal particles streaming from the Sun, can sometimes cascade all the way down to the surface. "On Mars, the solar wind can directly impinge on the Martian atmosphere," says Professor Bruce Jakosky, the project scientist for the MAVEN mission. "We wanted to know whether the regions of the magnetized crust might act like an umbrella and keep the solar wind from hitting the surface."

The answer is yes. The localized magnetism in the crust causes a bumpier flow of the solar wind around the Red Planet. "In some places, the solar wind can hit the atmosphere directly," Jakosky explains. "In other places, it can't." And that variation has produced something long suspected and finally confirmed in 2018 by MAVEN, that the natural light shows known as aurorae occur high above the poles of Mars.

Indeed, MAVEN has shown that where there is no localized magnetic field below, particles rush in where there are gaps. "We see discrete aurorae in the region between the magnetized crust where the particles can come in directly where the atmosphere isn't protected," says Professor Jakosky.

The outermost layer of the atmosphere, known as the ionosphere because of how the solar wind removes material escaping from Mars, is electrically conducting. *Ionization* refers to the way in which individual electrons are pulled off individual atoms. They leave behind electrically charged fragments (known as *ions*), which also disturb the magnetic fields.

As it interacts with this relentless solar activity, the Red Planet leaves behind a "magnetic tail" that extends away from the Sun. As a result, the solar wind has to flow around Mars on the side facing away from the Sun. "That produces a wake," Jakosky explains. "It's like the wake behind a boat, plowing through the water." As it does so, the energetic particles diffuse in from the sides "and fill in the wake behind it." Watching how that happens, MAVEN instruments have revealed how high-energy particles both escape and hit the planet.

"In other words, when the solar wind goes around Mars, it carries the magnetic field that is embedded in the solar wind," says Jakosky. "When the solar wind fills in the wake, the magnetic field lines come together to drive a process called *reconnection*." The magnetic field lines break and then recombine with the ones on the other side. "When that happens, it's like releasing a rubber band," Jakosky says. "It can snap the magnetic field lines. Those that snap outward away from the Sun drive the escape of material. The other one snaps back toward the planet and drives stuff back into the planet."

This phenomenon is unique to Mars. It propels some of the atmosphere into space; the "snapping" actively propels ions into space, twisting the tail as it does so. The interactions are complex, and MAVEN observations are ongoing. The aim is to work out how much this "reconnection" contributes to the wholesale removal of the outer layers of the Martian atmosphere, and how significant that might have been over geological time.

* * *

Around midnight, something else stirs in Homestead Hollow. It's not just the local wind but the local magnetic field, too, that is behaving in an unexpected manner. InSight carried with it the first magnetometer ever taken down to the Martian surface. And already it has found odd "pulses" of magnetic activity, technically known as "magnetic pulsations," in the middle of the night. Similar events occur on Earth, triggered by the solar wind. Here, they tend to occur at higher latitudes, linked to the aurorae of either pole.

On Mars, InSight is nearer the equator, so this phenomenon is an enigma, but possibly it is related to small-scale jiggles of activity deep within the planet's interior. One theory is that when Homestead Hollow is aligned with the magnetic tail—the wake described above—it could cause the fluctuations measured in the middle of the night by InSight.

This is another new chapter in research about the Red Planet. Maps of the crustal magnetization, observed from more than 60 to 160 miles (100 to 300 kilometers) up, have shown large-scale variations. From that altitude, image resolution is very poor, and no variations less than about fifty miles have been seen. "And even then, you have to iron out the variations due to changes in the spacecraft orbit which affect the measurements," Bruce Banerdt says.

InSight has also found there are small-scale variations in the magnetic field that are preserved in the crust on a much smaller scale than observed from orbit. Even stranger, the field strength in Homestead Hollow is completely out of whack with what has been mapped from above. Scientists use a unit called teslas, after the electrical pioneer, and on Mars, the readings are a millionth of that standard unit, known as nanoteslas. "The orbital measurements show about 250 to 300 nanoteslas," Bruce Banerdt explains. "We see about 2,000 nanoteslas from InSight. So it's almost ten times stronger than what was predicted from orbit."

For the seismologists, this unsuspected activity is yet another source to be eliminated from their own delicate measurements. The magnetometer helps "us to remove any noise from our extremely sensitive seismometers," in Sue Smrekar's telling phrase. "So we have a magnetometer that was really meant to measure electrical currents. That way, we can be sure that the solar rays powering up on the lander aren't creating signals in the seismometer that we mistake for marsquakes."

It could be that there is something in the vicinity creating these unexpected variations. They don't occur every day, and some seem to occur nearer to dawn. One theory is that over geological time, molten rock rose from below and cooled to form a new crust. "There's probably a lot more going on in the crust than we had anticipated," Smrekar explains.

Measuring how electrically conductive it is below the surface by modeling how the magnetic field behaves at depth will hint at the amounts of permafrost there may be. "How deep do you have to have to go to get to liquid water?" asks Smrekar.

Magnetic fields play an important role as a building block for a planet's evolution. As the planet solidified, the magnetic field that existed within these rocks and thus permeated whole swaths of the emerging crust might have been "frozen" into place. The surface rocks are too young to have been magnetized by the ancient magnetic field.

"We think [the signal] is coming from much older rocks that are buried anywhere from a couple hundred feet to ten kilometers below ground," said Professor Catherine Johnson of the University of British Columbia when the results were published in February 2020. "We wouldn't have been able to deduce this without the magnetic data and the geology and seismic information InSight has provided." Further analysis suggests some of these rocks were 3.9 billion years old—estimated from when they would have been buried—but this is a few hundred million years younger than when the core had effectively switched off. For now, more measurements are needed.

"As a lot of magnetic signatures are not being picked up from orbit, that has implications for how the crust was magnetized back when the Martian core was generating the magnetic field," Bruce Banerdt points out. "We are doing the analysis on it, but it is definitely providing new information about the magnetic history of Mars."

* * *

All these magnetic measurements, from the outermost reaches of Mars's atmosphere to its innermost internal layers, reveal a great deal about Martian evolution. When the Red Planet condensed out of the dust and gas from the primordial nebula of the solar system, its interior would have differentiated out. That is, the melting pot of planetary formation would see the heavier elements sink toward the core. It seems clear from the density and distribution of the magnetic field that Mars had a core, including iron, that was active and liquid.

"We have a fuzzy idea of what the core looks like," says Bruce Banerdt, "but InSight should really help us sharpen that up."

To date, the core's size has been estimated by comparing the strength of the observed gravity field of the planet with the average density of its mass. Even before InSight, it was known that there is just too much mass for there *not* to be an iron core. "But the exact size has large uncertainties," adds Banerdt. "InSight will measure the size much more precisely, using radio tracking and, if we are fortunate enough to get a large event, using seismology."

Within the first half billion years, the Martian core effectively "froze out"—that is, it started to solidify and was not available to churn and continue to fire up geological activity. The searing heat within would have been maintained by the natural radioactivity of the primordial elements as well as by the extreme conditions of compression. Although there wasn't enough heat to drive the convection within the mantle and break the surface into the jigsaw-like patterns of plate tectonics, there was sufficient left over to generate volcanic activity until comparatively recently.

"Volcanism results from the melting of the upper mantle rather than the core," explains Professor Bruce Jakosky. "That melting can still occur due to decay of radioactive materials. So it is not at all surprising to still get volcanism up to the present time."

Although the core probably solidified early on, the crust above it was much thicker, as it would have cooled for longer. That would have allowed volcanic activity to carry on unchecked, which buttresses the notion that the Red Planet may have been conducive to life early on. "It's not a stupid idea to ask whether there is life on Mars today or that there could have been in the past," says Jakosky. All the latest evidence is pointing in that general direction. "It doesn't mean that it's going to be there," he adds, "but it doesn't mean it won't be."

Indeed, other landers have found that the same basic ingredients in the biochemical recipe for life on Earth were present on Mars. It had a much denser atmosphere in its earliest history and possibly shallow oceans and volcanic heat sources, which support the notion that conditions were more likely to hold life.

Yet the Red Planet underwent the climatic equivalent of a sudden death. Understanding how this happened is nothing less than an astronomical detective story. Today's researchers are homicide detectives in a sense, each with their own pet theories about the strange death of the world next door. During the violent birth of the solar system, much debris was left behind. Mars, like Earth, was repeatedly struck by large asteroid-sized bodies. A greater proportion of the atmosphere that emerged in this earliest incarnation of the Red Planet would have been blasted off into space. The remaining atmosphere would have trickled away over time.

Where geological activity today continues on Earth, on Mars it did not. Over time, Mars could no longer sustain large-scale volcanic activity. The Red Planet would not have been able to effectively recycle carbon dioxide between the rocks and its atmosphere. When the volcanism ceased, the carbon dioxide would have become permanently locked up in the rocks. As conditions got progressively colder, the planet simply froze to death.

An emerging theory that is well supported by the latest evidence from NASA's MAVEN spacecraft suggests that the Martian atmosphere slowly bled away. Gaseous molecules were effectively knocked into cosmic oblivion as the Sun's energetic particles hit the atmosphere.

If Mars has always had as weak a magnetic field as it does now, then a gaseous blanket as thick as Earth's atmosphere may have been lost to space over the lifetime of the Red Planet. This will have had another considerable impact on the environment and conditions for life.

* * *

A year isn't long in Martian evolution. After InSight's first year of operations, tying in the findings from the InSight magnetometer with those of the broader MAVEN investigations is still in its infancy. The MAVEN orbiter, unfortunately, does not always fly over Homestead Hollow. Obtaining simultaneous observations from the ground up to orbit is a rare treat. When they are taken together, however, they show the flow in glorious detail. "We see the effects of the incoming solar wind that drives the changes in the magnetic field," Bruce Jakosky explains, "and how it is propagated down through the ground."

The aim is to tie what is happening within to what is happening without. These new observations show how the solar wind is driving the escape of the upper atmosphere. By looking at how the magnetic field is temporarily affected all the way down to the surface, the research teams can model what has triggered those changes. This will help them understand what kind of material is just below where InSight landed. "If we have a piece of ground and change the magnetic field it is sitting in, that will also propagate below the surface," Jakosky says. The observations of seismic and other activity are giving a clue as to how the magnetic field is changing immediately. Over time, observations of the field changes should allow the seismologists to probe down to about seventy miles' depth.

Similarly, the InSight teams say they will need at least two years of radio observations to narrow down the size of the core. In Sue Smrekar's judgment, "we haven't seen a big enough marsquake yet to give any supporting information."

Yet already a broader picture has emerged. "Mars is not as seismically active as the Earth," Bruce Banerdt explains. "And we're just starting, with the first few months of data, to get a rough estimate of the level of seismic activity on Mars."

One of the major conceptual difficulties is that they only have one seismometer location. Normally, a network would be used on Earth. "It's always easier with more," says Banerdt. Here seismologists have built up a three-dimensional picture of our own planet's interior. With several stations on Mars, they could pick up "local events" from many different locations. "And that's how we do seismology on the Earth," Banerdt explains.

So having one station entails a lot more waiting, especially on a planet that is clearly less seismically active. "That means we have to wait a long time, especially for the larger events. They happen very infrequently, and once a larger event occurs, we want to squeeze as much information out of that data as possible." Ultimately, seismology is a statistical science. "You put together your scientific conclusions based on an accumulation of seismic data," Banerdt says. "The more statistical information you get, the better your statistics are."

That greater haul will help pinpoint exactly where the marsquakes are coming from. In time, the hope is to trace greater numbers of "ray paths" from the way the seismic waves have moved through the planet to determine their origin more accurately. Bruce Banerdt is at pains to emphasize that it is very early days yet. The seismologists want to get everything working, including the mole, and then collect as much data as possible. "So we have a two-year mission planned," he says. "We still have the majority of our mission to go."

* * *

Taken together, these new findings show that all the ingredients for life were available on the youthful Mars: water, an internal dynamo, and extensive volcanism. Life requires energy, and it thrives in environments where energy is freely available. Volcanism would have provided that in plentiful supply. What that implies is something NASA's Christopher McKay has argued for years: the Red Planet really was a better planet upon which to incubate and sustain life.

Conditions on early Mars would have allowed life to form. There remain gaps in knowledge, caveats, and qualifications. Certainly, as the climate of Mars degraded, life might have hibernated far below the surface. There it will have remained, even if it only took a few faltering steps along the complex road that leads to biochemistry. If there are such things as "oases" for Martian life, subsurface permafrost will most likely sustain them—now, and in the distant past.

If life could evolve on Earth with all the various upheavals it had the indignity to suffer, then it could just as easily have survived on Mars. The Red Planet, where intense geological activity petered out far sooner, would have been that much more attractive.

"There is a dichotomy in the environments life survives in and where it can start," says Chris McKay. "Mars is a much better incubator as it was a much kinder, gentler place than the Earth was. The first billion years of the solar system's birth is one of the most unsolved riddles in astrobiology. On Earth, 4.5 billion years ago is the Moon-forming event, which really wipes out everything."

In a sense, Earth must have reinvented itself to have allowed life to flourish sometime later. Mars, however, was smaller, so there was insufficient heat to let that churning continue up till today. Lighter rock separated from heavier rock below, powering the dynamo that was its core. Mars probably had a similar history to Earth up to the point at which plate tectonics began to separate out.

In its earliest epochs, Earth would have been far more volcanically active than it is today. All over the surface of our world, molten lava poured out onto a steamy landscape still bearing the scar tissue of prolonged cratering. In places, that watery covering would have seeped deep into the crust and, superheated by the emerging lava, may have exploded upward and flowed out toward the surface. "Everything we know about life had been set in place by the late bombardment," says Chris McKay. "Mars may well be the missing link in the chain."

* * *

Talk to the older generation of Mars researchers, and some even today talk darkly about what was once called "the Martian horror story." For many veterans, the premature emphasis on biology in the early years of exploration, the headlong rush to detect life at the expense of other less glamorous investigations, was not very sensible. "Viking was predicated on the hope that the answer to life on Mars would be sitting there on the surface, ready for us to swoop in and pick it up," recalls Bruce Banerdt, who was a graduate student at the time. "But like most of the big important questions, the answers don't reveal themselves that easily. They are more often teased out by systematic, careful investigation covering many possible facets."

The fundamental mysteries that remain unanswered about the Martian interior are proof of this. Sometimes the more prosaic questions help frame the glamorous ones more effectively. So, until comparatively recently, information on the spread of minerals across the surface, the chemistry of the atmosphere, how that reacted with the soil, and, indeed, broader questions about the greater Martian environment were not forthcoming.

Now, with the first dedicated mission looking at the interior of Mars, questions of biology can be asked afresh. A new vision is coming into focus. Over the next decade, a host of missions will be scrutinizing the Red Planet in novel ways. Our view of Mars has changed and will no doubt change again. Thanks to InSight, we are examining the crucial links in the story of how life may have evolved on the Red Planet. Mars InSight is removing much of the mystery surrounding the role of the interior in the planet's evolution. Once that knowledge is in place, the larger question of life can then be addressed. That is the aim for the next American mission, due for launch in the summer of 2020. The Martian horror story has ended.

3

CURIOSITY

I f ever there was an unlikely face for exploring Mars, surely it belongs to Adam Diedrich Steltzner. With his Elvis quiff, pierced ears, and snakeskin boots, he doesn't exactly look like a rocket scientist. But as the lead engineer for safely landing the heaviest machine ever sent to another world, he had his work cut out for him. When the Curiosity rover dropped out of the sky on Mars, Steltzner became synonymous with "the seven minutes of terror"—as it became known—he had pioneered. Only one question remained. Would it work?

On the day of the landing in August 2012, on a live feed from the Jet Propulsion Laboratory, Adam Steltzner was clearly in his element. He describes himself as having been a "listless, restless wannabe rock star" when he was younger, but one night when he was driving home from a gig over the Golden Gate Bridge, he noticed the constellation of Orion. He wondered why it appeared in different positions in the sky on subsequent nights. It started him on a course at community college, engineering excellence, NASA's Jet Propulsion Laboratory, and ultimately, the Red Planet.

With the cost of the Curiosity mission totalling $2.5 billion, the prospect of potentially losing the mission had reportedly caused President Obama's science adviser sleepless nights and to become physically ill with worry. Though Adam Steltzner had complete faith in his and his team's abilities, there was constant anxiety during the rover's development and a sinking feeling they might have missed something. "All of that was still churning through our minds," he says of the night before the landing. "And so, I didn't expect to sleep a wink, and I slept like a baby. I'm like, 'Wow.'"

To prepare for the landing, the engineers had turned off the "ears" of Curiosity's telecommunications system, essentially locking it down. "We command the spacecraft to stop listening to Earth," Steltzner explains, "because we don't want an errant piece of uploaded code or somebody's transmission to confuse it." And then, with his trademark grin, he adds, "It's either going to successfully land or not. We can't change anything."

* * *

The Curiosity rover was hardier and more capable than anything ever before dispatched to the Red Planet. One NASA higher-up had called it a rover on steroids. Its sophisticated instruments—cameras, lasers, and spectrometers—along with a self-contained chemistry laboratory were designed to take scientific investigation to a new level. But all those new instruments ushered in a fundamental, intractable problem. Curiosity weighed just over a ton. That made it five times heavier than its immediate predecessors (and, indeed, than the more recent InSight lander). With that amount of mass, there was no way Curiosity could simply come in on a parachute alone. Airbags were out, even though they had been successfully used on many JPL-built Mars landers. They were too heavy, and the launcher wasn't powerful enough to accommodate them.

They *had* to do something new.

The entry, descent, and landing sequence, over which Adam Steltzner was in charge, was so complex and daring that it seemed too bizarre to be practical. With words that are richly ironic for someone who had flunked geometry at school, he would shrug and claim, "We look for the simplest, lowest-risk way of solving a problem." There were, he says, no sacred cows and certainly no dogma. Ripping up the rule book was expected. Steltzner thrived on the sheer impossibility of it all.

So what became known as a "sky crane" was developed to bring the rover, its six wheels already extended, to a soft landing on the surface. As it descended, Curiosity jettisoned its heat shield, reeled out a cable, and the rover was pulled out below, ready for touchdown. This "sky crane," similar to the gantries in work yards, would lower the rover gently to the ground. The crane would then automatically cut its landing cords once the rover had come to rest. The crane mechanism would then smash to the ground some distance away.

Several people thought it was crazy. Yet Steltzner maintains that when the team knew what to really focus on, the chances for success shot up. "Once everybody understands what is required and essential and must-do," he reflects today, "it's pretty easy to have them keep focused on the most important elements. That means not being distracted by things that are not that important."

Late on the evening of Tuesday, August 5, 2012, Adam Steltzner paced the floor in a tiny mission operations room at the Jet Propulsion Laboratory. Everything relied on automatic sequencing. If something went wrong, "there was nothing we could do," he says. "All of us were just watching."

The Red Planet is so far away that a single, one-way transmission takes almost fourteen minutes to reach Earth. Rationally, if not emotionally, Steltzner had accepted they were just spectators. "We're already alive or dead on the surface of Mars. It's already happened, and we're just watching the tape delay of the game, as it were." It was a very weird moment.

"It was the culmination of a decade of investment by over a thousand people. There's a lot of pressure, but again at that point, you have nothing that you can personally do about it."

* * *

Adam Steltzner's coping mechanism was on display for the rest of the world to see. Reports from his team filtered through his headphones; his work pass, dangling around his neck on a red lanyard, jangled with each step. "I was walking between three screens showing three different elements of data," he says today. "Mostly I'm walking, because when I get nervous I can't sit still. We were supposed to be sitting still, but I am a pacer, and I needed to see those things."

Squinting at this digital triptych, Steltzner picked up the pace as the raw engineering data came through. With the atmosphere of Mars perilously thin—at the surface the pressure is the same as 130,000 feet above terrestrial sea level—even minor changes could affect the incoming trajectory. Pressure changes are also unpredictable. The atmosphere balloons outward and condenses back down with alarming speed (also reflected in pressure drops measured at the surface). A cold snap, a faulty reading, or a miscalculation would have already meant the difference between success or failure.

When Curiosity was low enough, roughly 1.2 miles or two kilometers from the surface, it popped out a parachute that was 10 percent larger than those used on previous landers, an innovation in itself. Exactly like Buzz Lightyear, Curiosity was falling through the sky with grace and ease. Broadcast live, Steltzner seemed to be reacting to a foreign language.

"Tango delta nominal."

"RIMU stable."

"UHF good."

The clipped acronyms belied their significance in this precisely choreographed sequence. The cable had unreeled. The rover was dangling above the ground. Then there was a moment of disbelief, when Curiosity had either crashed or made it. Steltzner pointed toward the back of the room, a question framing his features.

And then it came. The room of controllers, all dressed in identical blue polo shirts, erupted in cheers and whoops in the soft Californian evening. Steltzner wrapped his arms around the nearest teammate, enjoying a moment of elation. They had done it. They had pulled off the impossible.

Now the rover was ready to start its work and live up to the promise of its name. Adam Steltzner's simple appreciation—"I got curious and I followed that curiosity"—has guided both him and the machine he helped build in all their subsequent work. "The outgrowth of that curiosity is our desire to explore that which we do not know. We hope that all of our missions are filled with those twists and turns where we discover that we don't know really what's going on, and that there is something new to be understood and to unfold in front of us."

* * *

In the current search for life on Mars, Curiosity's landing site represents ground zero. Within a few hours of landing, the rover was tootling off and exploring the floor of Gale Crater. Located in Elysium Planitia, it is roughly four hundred miles south of where InSight later came to rest. "We can actually see the same clouds, which is kind of fun," says one scientist involved in both missions.

The crater was gouged out by a fiery cataclysm in the ancient past, an impact so powerful that it excavated, at its deepest, a hole three times the depth of the Grand Canyon. The immediate shock wave formed a peak at its center. Informally christened Mount Sharp, after a pioneering geologist who studied Martian landforms, it is unique. It rises out of the crater floor to just short of the commanding heights of Mount Rainier, which towers above Seattle. Mount Sharp is 18,000 feet (roughly 5,500 meters) high. It has acted as a repository for all the geological activity that has subsequently taken place inside the crater.

Gale is what geologists call "a complex crater." From rim to rim, it is ninety-five miles (154 kilometers) wide, which would take the average family car a couple of hours to cross. For Curiosity, "it's probably an eighteen-month drive if we just turned on the afterburners and had gone for it," Adam Steltzner notes.

Water has played a key part in the crater's evolution. Lakes have repeatedly filled and left sediments behind. The biggest puzzle was how the central peak has been sculpted by the wind to "get the shape of the mountain somehow," in the words of one baffled geologist. Its formal name is, significantly, Aeolis Mons, after the ancient Greek word for wind.

Curiosity landed on the crater plains to the north of the mountain. For the first half of the mission, the rover drove parallel to the extensive dunes, which were potentially dangerous as sand traps, before skirting up through the layers that form the foothills. It has since been exploring layers of material deposited by water over geological time.

Almost a day after the landing, the engineers ceremoniously handed over the keys to the rover. Adam Steltzner marvels at how the scientists, too, have broken the rules. "What is the first thing that they do?" he asks rhetorically. "They go ninety degrees to the plan. The plan was to go due south, and they go due east as they saw something interesting." A few weeks later, as Curiosity headed east, there came one of many exciting single finds from the mission. The rover stumbled across evidence that an ancient stream had flowed, complete with rounded rocks.

Flexibility has informed their subsequent exploring. Steltzner uses the famous Dwight D. Eisenhower wartime quote: "Planning is everything. The plan is nothing." Heading east into the foothills of Mount Sharp was exactly the right thing to do. The scientists were not enslaved to the plan itself. If it sounds a lot like luck, not so. The team should always be trusted to make the right decisions, he says: "If you're not awake to what the future really is, if you force yourself to believe that the future is what you expected it to be, you miss an opportunity to be competitive, to be innovative."

And that, in a nutshell, is what the mission has been doing. *Exploring.*

* * *

Dr. Ashwin Vasavada also marvels at the ingenuity that has made Curiosity's journey into the unknown possible. "We really are just tiny human beings using all our intellectual strength to sling ourselves virtually to other worlds," he says. As project scientist for the rover, he has had a ringside seat for what is without a doubt the greatest journey of exploration in his lifetime. It was the Red Planet that first got him interested in science when he saw a Viking photograph in a book. "For the first time, I really noticed planets were other worlds," he says. "It blew me away."

After studies at UCLA set him on what he calls a winding road, Vasavada slowly inched his way toward the Jet Propulsion Laboratory. He worked on Curiosity for seven years until the launch in 2011. That event made the whole thing "much more real." By that point, NASA had successfully landed a trio of rovers on Mars. The first, Sojourner, was a small demonstrator in 1997 that gave NASA the confidence to plan ever more complex machines. Seven years later, two separate rovers named Spirit and Opportunity were dispatched. They were the scouts that lasted way beyond their planned lifetimes. Though their warranty was for ninety days, they remained active for six and fourteen years respectively. Since 2004, thanks also to Curiosity's subsequent landing, there hasn't been a time without a working rover on Mars.

Dr. Vasavada's hometown of Stockton, California, also produced the acclaimed musician Chris Isaak. "He's sort of the biggest thing to come out of our little high school," Vasavada acknowledges, but the two never crossed paths. Neither did Spirit and Opportunity. After assembly in the same building at JPL, they departed at separate times. They landed thousands of miles apart on opposite hemispheres of the Red Planet. As they moved around, they literally shook, rattled, and scratched any interesting rocks they rolled up to with their mechanical arms. They took detailed images with panoramic cameras and the first-ever close-up, magnifying lenses sent to Mars. Each had three spectrometers that were able to tease out the mineral composition of rocks and soil. They used a rock abrasion tool that ground out the inside of stones for analysis. They also employed magnet arrays to pick up airborne dust.

What Spirit and Opportunity found was the surface expression of how water had pooled on Mars in the ancient past. The nuances and details were different on the respective hemispheres where they had come to rest, but taken together they spelled out one irrefutable message. The Red Planet had undergone wet periods that had endured for considerable time. "And then once you've determined that liquid water was present and persistent on ancient Mars," Vasavada explains, "you can follow that up with a broader payload designed to look at habitability"—whether conditions would have sustained life—"and you could jump to look for signs of life, but at the time, in the early 2000s, it was still thought to be premature in terms of technology."

Looking for where life might have formed was Curiosity's mission. But where to land? It is the dilemma for every mission sent to the Red Planet. Though smaller than Earth, Mars presents about the same landmass as our world, whose surface is for the most part covered by ocean. There are an awful lot of places worthy of investigation.

Determining where to land begins with scouting from above. The powerful cameras aboard the most recent spacecraft, in particular the Mars Reconnaissance Orbiter, have discerned features as small as ten inches (twenty-five centimeters) across. Where once astronomers had to attach spectrometers to telescope eyepieces, today vastly improved instruments in the vicinity of the planet can pick out the telltale signs of individual minerals that cover proposed landing sites.

If persistent water sat in any region, it is clear from even a couple of hundred miles above, especially from the signature of hematite, an oxide of iron formed in the presence of water. Hematite has a unique spectral fingerprint and is a significant component of the "rusty dusts" that give Mars its characteristic red color. In 2010, the Mars Reconnaissance Orbiter detected hematite within Gale Crater. Four years later, Curiosity confirmed it in situ by exploring an entire layer of hematite a quarter mile up the slopes of Mount Sharp.

Hematite is not the only evidence that water had pooled in Gale Crater. From orbit, layers of clay have also been spotted. There appear to be channels both in the wall of the crater and all along the edges of Mount Sharp itself where water flowed. Because of this, Curiosity has slowly but surely been making its way up the first fifth (just over 3,000 feet) of that three-mile-high mountain. Vasavada and his colleagues back on Earth are attempting to work out when these layers were deposited and how much water was involved.

As it has moved on up toward the central peak, Curiosity's instruments have identified a collection of minerals that *must* have formed in water that flowed through Gale Crater. The rover has subsequently observed how the layers of sediments were laid down. Ashwin Vasavada notes, "What we've now found is that the record of lakes is relatively continuous and has not stopped for a thousand feet or more."

* * *

Follow the water.

If there was a simple way to describe why Curiosity had come to rest in Gale Crater—and, indeed, the overall strategy for JPL's current Mars exploration—it was to examine, in great detail, evidence for how water had once flowed across the surface. Where there was water, life might have evolved. It could support the complicated biochemistry needed to incubate life. The nearby volcanism showed that energy was plentiful enough to sustain metabolism. Within months of landing, Curiosity came across an area that was christened Yellowknife Bay, which, at least in pictures taken from orbit, appeared to form the end of a river system fanning out into what looked like lake deposits.

"When we got there, we found a landscape that was unlike anything ever seen by a Mars lander," Ashwin Vasavada recalls. "It was not the typical sand and gravel and boulders but a huge flat expanse of bedrock, which immediately strikes you as something that looks like a dry lakebed on Earth."

When they got a closer look, the rover found the surface was made of very fine-grained rock, the kind of sediment that forms as mud collects at the bottom of a lake. Even more revealing, it seemed to have formed in a stable, standing body of water. Adam Steltzner spells out what that actually meant. "The scientists had enough information within six months to know that the early wet Mars was habitable for life," he says. "That's pretty incredible."

That was where Curiosity first deployed its drill, making two small holes in the Martian soil, an achievement in itself, as it was the first time a rover had ever drilled into the surface of Mars. The rover determined that the rock was made up in no small part of clay minerals, around one part in five. The sediment was rock that had been eroded elsewhere, then delivered by the water that had flowed into that lake. As the sediment was buried, its temperature would have increased, "cooking" the rocks to form clays. "We saw fine-grained mudstones which came from some sort of standing bodies of water," says Professor John Bridges of Leicester University in the UK, one of the many hundreds of scientists participating in the Curiosity mission. "After that there was no debate, it was a significant find."

Since then, much more has been learned about what happened to the water in the ancient past. "The streambed and the weathering of the pebbles within the streambed gave us a hint that the stream went back to the length of time that at least that portion of Gale Crater was wet," says Steltzner. "The chemistry in the deposits at Yellowknife Bay told them the pH and salinity were acceptable to support life."

Over the last five years, Curiosity found that the bottom half-mile of the mountain lays out a rich geological history. What Vasavada terms "an enigmatic layer at the base"—"enigmatic in the sense that it just didn't have any actual signature from orbit"—took them by complete surprise. Next they saw a hematite ridge, "with a strong signature of iron oxide," then areas that started to contain clays. Over the year 2020, the landscape will change to more sulfate minerals. And that, Vasavada explains, "will close out everything we ever hoped to get out of Mount Sharp."

* * *

Driving on Mars isn't just a case of turning on the engine, putting the clutch in, and letting the interplanetary jalopy go off for a spin. Roving across the Martian surface comes with considerable challenges. It is akin to playing a complex form of chess. If controllers don't think carefully about their next move, the rover could stumble blindly into a sand trap or some other situation that puts it in danger.

They have already learned the hard way. Both the Spirit and Opportunity rovers got stuck in the sand. The Jet Propulsion Laboratory, naturally, has a backup for when such situations arise. A team of engineers is always available to look at how to free the rover, through modeling and re-creating the exact same situation in JPL's Mars Yard, a kind of sandbox for adults where real-life backup rovers roam. These identical twins of the rovers doing the real work on Mars are crucial.

In 2005, engineers freed Opportunity from a dangerous sand trap after several weeks of planning wheel turns and other maneuvers in the Mars Yard. Spirit, however, got permanently stranded inside a soft area of sand in 2009. It effectively died there a year later, unable to move to a more favorable spot to recharge its solar panels during the winter. As a result, JPL engineers became smarter in operating Opportunity; indeed, the rover exceeded a marathon's distance of driving on April 15, 2015.

And it was still in rude good health when the fiercest global dust storm on Mars in recent memory rolled through three years later. Opportunity shut down into hibernation mode. Because the dust restricted the amount of sunlight filtering down to the surface, not enough power was generated by its solar panels. Despite repeated wake-up calls, Opportunity never responded. Though one project scientist called it "a very tough vehicle," by February 2019, NASA pronounced the rover dead, a sad and untimely fate for such a plucky pioneer.

Clearly, driving on Mars isn't as easy as just telling the rover where to go. There's the time delay, for one thing. There are several minutes of lag between a command being sent from Earth and when it is received on the Red Planet, depending on the relative positions of both planets. By the time controllers realize there is a cliff on Mars, it could be too late to do anything about it. And if the $2.5 billion investment in the rover technology is crippled forever, it won't just be the presidential science adviser who can't sleep at night.

So Curiosity is equipped with some artificial intelligence to keep it out of trouble. It has a fault protection system that senses when the rover may be falling into danger—for example, when it is slipping too much compared to how fast it *should* be moving. It can also tell if the wheels are stuck from the amount of electric current running through its motors. That electrical flow is a useful indicator for the amount of resistance the rover is encountering. If it is getting stuck, the electric current will fluctuate unexpectedly.

In case of dire emergencies, Curiosity can just sit it out and wait for help from Earth. If the problem is more minor, the rover can skip the problem and move on. Controllers don't want the rover idling just because one of its instruments isn't working properly. A problem with the drill back in December 2017 meant there followed what is termed "forced restrictive planning." Traverses were purposely outlined in two-day increments, instead of the usual one day. That gave the ground-based designers greater time to troubleshoot the drill.

In the JPL Mars Yard, engineers worked out how to fix the problems using an Earth-bound Curiosity twin. Unfortunately, the drill proved balky for much of 2017 and part of 2018 due to a failed motor. Fixing it required almost two years of troubleshooting from Earth, which limited scientific inquiry.

A number of instruments require the drill to "feed" them with ground-up samples for analysis. When the drill worked, some of the first samples were delivered to the Chemistry and Mineralogy (CheMin) instrument, an analytical laboratory technically known as an X-ray diffractometer. What that means is it essentially wallops the sample with X-rays to force its contents to fluoresce. Measured accurately, the patterns reveal the energies that are bound up within the atoms inside, allowing the scientists to determine the chemical composition of where the sample was taken.

The minerals observed show what the environment used to be like.

For example, olivine and pyroxene (from basalt, essentially fine-grained volcanic rock) are components of lava. Jarosite comes from sedimentary rocks, which are precipitated from water. All have been detected by Curiosity as it has trundled around Gale Crater. The extent of these and other minerals reveals that the lakes in which they were deposited "weren't just ephemeral or temporary," in Ashwin Vasavada's estimation. "There were lakes for tens—perhaps, hundreds—of millions of years."

* * *

Abigail Fraeman is typical of today's professional Martians. One day in suburban Maryland, where she grew up, her father brought home a telescope. Like so many others, she was quickly hooked. At seventeen, she won a prestigious student competition to design a rover at JPL and later worked for a summer at the Carnegie Institution for Science in Washington. There she looked at what she has called "three squiggly lines" in the spectrum of a star suspected of having a large water-rich planet orbiting it.

Fraeman also met an extraordinary pioneer whose name now lives on Mars. Vera Rubin was a pioneering astronomer who first proposed the existence of dark matter and worked in the same building. Rubin died in 2016 at the age of eighty-eight, but not before a ridge of hematite ("I was the campaign lead," Fraeman says with a smile) in Gale Crater was named for her due to singular contributions to science. In the meantime, Abigail Fraeman is carrying on Rubin's legacy, blazing the trail for other explorers to follow on the frontiers of knowledge. After studying geology in college, Fraeman has wandered virtually around Gale Crater as part of her doctoral program. While many PhD students end up stuck in libraries, Fraeman is now on staff at JPL, where she works with the planning team to figure out what Curiosity will do next. To her delight, she has played an important role in piecing together the crater's rich geological history.

JPL sends sets of commands up to Curiosity every day or few days. Before they do, Fraeman and others have a strategic planning meeting to decide what they want to explore. There is always a trade-off. Time is precious on the Red Planet. The team must decide what is most worthy of examination. The rover could spend years at a single location, but its mission is to survey as much variety in the Martian terrain as possible. That means capturing the main features of a single site, then moving on to the next.

"We have people called long-term planners who work on a long-term traverse," Fraeman says of these extended road trips. "You have knowledge of the terrain, physical properties, and then figure out, 'Okay, in the next few weeks and months, where do we think we want the rover to go? What units did we identify from orbit? Do we think we want to visit on the ground?' And how is that all going to fit in terms of what we do in the next few months."

One of the greater challenges is reaching a consensus. The rover cannot move in opposite directions at once. They have to pick the target of interest well, based on the data gathered so far and what they want to do next, so as to better investigate a particular hypothesis. Is it worth staying at one particular location longer to gather more information, or is it better to move to another place? These questions must be figured out day by day, week by week, year by year. Otherwise, the rover would just stand still while the scientists argue among themselves.

"For example, if there's some bedrock in front of us, but we've been studying bedrock every day for the last two months and it's all the same, but [we see] this really interesting vein," Fraeman explains, "that day we might say, 'We're gonna want to look at the vein [instead of] looking at the bedrock.' Of course, if there's anything super-compelling, we'll change our plan. 'Tomorrow we wanted to drive, but this is too cool. Let's stick around here for a little bit longer, just to make sure we really do a thorough job collecting all the data we need.'"

Older rover traverses were planned with help from above. That still takes place—the Mars Reconnaissance Orbiter passes less than 125 miles (200 kilometers) overhead—and remains a key part of identifying where Curiosity has been traveling across the surface. But computer technology is finally catching up with what the movies have been suggesting for years. Using pictures from Curiosity, JPL can place a virtual rover inside a virtual landscape that exactly replicates what the rover can see. Then, traverse planners like Fraeman can put on a virtual reality headset and, in an instant, be transported to Mars. They see what the rover sees. They can walk around to observe the most promising targets, as well as the hazardous areas to avoid.

Even as a PhD-level geologist, Abby Fraeman has her work cut out for her. She works with the engineers to assess the potential hazards in the local terrain. Foremost among them are slopes: the rover can only traverse 20°, depending on the surface type underneath. The sandier the terrain is, the tougher it's going to be for the rover's wheels to get a grip, as occurred in the summer of 2014. Unexpectedly, the rover came across what was christened Hidden Valley, an area of surprisingly slippery sand that covered the extent of a football field. Rather than risk problems with the wheels, the rover was ordered to turn on its heels. "That was way sandier than the rover could handle," Fraeman says, "so we learned quickly the kinds of terrains and the sandiness the rover can drive over and what might be a little bit dangerous."

Curiosity also needs to worry about rock hazards. Technically, it can crawl over obstacles as high as twenty inches (fifty centimeters), or about the height of the seat on a standard dining-room chair. In practice, however, early on in its mission, Curiosity developed holes and punctures in its wheels at a rate that worried JPL controllers. The rover had been spending more time on "harder" bedrock than had been anticipated. Further analysis shows that the wheels are suffering from a combination of metal fatigue and punctures from sharp rocks. Between the bedrock and the pointy stones, the JPL team are now looking for the smoothest places possible for the rover to roam.

"There are some algorithms that the rover drivers put on board the rover that do the traction control that mitigate the forces between the wheels and the rocks," Fraeman says, "but then as geologists too, we also look at the rocks and the field around them. We try to identify the ones that are potentially the most hazardous, so these would be sharp, hard rocks that are really well-embedded in the terrain. Those are the worst types of rocks for the tires in terms of wheel damage."

Ironically, the tire problems have been exacerbated by a typically JPL element of subversion. Before Curiosity's launch, NASA Headquarters declared that the only decals to be shown on the rover were to be those of NASA and not JPL. A few weeks after the landing, however, a few higher-ups in Washington, DC, started to have conniptions when they realized the tire tracks spelled out in Morse letters, as the rover trundled along, the symbols for J, P, and L. Those Morse indents embedded in the wheel track have added to the wheel damage. For now, the controllers don't want to tempt fate, so they tend to hedge their bets on where they're sending the rover. They will be taking radical measures to improve the wheels on the next JPL rover to be launched in the summer of 2020.

* * *

Curiosity allows controllers to do what they would do themselves if they were standing on the surface of Mars: scan, poke, prod, and look for interesting features. The rover carries a remarkable tally of seventeen cameras; most are used to avoid hazards and for navigation. While one camera was used for the descent down to the surface (to determine where exactly the rover had landed), another is used for extreme close-ups, and there is a dedicated camera for panoramic views.

The rover has steadily been gathering information about the local environment. One of its radiation instruments has shown that humans who one day will travel to the Red Planet won't receive a devastating total dose of radiation, at least at first blush. Another radiation detector looks for how much water lies immediately under the surface. This instrument, called Dynamic Albedo of Neutrons (DAN), measures the escape of neutrons, which form the heart of atomic nuclei, from the surface of Mars. By analogy, it shines a light on what is happening directly underneath the surface. When neutrons hit a hydrogen atom, they slow down by a distinct amount of deceleration. As water contains two hydrogen atoms for every oxygen atom, looking at how the neutrons react is revealing. Subsurface water on Mars can be estimated from exactly how the neutrons slow down.

The DAN instrument is directed downward and sends ten million neutrons with each pulse. It then looks for the "slow movers," which reveal water-borne minerals as well as the presence of water ice. How they are reflected is useful in measuring not only hydrogen but chlorine and water content in the soil all along the rover's path. The chlorine is there in trace amounts, delivered to the Red Planet over the eons by meteorites from space that have gouged out craters like Gale.

More data on everyday conditions comes from a Spanish weather experiment called Rover Environmental Monitoring Station (REMS). This measures conditions in the surrounding atmosphere and sends regular weather reports back to Earth. Wind speed and direction are measured via a pair of small booms mounted on the rover's mast, which also measure air temperature. Another sensor on one of the booms picks up infrared radiation (a proxy for ground temperature), while a sensor on the other boom measures atmospheric humidity. The rover's electronic box can pick up atmospheric pressure changes as well as ultraviolet radiation from the Sun.

Taken together, all this information—the pictures, the radiation data, and the weather—reveals details of the here and now in Gale Crater. To piece together the story of what happened in the ancient past requires a more sophisticated form of investigation, so there are a series of several innovative spectrometers that work inside and outside the rover. Their aim is to unravel the mysteries concerning the minerals of rocks and soil that the rover comes across.

The most famous of these is Curiosity's laser instrument, which often prompts usually serious journalists to gleefully use "Pew! Pew!" in their headlines. The more stately and staid name for this instrument is the Chemistry and Camera tool (ChemCam). As the rover approaches a target rock, it aims and fires the laser, which vaporizes the minerals into electrons and ions. Then ChemCam's spectrometer observes the cloud of free-floating particles that results, which allows the constituent elements inside the rock to be estimated.

Another instrument, built in Canada, is used to disturb the ancient soils of Gale Crater. Known as the Alpha Particle X-Ray Spectrometer (APXS), it bombards samples with X-rays and alpha particles that result from the natural radioactive decay of an element called curium. This creates an energy release. The resulting shower of energetic particles that it unearths reveals what the rocks are made of.

Most of the world's attention, however, has been focused on a sophisticated, self-contained chemical laboratory that has homed in on the complex chemistry that characterized the earliest epochs on Mars.

*　*　*

The centerpiece of Curiosity's scientific investigations is the Sample Analysis at Mars (SAM), an instrument suite that comprises state-of-the-art spectrometers that truly make it stand apart. SAM is a sort of science oven that takes up about a third of the scientific payload on the rover. Sometimes referred to as "Samantha" by its development team, this self-contained chemistry laboratory has opened a whole new vista in research.

SAM uses the rover drill to deliver up powdered samples, which are dropped through a funnel into a series of sample cups that sit on a miniaturized carousel. This material is then heated to very high temperatures. Portions of the vapor that results are released into a trio of instruments that analyze molecules based on their constituent chemical composition. This includes a mass spectrometer, a gas chromatograph, and a tunable laser spectrometer. All have slightly different ways of measuring the abundance of certain elements within the samples taken by the rover's arm.

The mass spectrometer separates out elements and compounds by mass, while the chromatograph heats them up for vaporization, measuring the rainbow spread of spectral "colors" that result. The samples release gases at different temperatures, which also provides a clue as to how simple or complex their chemical constituents are. Finally, the laser spectrometer steps in to look for isotopes—different versions of elements that are chemically the same but have slightly different masses—of oxygen, hydrogen, carbon, and other essential constituents.

To date, the SAM instrument has not reached its full potential, because there was a nightmare scenario when the instrument risked being little more than dead weight. During the early days in Gale Crater, it suffered from a contamination incident. This involved leaking "wet chemistry" cells filled with the wonderfully named N-(tert-butyldimethylsilyl)-N-methyltrifluoroacetamide (MTBSTFA), which plays a key role in identifying organics. MTBSTFA is an excellent reactant and "derivatization agent" that chemically modifies any given compound in a known way. Essentially a very special chemical solvent, MTBSTFA has properties that are suitable for analysis using a gas chromatograph.

Organics can be tricky to detect, as they are invariably delicate and often decompose into simpler molecules. But if they are "derivatized," these products will, to borrow Mr. Spock's phrase, live longer and prosper, certainly on the timescales needed to work out where they came from. Due to the leak, NASA had to take some extra care with its testing. As part of the process of identification, samples in one particular cup were heated to high temperatures. They dropped three samples into the cup, which reduced the contamination to much lower levels. This has allowed experimenters to report their findings with greater confidence.

In total, SAM has seventy-four thumbnail-sized sample ovens. Only nine contain various "derivatization agents" for wet chemistry. In 2019, after more than seven years of operation, they are finally being used. "They're a little riskier to run in the sense they might change the character of the instrument a little bit," says Ashwin Vasavada. "But this derivatization experiment is a gentler way of extracting the information."

* * *

Researchers involved in the Curiosity mission are scientific detectives. With a nod to one of the greater detectives in fiction, the discovery of ancient organics by the SAM instrument was worthy of Sherlock Holmes himself. One of the more famous Baker Street stories concerns the dog that did not bark in the night. In this case, its Martian equivalent is the riddle of whether there were organics on the surface.

The word *organic* means "having to do with the chemistry of carbon," most particularly as a highly flexible molecule that provides the biochemical backbone for microbial life. The great strength of organic molecules is that they are remarkably stable. They have been found on many objects within our solar system. Space missions have shown, for example, that as much as a third of Halley's Comet's weight was composed of organics.

On Mars, organics would certainly have been brought from elsewhere by meteorites. When the *Viking* landers sifted the Martian soils in the 1970s, however, no organics were discovered. It was believed that their absence was the result of strange chemistry taking place on the surface. Regardless, the lack of organics limited the possibility for any biological activity. As we will see, subsequent reanalysis of the Viking data has revealed that their absence could have been because of unusual, and hitherto unsuspected, reactions that took place when they were being analyzed.

Complex and tough organics—those that have survived over geo-logic time—were discovered by SAM in the summer of 2018. They were preserved in sediments from the ancient past, and their persistence and the clear longevity of water on the surface are an important indication toward biochemistry. What the SAM instrument found was a small but significant amount of organic material in the form of complex chains and rings of carbon. These organics were ancient, from what Curiosity could determine. It was believed dust had settled on to the bottom of Gale Crater and become sandwiched together in layers as more dust fell on top of it. Geologists estimated the rocks to be three billion years old.

The organics also had a lot of sulfur in them—no surprise, since the Martian crust is, like Earth, rich in sulfur, the result of extensive volcanic activity. Sulfur is bound by strong chemical bonds that have allowed the organics to latch on to them for greater lengths of time. "That would be a way of having the organics become more robust against the things that can destroy them," Ashwin Vasavada says.

Taken together, all these elements observed in Gale Crater are the raw ingredients for biology. In both clay-rich minerals and sedimentary layers, organics are likely to have survived for much longer. On Mars, they have increased the "habitability envelope," the time when condi-tions were right for life to have formed. "It is not telling us that life was there," one of the SAM team scientists cautioned when announcing the details in the summer of 2018, "but it is saying that everything organisms really needed to live in that kind of environment, all of that is there."

The presence of these organics implies that life-friendly elements were available on the Martian surface. Curiosity also found simple ben-zene rings with chlorine atoms attached to them, known as chloroben-zene. Chlorine is a fairly obvious constituent of many cleaning fluids and herbicides, and all instruments sent to Mars require cleanliness to an extraordinary degree. The science team was very careful to check that there was no contamination that might also have produced this result, especially given the earlier problems.

The finding was that the chlorobenzene was real: it was not the result of a leak or any latent contamination. The measurements still stand today as the first organics found on the surface of Mars. And the best is yet to come. "Our organics story is not over yet," Ashwin Vasavada said in late

2019. The SAM science team is "attempting to examine mudstones with more sophisticated uses of the instrument."

In the same way that Curiosity built on earlier results from Spirit and Opportunity, the next missions to Mars will follow in these impressive footsteps. They will search for the signatures of biological activity that may have been left behind in ancient material elsewhere on Mars. Curiosity has made that leap possible, not least with some other findings concerning the most elusive will o' the wisp in today's Martian atmosphere.

* * *

As it keeps heading upward, Curiosity has found ever muddier layers, a measure of a watery past that has been surprising in its extent. The geologists want to see how high these mudstone layers extend along with a "bonanza" of other minerals associated with them. "That's where all these interesting minerals are that might have evidence of habitability in the past," Vasavada says.

The key finding is that Mount Sharp is stacked and packed with sedimentary layers. The "enigmatic bottom layer" where Curiosity started out was extraordinarily thick, Ashwin Vasavada says. "That is why it took around five years of climbing to pass through it." Even now, there is no race to the top, as they have been able "to explore this mound of material to help us understand exactly what those environmental changes might have been," in the words of Abby Fraeman.

As Curiosity has ascended Mount Sharp, its measurements have, in Fraeman's vivid expression, put a pin in the past climate of Mars. What she terms "constraining the climate over time" is crucial for understanding whether conditions might have been able to sustain the complex biochemistry needed to metabolize nascent cellular life-forms. The surface of Mars was warm enough, at least in this local area, to support liquid water for a significant amount of time. "So there's at least one period at some point around 3.5 billion years ago," Fraeman says, "where it must've been warm enough for at least a million plus years to have been wet on the surface."

There have been many episodes of flooding from lakes. "We've been finding a lot of evidence that liquids came in and further altered the rock," she says. This not only makes the rock record more complicated to read but shows that liquid water was persistent, either on the surface or bubbling up from below. But what can such findings reveal about any indigenous life-forms on ancient Mars, if they ever existed?

Ashwin Vasavada cautions that on Earth, geologists would have trouble picking up an ancient rock and figuring out if there is life contained within it. Erosion from wind, rain, and other geologic processes can easily mottle the signatures of life, especially on the microscopic scale of microbes or bacteria. That is why some of the recent findings of so-called ancient terrestrial life, fossilized biochemical remnants known as stromatolites, remain controversial. Some are believed to be as old as 3.8 billion years, although there is greater consensus that they may be as old as 3.5 to 3 billion years, around the time when life first evolved on Earth and, possibly, Mars as well.

From Curiosity's measurements in Gale Crater of the minerals and the surface chemistry, the water involved wasn't salty, and in the layers examined so far, the acidity was right, too. Such deposits are preserved in rock chemistry for eons after the water has evaporated. "My feeling is that, if anything, Curiosity has shown us that Mars was even more friendly to life than a lot of people on our team would ever have ever imagined prior to getting there," Vasavada concludes.

* * *

On the first weekend of June 2019, Curiosity scientists awoke to surprising news. NASA ROVER ON MARS DETECTS PUFF OF GAS THAT HINTS AT POSSIBILITY OF LIFE was the banner headline from a *New York Times* story alert. A leak of an email intended for the Curiosity science teams meant there was a great deal of media speculation. Such internal communications routinely summarize the findings of their weekly "happy hour" teleconference, as participants call them, so was hardly out of the ordinary.

The "puff of gas" referred to methane, which is a key indicator of life on Earth. In the search for life on the Red Planet, its presence, even in insignificant quantities, is for once a dog that has actually barked in the night. Overnight measurements taken by Curiosity in Gale Crater have detected minuscule outbursts of methane a number of times. It is present in trifling quantities that seem to ebb and flow in a seasonal cycle. This latest finding was a new spike from earlier that same week three times greater than previous measurements. Intriguingly, it faded far more quickly.

The yin and yang of the Curiosity measurements imply short, small events rather than large "exhalations"—something that InSight might narrow down, as that kind of activity will depend on how geologically active the interior is. When it was originally observed using the SAM instrument to take in the local nighttime air, one participant deadpanned that "the people who did the work thought they were going to get Nobel prizes."

The greater truth is that the amounts detected have been so small that it is difficult to know what they actually mean, if anything at all. It was only in 2004 that Martian methane was first detected from Earth with very sensitive detectors attached to large telescopes. A year later, Europe's Mars Express, still working today after more than fifteen years in orbit, had found that methane was present in at least ten parts per billion. Some researchers have questioned both these findings. The telescopes on Earth, they say, were not nearly sensitive enough to detect such small quantities. A similar claim has been made for the Mars Express measurements that, in 2012–2013, mapped methane across the Curiosity landing site.

Using a technique called "patch tracking mode" as it flew overhead, Mars Express trained its spectrometer to observe light absorbed by atmospheric molecules and how that was then emitted as a different form of heat. Methane has a very specific infrared fingerprint, which Mars Express pinpointed to an area to the east of Gale Crater, where Curiosity had also detected it. A year later, the NASA rover itself measured a methane spike at seven parts per billion. This lasted for several months, and since then, Curiosity has observed a small seasonal cycle.

* * *

Into this debate has come the European Space Agency's Trace Gas Orbiter, which was launched by Russia's Roscosmos in March 2016. Its aim, as its name suggests, is to search out the rarest, most elusive atmospheric molecules, and it has a greater sensitivity to these poorly understood concentrations. In short, it was designed to resolve the matter of Martian methane.

Methane, along with other minor atmospheric constituents like ozone, nitrous oxide, and formaldehyde, is short-lived. The earlier observations imply the existence of an active, current source that seems to be replenishing every few months. Yet seeking confirmation of this has been akin to hunting down the Snark.

The Trace Gas Orbiter is capable of detecting a few tens of molecules in any given trillion. Two onboard spectrometers have, between them, extended highly sensitive measurements from the infrared into the ultraviolet. Twice each orbit—when it rises or sets as seen from the Orbiter—the Sun acts like a very bright infrared lamp that shines through the Martian atmosphere. Because the characteristics of sunlight are well known, the spectrometers can determine what gases are present. Any telltale signals of trace gases from within that localized slice of illuminated atmosphere are very strong.

The two European instruments essentially check on each other to make sure they are observing real phenomena. They can also scan the surface directly below, observing the infrared portion of sunlight that bounces off the rocks and plains. Such signals are very weak, however, and the instruments themselves have to be cooled to ensure they can sense anything. But it does allow maps of methane to be made across whole swaths of the Martian surface as the Trace Gas Orbiter flies overhead. Its instruments are sensitive enough to have seen the Curiosity exhalations.

To add to the authentic sense of mystery, in the middle of June 2018, Curiosity observed five hundred parts per billion of methane. When the Trace Gas Orbiter passed over at the same time, it didn't actually detect anything. "We've clearly got some pieces of the jigsaw," acknowledges Professor John Bridges. "We can't quite put it together."

* * *

Many researchers quietly acknowledge they are baffled with regard to Martian methane. Something is creating methane close to the Martian surface, and another mechanism is then destroying it before it reaches altitude. Exactly what these mechanisms are remains an enigma. One school of thought suggests that the sublimation of ice in which bubbles of methane are trapped may explain the observations. Subsurface permafrost might be melting, releasing methane along geological faults. Though the release is taking place today, the gas involved might well be ancient.

If the researchers don't know the source, they also don't know the "sink": how and where it goes when it disappears. Methane should float upward when released. It would then take a few months to mix in with the rest of the Martian atmosphere. Exactly how it would mix remains elusive, though unfiltered ultraviolet radiation from the Sun would eventually break it down. The fact that incessant radiation breaks down the gas in the atmosphere implies that this freshly observed methane was created recently. All the chemical reactions that explain its eventual destruction occur over what one researcher has called "relatively long time scales," meaning a few hundred years.

At the surface, Curiosity sniffs out the traces of gaseous molecules three feet (one meter) above the local terrain; the Trace Gas Orbiter instruments have their greatest sensitivity at an altitude of three miles (five kilometers). The closer to the surface they scan, the more the atmospheric signal is swamped by thermal "noise," thanks to the local rocks and terrain reflecting sunlight back into space.

Using its SAM "oven," Curiosity takes in the nighttime air. Its first task is to remove all the carbon dioxide from the sample. The signals from whatever gaseous molecules remain are then amplified. This air, as might be expected, is very, very cold. One theory is that the boundary layer—the part of the lower atmosphere that directly feels the effect of the surface—acts as a lid to keep the methane close. Once the Sun comes up again, turbulent mixing starts again. The boundary layer pushes up and the methane can rise upward.

Given how tiny they are compared to the vast bulk of the Martian atmosphere, these methane releases may not be significant at all. Others have suggested that the observations are "false positives" or artifacts of how the data has been processed. Most controversial of all is the accusation that the findings stem from something outgassing from inside the chambers that feed Curiosity's SAM instrument, a charge the NASA experimenters vehemently deny.

A complication has been the effect of enhanced levels of dust in the atmosphere. The best guess is that some of the observations result from of the action of unfettered ultraviolet light, possibly aided by electrostatic reactions taking place on atmospheric dust suspended in the air. When the Trace Gas Orbiter began its measurements in 2018, the whole of Mars was obscured from the fiercest global dust storm seen in recent years. Several scientists think the dust storm may have obscured the underlying picture of what was happening to the methane.

Instead of solving the mystery, the Trace Gas Orbiter has added to it. In the spring of 2019, the principal investigator for the orbiter's Belgian-built spectrometer said that, although there was a signal from the dust, "we already know we can't see any methane." Two months later came the spike seen by Curiosity that caused so much excitement. Once again, when the Trace Gas Orbiter flew over Gale Crater at the same time in June 2019, it saw no signs of that very same methane.

Five months later, it was reported that there are seasonal cycles with other atmospheric gases close to the surface as measured by Curiosity. Most seem related to the way atmospheric carbon dioxide freezes out onto the winter pole and is later released with the onset of summer. Nitrogen and argon seem to behave in the same way. Oxygen, however, does not. Once again, this has caused headaches for the scientists who have made the observations.

While 95 percent of the Martian atmosphere is composed of CO_2, molecular oxygen is present as just 0.16 percent, more than a trace but hardly significant and probably not the result of biology (its presence and patterns are not characteristic of any known life mechanism). The new results show that over the spring and summer, oxygen rises by as much as 30 percent in Gale Crater as observed by the SAM instrument. The researchers have pronounced themselves completely stumped. "We're struggling to explain this," said one when their results were published in November 2019.

The amount of oxygen added to the atmosphere was much more variable than expected. The variations cannot be explained from oxygen in near-surface water being physically pulled apart by sunlight or from carbon dioxide being chemically broken down by the same process. As with methane, there appears to be another chemical source and sink, as well as synchronicity, since the changes seen with oxygen seem to be related to the seasonal cycle of methane itself. "We're beginning to see this tantalizing correlation between methane and oxygen for a good part of the Mars year," said Melissa Trainer of NASA's Goddard Space Flight Center.

The oxygen could be produced biologically, but there were no telltale indicators that would have been associated with such a large-scale variation. Similarly, if the associated observations of methane were the result of microbes, it is not possible to say so with any degree of accuracy at the moment. The only way to confirm that would be to observe in detail if any take-up of the isotopes in the methane mimics what happens on Earth when biology is involved. The very scarcity of the methane makes it difficult to obtain an unequivocal reading, although the tunable laser within SAM should be able to do that over time. "We really would like to get the isotopes," says Professor Bridges, "but for that we will need to look at the material in our laboratories."

For now, the mystery of Martian methane remains. The general consensus is that whatever is causing it is probably not as active as the level of debate within the scientific community.

* * *

Finally, in the fall of 2019, Curiosity reached what is known as the clay unit within Mount Sharp for the first time. The forensic examination of such clays was one of the major justifications for landing in the Gale Crater. According to Ashwin Vasavada, this was "the only place you could actually see clay minerals from orbit." Significantly, clay minerals are associated with "the kind of water that's good for life, freshwater, not too salty, at cooler temperatures." The clays will have preserved within them the minerals from the most life-friendly epochs in the Martian past.

"We have just done the experiment that we've been wanting to do for years," says Vasavada. This involves the "wet chemistry" ovens within SAM whose derivatization agents "help pull out things that might be hidden." Already, a trend has emerged from the kind of minerals seen as they have headed up the mountain. In simple terms, conditions have been changing from wet to dry. Two years earlier, the rover passed through a region the geologists christened Sutton Island, where there were distinct mud cracks in the surface. Curiosity trained its instruments on a feature the geologists nicknamed "Old Soaker," as it looked like a salt bed on Earth. What the rover found was various mineral salts mixed in with sediments that suggested they had crystallized in a wet environment that had then subsequently dried out.

In research published in October 2019 in *Nature*, what the scientists termed "lacustrine deposits with sporadic occurrences of sulfate minerals" are the result. In plain English, that means salty brines that were further concentrated as they evaporated when the local atmosphere deteriorated. These features in the Martian mud looked similar to the dried salt fields found in the Altiplano region of the Andes. In South America, they are fed by streams that fill plateau basins but do not drain into any seas. On ancient Mars, the flowing water would have similarly filled shallow briny ponds, undergoing repeated dryings and then overflowing again.

This latest study on Mount Sharp has shown that at least one episode of drying out, and probably more, took place through evaporation. As the atmospheric conditions declined, the salts left behind would have been sufficient to form the concentrations measured by Curiosity. "This further suggests that [the conditions] were more arid—and had more sulfate in the environment—than previously imagined," the team said in their paper.

In that sense, Old Soaker is a preview of what comes next. Over the next year, Curiosity is finally expected to reach the "sulfate unit" itself. This transition from the clays seen to date will provide some important clues about how Mars went from wet to dry, "which," in Ashwin Vasavada's estimation, "is one of the biggest mysteries about the planet." When the deposited minerals changed into greater amounts of sulfate, there would have been less water around; in any case, it was likely to have been much more acidic. "And if it gets too acidic," says Vasavada, "then you can cross some limits that we've never seen life able to cross."

If the results to date have shown that Mars could well have supported life—"and we can say that it was habitable for tens of millions of years"— then the sulfate unit will draw that period of possibility to a close. As to when the rover will reach there, Vasavada smiles. "We don't know where the sulfate unit is. We didn't know where the clay unit was. You don't know until you get there."

As Curiosity continues its extraordinary climb, one limitation looming on the horizon is that its nuclear power source will last for only so long. Certain components are already showing their age. "We believe that we still have a few good years left," Vasavada says. "We can plot the energy coming out of radioisotopes through the electric generator. And that's very predictable."

Much less so are mechanical failures. "We've had issues with the drill, with the wheels," Vasavada says. "We're always one catastrophic failure away from never getting to the sulfate unit. So we're aiming to get to the sulfate unit within the next year."

* * *

In the eight years Curiosity has been working on Mars, it has revealed fundamentally important information concerning how conditions might have supported life in the ancient past. Now, the next generation of rovers is poised to look for signs of ancient life directly. They are in capable hands. Adam Steltzner has stepped up to become chief engineer on JPL's follow-up mission, Perseverance. Sometimes, he will say with a knowing smile, "We're having too much fun here."

Using the same kind of hardware as Curiosity, Perseverance will, for the first time in more than forty years, consider "key questions about the potential for life on Mars." This potential will show up in the form of markers like hydrocarbon chains, amino acids, or lipids (fatty acids found in cell walls) that are more closely associated with life. Such biosignatures should, if they are present, make themselves known.

To do this, the Perseverance rover will carry seven instruments. Many are either new or improved designs of the ones carried on Curiosity. Perseverance will come to rest in a crater in what some think may have been a sea that covered a vast area of the northern hemisphere. Jezero Crater has a breached wall where a delta flowed through it in the past. That means an irresistible mix of water, minerals, and conditions that may have been conducive to life coming together in the form of telltale organics. "All the sedimentologists are drooling over these kind of deposits," says one amused observer.

For now, the scientists have to steel themselves a little longer. From the very beginning, JPL's founding engineers were the original space cowboys—"spoiled, able brats" in one estimation—in the earliest years of the space age. Like Adam Steltzner, they were brilliant. From the outset, they could always be relied on to come up with crazy ideas, not least when they came under the direction of a bona fide genius who many were convinced was actually a Martian.

4

THE ROAD TO UTOPIA

Take the 210 Freeway that snakes along the foothills of the San Gabriels of southern California and just beyond the Rose Bowl, north of Pasadena, and you will find the spiritual home for America's exploration of Mars. The Jet Propulsion Laboratory (JPL) announces itself only with a few undistinguished road signs. Unless you have the right credentials—in this post-9/11 world, security is tighter than ever—the farthest you can get is the Visitor Center, known as the von Kármán auditorium. This is where the press assembles, in the large lecture theater next door, for the latest results from the Red Planet.

Without realizing it, today's army of social media flunkies, journalists, and wannabes who congregate there are following in the footsteps of Theodor von Kármán himself. Few know who he was, where he came from, or less still why the auditorium is named after him. Yet he was crucial to the story of America's entry into space and its confidence in sending ever more complicated vehicles to Mars, and other planets, too.

Theodor von Kármán was born in Hungary and was a child prodigy. At age six, he could multiply six-figure numbers in his head. He and many of his compatriots who ended up in the United States "were really visitors from Mars." Their accents, so it was said by one who knew them all, made it difficult for them to pretend they were anything but Hungarian. In the 1920s, violence and anti-Semitism in their native land propelled them across the Atlantic. Von Kármán joined the Caltech campus to help establish its name in the fledgling science of aeronautics. Suave and cultured, von Kármán encouraged his students to think in outlandish, seemingly surreal ways. "Often impatient with engineers who seemed to him tied to conventional thinking," notes one academic historian, "the strongly theoretical von Kármán was receptive to ideas that struck others as bizarre."

Engineers like Adam Steltzner are the latest in a long line of mavericks who have flowered in the foothills above Pasadena. The first were a handful of pioneers who began to experiment with rocket engines in earnest in the mid-1930s. One day, they accidentally blew up a tank of red-fuming nitric acid in the aeronautics building. Everything in sight was covered with a rusty brown fog. They were soon ostracized and banished to the Devil's Gate Dam up in the Arroyo Seco.

Under von Kármán's kindly indulgence, they started to call themselves the Suicide Club. They, and others who followed, established the first generation of US missiles, which eventually begat the ones that went into space. Within a decade, they eventually named their establishment the Jet Propulsion Laboratory, even though jet engines were never built or even researched there. Rockets were disreputable and hardly carried the patina of respectability. The result was that then, as now, JPL was a misnomer that somehow stuck.

Ever more powerful missiles were built under the aegis of the US Army, until JPL came to launch America's first successful satellite, Explorer 1, on the last day of January 1958. When the National Aeronautics and Space Administration (NASA) was formed eight months later, it took the JPL mavericks under its wing. Managed by Caltech, they were not hidebound by civil service rules, nor did they want any part of the billion-dollar boondoggle that soon became human spaceflight. They were outsiders from the very beginning.

JPL wanted to claim the solar system as its turf, propelling legions of robotic spacecraft out into interplanetary space. In 1959, NASA Headquarters in Washington, DC, reluctantly agreed to fund such an enterprise, while being more preoccupied with the details of launching humans into space. President Kennedy's decision to send astronauts to the Moon meant the laboratory was given the mandate to fire a series of probes called Ranger to scrutinize the surface ahead of them. That the first six failed led to shaky relations between NASA Headquarters and

JPL, but thankfully, the first JPL missions to the planets were more successful. They were named Mariner in honor of their nautical antecedents dispatched across the oceans.

Today, the early *Mariner* spacecraft are remembered as the Model T Fords of the space age. Into hexagonal boxes, engineers crammed early electronics to remotely control the craft in interplanetary space, with sufficient volume left to carry suites of scientific instruments. They were powered by the Sun, using solar panels. Their basic design was easily adaptable: they could be launched inward to Venus, with two solar panels, or outward to Mars, where four were needed.

In those early days, however, the pioneering spirit required more than a little good luck. "John F. Kennedy was president, and there was a Camelot feeling that better things were going to happen," recalled one JPL veteran. "Elvis provided the music of the times at Cape Canaveral, which was called Cape Carnival then."

The funfair atmosphere mostly came from the vagaries of launching probes to the planets. If you see a planet in the sky, you cannot fire a rocket at will and head off. Long, looping orbits that intersect where the planet will be months hence are the only way to reach them. Known as launch windows, such opportunities to launch to Mars occur every twenty-six months or so.

In the autumn of 1962, JPL was ready, and its first target was Venus. Yet *Mariner 1* got no farther than the bottom of the Atlantic Ocean. Its Atlas booster malfunctioned because, among the thousands of lines of computer code guiding its ascent, a hyphen instead of a minus sign had been typed in.

Two weeks later, it was *Mariner 2*'s turn. The Atlas rose into the sky and suddenly spun around on itself. Its orientation sensors went haywire. The onboard navigation system was so disturbed that its Agena upper stage also rotated but in the opposite direction—incredibly, by just the amount needed to cancel the original error. A US Air Force launch official at the Cape remarked to the JPL project team, "We don't know how, but you're on your way to Venus."

Further vexations were to occur in flight—at JPL it became known as the five-miracle mission—but eighty-eight days later, in December 1962, *Mariner 2* flew past Venus. Its instruments revealed that the planet's atmosphere was composed predominantly of carbon dioxide and its surface was so hot lead could melt. But it was a triumph and one that would, with similar mishaps, be replicated on the road to Mars.

* * *

In the fall of 1964, JPL was ready to shoot for the Red Planet.

Mariner 3 successfully lifted off on November 5, but the aerodynamic shroud protecting it from drag and buffeting during the ascent through our atmosphere failed to deploy properly. *Mariner 3* was left stranded inside its fiberglass covering, which refused to open. The spacecraft was left in orbit around Earth.

Three weeks later, they tried again with *Mariner 4*. This time, the spacecraft successfully separated from its own hastily redesigned shroud, but then *its* problems began. Like their nautical predecessors, spacecraft navigate by the stars. They orient themselves in space and head in the right direction by the use of sensors that "lock" onto specific stars. The early *Mariner*s navigated by Canopus, a particularly bright star seen to best advantage in the southern hemisphere on Earth.

Within hours of leaving Earth, however, *Mariner 4* could not find Canopus. Its tracking system literally wandered across the sky, locking on to the wrong stars. The sensor was confused by reflections from paint flakes that gathered around the craft long after it had separated from its booster. Only by sending up repeated commands did JPL mission controllers find Canopus in time so that *Mariner 4* would get on the right track toward Mars.

On December 5, *Mariner 4* made another navigational correction and promptly lost Canopus again. By this time, one of its instruments, measuring the flow of energetic particles streaming from the Sun, was returning nonsense data. The instrument was reluctantly abandoned. By the following spring, however, it mysteriously switched itself back on again and worked perfectly.

Such aggravations were par for the course in those days. Thankfully, *Mariner 4*'s onboard systems behaved well enough to actually reach Mars on the evening of July 14, 1965. The Red Planet came under scrutiny and yielded some of its secrets. As *Mariner 4* disappeared from view behind the planet, its radio signals were attenuated as they passed through the Martian atmosphere. The way in which they were cut off gave important clues to the structure and the density of its atmosphere.

Mariner 4 provided scant hope for the possibility of life. The atmospheric pressure at the Martian surface was found to be about a hundredth of that on Earth—a measly 10 millibars—and the temperature on the surface was −100°C (−150°F). Only the hardiest of microbes might survive, but, for most people, the evidence from *Mariner 4*'s cameras provided the clincher.

Under the guidance of Professor Robert Leighton at Caltech, *Mariner 4* carried what was then a state-of-the-art television camera. It built up pictures from individual dots rather than lines, as was common with domestic television sets of the time, due to limitations of power. Each of these dots, known as picture elements or pixels, was recorded in sequence. One image of Mars contained 60,000 pixels. Their brightness was stored as an individual bit of electronic data by a tape recorder. It was space-age painting by numbers, cumbersome by today's standards yet revolutionary at the time.

Mariner 4 had only ten watts of power to play with. It could transmit the data at only eight bits per second through space, for the listening ears of JPL's Deep Space Network (three dedicated radio telescopes in California, Australia, and Spain) to pick up the signal. The plan was that *Mariner 4* would take a strip of twenty-two pictures as it swept over the southern hemisphere of the planet from a range of 6,000 miles (9,600 kilometers). This precious data would then take ten days to replay and build up into photographs at JPL.

Improbably, it worked.

In all, eleven useful frames were returned, some of which revealed a surprise: craters. That Mars would be peppered with craters is, in hindsight, no big deal: from innermost Mercury to Pluto at its outer extremities, all the solid bodies in our solar system show signs of cratering, a record of the violence of their birth.

But in 1965, craters were a shock to geologists almost expecting a windswept desertscape and a general populace half-believing there might be canals. Taken altogether, the *Mariner 4* data implied that Mars was dead in the geological sense and quite probably biologically, too. Craters have been erased on Earth due to its active geology, while on the Moon they have remained in place because geological processes ended early in lunar history. Mars seemed to be as dead as the Moon.

Mariner 4 also found no evidence of a Martian magnetic field, which meant the surface was blasted by the often-lethal radiation from the Sun. Unprotected by anything approaching an ozone layer, any living organisms would have the handicap of having to survive this onslaught of deadly ultraviolet radiation. Thanks to *Mariner 4*, the planet itself seemed rather more dead than red.

* * *

At the same time that JPL began building the first *Mariner*s, an exotic new species of scholar flowered in the halls of academia. Calling themselves "exobiologists," in the early 1960s they were very much the new kids on the block. Their more staid colleagues were wont to call them "ex-biologists," for many dismissed the search for life elsewhere in space as a science without a subject, a field that had more to do with wishful thinking than real research.

Yet as NASA began to draw together the expertise to consider the complexities surrounding the question of life on the Red Planet, exobiology quickly gained pace, and its practitioners became preeminent in this exciting new field. Gerald A. Soffen's background was typical. Originally, he had studied zoology at the University of California, Los Angeles, but as a postdoctoral student at Princeton he became fascinated by the philosophical implications of life in the universe. Like many of his contemporaries, when Soffen joined JPL in 1961, he was certain there must be life on Mars. "The feeling was that there would be microbial-type stuff on the surface," he says. "Maybe it came from comets or some other mechanism, but it would be there for us to find."

Until *Mariner 4*, just getting a coherent picture of conditions at the surface was difficult. Even with the best telescopes available, research was stymied because Earth was so very far away from Mars. Spectroscopy gave science the wherewithal to remotely probe the innermost chemical workings of objects elsewhere in the universe, yet it was as much art form as science, because its practitioners had to work around distinct handicaps. Astronomers had to learn which spectral lines resulted from our own atmosphere and appreciate how much they would be shifted due to the movement of the planets themselves.

Worse, Martian spectral lines often ended up playing peekaboo with well-established ones in Earth's atmosphere. Because our atmosphere is very much denser than that of Mars, any spectral signatures characteristic of the Martian atmosphere would be faint. They were often swamped by the stronger and broader signals from Earth's own atmospheric molecules.

When Sir William Huggins, a pioneer of spectroscopic science, tried to find water vapor in the Martian atmosphere in 1867, he failed miserably. It was only in 1963 that water vapor was actually found on Mars, and only then in trifling quantities. Carbon dioxide had been discovered in 1947, but most intriguing of all was the possible presence of organic materials detected in 1956 by William Sinton of Harvard University. When he suggested they might be from chlorophyll from lichen, there was a great deal of speculation. (His findings were later shown to be wrong.)

In the early 1960s, the fledgling science of exobiology was rife with controversy and very much frontier territory. It would have helped had astronomers been able to accurately ascertain the composition of the Martian atmosphere. The pressure at its surface was another vexed question. By then, many had already deduced the same figure of 85 millibars that astronomer Percival Lowell, who had confidently claimed there were canals on Mars, had suggested at the turn of the twentieth century. Compared to the thousand or so millibars at Earth's surface, this meant the Red Planet would be a forbiddingly frozen and arid world. As to the chemical composition, the feeling was that the Martian atmosphere would be 1 percent carbon dioxide, 1 percent argon, and the rest—98 percent—nitrogen. Both argon and nitrogen were difficult to detect spectroscopically, but that didn't deter many exobiologists from claiming it would be found on Mars. Nitrogen is a key element in terrestrial life, and, given that it forms 78 percent of our own atmosphere, the larger figure on Mars didn't seem too wild.

A single spectroscopic measurement from Mount Wilson in 1963 not only uncovered water but revealed several new absorption bands characteristic of carbon dioxide. By adding up how each individual line in the spectrum absorbed light, it was possible to get a better estimate of the surface pressure: 25 millibars.

Mars was perceived as a positively parched planet when *Mariner 4* was being prepared for launch. And there was scant comfort that, under such conditions of pressure, water would evaporate at the same temperature at which it froze on Earth, 0°C (32°F). Such appreciably minute amounts of water in the atmosphere meant there might not be enough available for it to exist as a liquid on the surface. So Martian life-forms, if they existed at all, would have to be remarkably well-adapted to survive on such a self-evidently hostile world.

In February 1965, just five months before *Mariner 4*'s arrival, the influential journal *Science* weighed in with an editorial warning that by concentrating on life on Mars at the expense of the rest of the universe, "we could establish for ourselves the reputation of being the greatest Simple Simons of all time."

Another devastating argument came from a particularly important doubting Thomas, an independent British scientist who was consulting at JPL around this time and sharing an office with Carl Sagan. James Lovelock was later to formulate his famous Gaia theory with Sagan's first wife, the biologist Lynn Margulis. But in the early 1960s, Lovelock was content to pore over spectra and pronounce something that his colleagues often didn't want to hear. "You could tell there would be no life on Mars," he says today. "The atmospheric gases were in equilibrium."

A subtle point but a powerful one: all biological entities delicately affect the chemical makeup of the air around them. By measuring which gases are present, you can calculate whether they could continue to react with each other. The presence of life would mean that they would alter the chemistry of its localized environment and the gases would not be in balance with each other. The fact that they were balanced in the Martian atmosphere suggested to Lovelock that it could not support life.

That certainly seemed borne out by the results of *Mariner 4*. In vain did Bob Leighton's camera team declare in an official summary of their results that as they had anticipated, "Mariner photos neither demonstrate nor preclude the possible existence of life on Mars."

Yet exobiologists were not quite ready to throw in the towel. In the heady days of the space race in the 1960s, an opportunity to scour the surface of Mars for life presented itself in the form of a veritable behemoth. This was Wernher von Braun's mighty Saturn V booster, which propelled the Apollo astronauts to the Moon. NASA Headquarters directed all its research centers to work up a complicated and ambitious mission. It would carry not one but two fully automated laboratories to sift through the soils to find evidence of life on Mars.

Known as *Voyager*—not to be confused with the later JPL mission that explored the outer planets in a grand celestial tour—it would have weighed some eighteen tons at a time when the *Mariner*s were at most one or two. *Voyager*'s payload would carry as many experiments as possible to solve the riddles of Martian biology.

Jim Lovelock was astounded that a flea trap was seriously considered, in the belief that the sands of Mars might be like their terrestrial counterparts and harbor fleas. Other experiments were less exotic. They consisted of a microscope attached to a TV camera to magnify any microbes present, a chamber to detect the growth of microorganisms, and an air sampler to look for enzyme activity. There also was an instrument named Gulliver that would detect life by feeding Martian microbes with a "broth" of nutrients and observing their metabolic reactions.

By 1967, the projected cost of this complicated mission had swelled from $700 million to $2.2 billion ($13.5 billion in 2018 dollars—even worse than today's ballooning budget of NASA's much-delayed James Webb Space Telescope in comparable dollars). Though everybody within the space agency wanted a piece of the pie, something had to give. *Voyager*'s prospects faded as NASA's Vietnam-era budget was severely pruned by Congress. The original *Voyager* was canceled.

In any case, it was already clear from *Mariner 4* that many of the experiments considered originally would not have worked on Mars. Yet the lure of Martian life was too great. So there came, from an entirely unexpected quarter, a much more modest assault on the Red Planet.

* * *

Ten days after Neil Armstrong made his giant leap for mankind in July

1969, reporters flocked to Pasadena to await results from the next JPL missions to Mars. They would not be disappointed.

*Mariner*s 6 and 7 would track across the southern hemisphere of the Red Planet, mapping about a fifth of the planet's surface. Together, they returned a million times more data than *Mariner 4* had done five years earlier. Advances in electronics meant greater numbers of instruments could be packed into the same sort of hexagonal box that formed the main body of the spacecraft. They carried more advanced cameras capable of better resolution and spectrometers that could sift spectral lines with greater acuity as they sped past.

Ninety-nine percent of the Martian atmosphere seemed to be composed of carbon dioxide. There was little room for the all-important nitrogen on which biology was dependent. There was, however, a moment of genuine excitement when results from the infrared spectrometers were unveiled. Tracking across the southern hemisphere, the *Mariner*s found that near the south pole, temperatures were around −160°C (−256°F). That implied that the south polar cap was made of frozen carbon dioxide. But at its extremities, temperatures rose appreciably. They seemed too high to let carbon dioxide alone freeze out, which hinted at the presence of water ice—and more besides.

From a hasty overnight analysis of spectroscopic data from *Mariner*s 6 and 7, it appeared there were signatures of both ammonia and methane. "We can't refrain from speculation that they might be of biological origin," said the lead scientist at a press conference. Even the *New York Times* joined in the excitement by reporting on its front page that there was "a general gasp among scientists and newsmen." In the event, it was a gasp too soon, for the spectroscopic signals were actually caused by solid carbon dioxide.

There was a moment of equal excitement for the JPL engineers who steered the *Mariner*s to Mars. Just hours before reaching the Red Planet, *Mariner 7* suffered a power loss that initially was blamed on a meteorite hitting the spacecraft but more likely resulted from a battery that exploded. So began the myth of a curse upon Mars missions, which one JPL manager jokingly christened the Great Galactic Ghoul, an all-too-believable modern-day manifestation of Martian malevolence. It would strike capriciously over the next few years, more often at Soviet Mars missions.

The next launch window in 1971 promised to be the best yet, for Mars would make a particularly close opposition to Earth. The margins in mass, meaning more hardware for less fuel, would allow more sophisticated spacecraft to actually brake into orbit around the Red Planet for the first time. To do this, they would still need to carry more than half their weight in fuel. Nevertheless, it would be worth it, as an orbiting spacecraft would be able to complete the first high-resolution mapping of the whole of the Martian surface.

A new version of the *Mariner* spacecraft would carry an onboard minicomputer that could be reprogrammed from Earth with new instructions. Such flexibility would be needed as scientists reacted to the discoveries that would be made as the spacecraft unraveled the surface frame by frame. In tandem, *Mariners* 8 and 9 were designed to map the whole of the surface down to a resolution of 330 feet (100 meters). Unfortunately, *Mariner 8*'s launcher malfunctioned and crashed into the Atlantic. Not even a Great Galactic Ghoul lurking in the ether was needed to foul up the start of so hopeful an enterprise.

On May 30, 1971, *Mariner 9* successfully left Earth, with JPL controllers knowing that it would have its work cut out for it. As it journeyed toward Mars, its target became engulfed by a global dust storm. This in itself was hardly a surprise. The occurrence of Martian dust storms had been remarked as far back as the eighteenth century. Virtually every Martian year, individual storms grew together to cover the whole of the planet in a featureless ocher-colored haze.

And that was what awaited in the autumn of 1971. *Mariner 9* raced two Soviet probes, *Mars 2* and *3*, toward the Red Planet. While *Mariner 9* could be reprogrammed, the Russian spacecraft could not. Both were consumed by the swirling dust as they attempted to land, with *Mars 3* transmitting garbled information for all of twenty seconds before it disappeared that December.

In the meantime, *Mariner 9* successfully entered orbit and waited for the dust to clear. When it did, geologists on the *Mariner 9* television camera team were presented with the bizarre spectacle of a series of black spots poking out of the murk. These were massive volcanoes, the largest of which was greater in extent than the island of Hawaii. It towered some nine miles high, including cliffs at its base that were nearly a mile deep. It was named Olympus Mons, after the home of the ancient Greek gods.

Nearby were three smaller but nevertheless enormous volcanic piles that would dwarf Mount Everest on Earth. In time, other less spectacular volcanoes were found dotted around the planet. Improbably, the earlier *Mariner* flybys had missed all of the most interesting features on Mars. Thus, the fact *Mariner 9* made repeated orbits around the Red Planet showed how valuable its mission was.

Our picture of Mars was redrawn as a world that had been geologically active in the past. The largest canyon on Mars was another entry into the record books. Valles Marineris, as it was soon named in honor of its discoverer, stretched some 4,000 miles (6,500 kilometers) in length, a fifth of the way around the planet's circumference, and in places was a few miles deep.

Surrounding features hinted at catastrophic changes. At the eastern extremities of the Valles Marineris, the ground had collapsed at around the same time as a release of vast water floods. These outflows covered much of the northern hemisphere and smoothed out the surface, creating extensive plains in their wake. There must have been sufficient liquid water available to have flowed across the surface unhindered.

Even more importantly, *Mariner 9* detected fogs and frosts on the surface as well as clouds. That meant that there was still enough water vapor around to be biologically significant. So, scientific opinion swung to favor Martian life-forms, however remote and unlikely they might be. The evidence for water and volcanism in the Martian past was irresistible to those hoping to find life.

That water had flowed across the surface meant that the climate must have been warmer for considerable periods. Though the water flows didn't seem recent, the inference was clear. If liquid water was once freely available, it was more likely that life could have evolved. By adapting to the conditions that had obviously deteriorated since then, Martian life could have survived. That was what many exobiologists clung to. After *Mariner 9*, microbes on Mars no longer seemed like so much wishful thinking.

When *Mariner 9*'s mission ended in November 1972, the Red Planet had been revealed in all its full and spectacular glory. The official NASA summary was titled *The New Mars*. It carried a choice selection of the 7,329 pictures obtained throughout the course of the *Mariner 9* mission. Potential landing sites were chosen from the photographs that showed flat and featureless terrain.

Landings would not be long in coming, thanks to NASA's next missions to the Red Planet. The first American attempt to land on Mars was to take place, ordained by considerations more political than scientific, on July 4, 1976. If all went according to plan, Mars would give American taxpayers a bicentennial present they would never be allowed to forget.

* * *

Langley, Virginia, is best known for America's most famous, yet shadowy governmental bureaucracy, which resides there. The Central Intelligence Agency shares the pleasant, wooded landscape with a NASA research center, where aircraft crashes are deliberately staged for an agency that is still charged by Congress to investigate the issue of aeronautical safety. So NASA Langley was better known for its work in the often laborious—and to many, unimaginably dull—science of understanding how air flowed around engines, wings, and the first ballistic capsules returned from space.

What happened to allow Langley to send a mission to the Red Planet sounds suspiciously like a scene from an old Andy Hardy movie—*"Hey kids, let's build a Mars probe right here!"* It really was based on a curious coincidence. By 1967, the newly appointed director of the Langley Research Center was Ed Cortright, a friend of Deputy NASA Administrator George Low, who had been in the same fraternity at college. One day that summer, Low asked his newest center director what mission he would most like to tackle. The plum job, Cortright replied without missing a beat, would be a Mars lander. Langley, he insisted, really should be involved. In due course, it was. But it was hardly through an act of cozy cronyism.

NASA Langley was then riding high on the success of the Lunar Orbiter spacecraft, which mapped the whole of the Moon to select landing sites for the Apollo astronauts. Its expertise in what was known as "site certification" weighed in Langley's favor. As ever, there was the usual internecine fighting as to which center should run the first landing on the Red Planet. NASA Langley, disparaged by one critic as "just a bunch of plumbers," came out top. Its engineers had been looking at Mars missions for more than a decade by this point.

Gerry Soffen for one was glad to escape the post-Voyager doldrums in Pasadena. He and a handful of colleagues were assigned to Langley, with himself as project scientist. "Getting Langley involved was perfectly right," he says. "They had the experience in aerodynamics that JPL hadn't. Up until then, we had only landed the Surveyors on the Moon, where there is no atmosphere."

What became the most ambitious assault ever attempted on the Red Planet was eventually given the name Viking. In a time before political correctness, any unwarranted associations with invasions, rape, and pillage were deemed perfectly acceptable. The mission eventually crystallized as a concept involving pairs of identical spacecraft launched atop Titan-Centaur boosters that would take nine months to journey to Mars.

Once in orbit, the landers would separate from the orbiters. They would endure the fiery heat of entry into the Martian atmosphere within an aerodynamic shell. After separating from this kindly cocoon, each lander would parachute down to the surface and fire rockets to slow their speed to zero.

NASA Langley led the project, with JPL building the *Viking* orbiter, which was essentially a *Mariner 9* with knobs on (it had improved, higher-resolution cameras and a brand-new scan platform to scrutinize the surface). The lander, however, was quite unlike anything that had ever been built before. At a time when the television series *Mission: Impossible* was wowing viewers, NASA Langley had to do it for real.

The *Viking* landers required the safe dispatch of a fully automated laboratory, smaller than a VW Beetle, that could scoop up samples of the surface and work for many months at a time. Biology was the focus. There would be three ways of looking for life: an imaging system, organic analysis, and direct life detection. If there were anything remotely resembling Martian animals, their movements could be detected by a facsimile camera that slowly scanned the horizon. It built up its portrait pixel by pixel in elongated scans and was truly revolutionary for its time.

The direct search for life would make use of a sampler arm that would scoop up soil and drop it through a hopper into a self-contained laboratory. In time, additional experiments were added, including a self-contained weather station and a seismometer, among others. Throughout the Viking project, the marshaling of unknown numbers of personnel in an era without faxes or emails was supremely successful. For the many hundreds of contractors, their participation meant endless journeys between manufacturing plants and meetings around the United States.

"I recall shuffling out of an airplane late at night after a few hours of half sleep," recorded the geologist in charge of the camera, the late Thomas "Tim" Mutch, in an official NASA reminiscence published in 1978. "[I failed] to recognize either where I was, or to what end I was traveling. For several moments, I had the Kafkaesque feeling that I had somehow lost my identity, that I had become separated from the real world."

His experience was typical. The complexities of *Viking* never unraveled due in large part to the person whose shoulders bore ultimate responsibility, project manager James R. Martin Jr. Martin was a true colossus of the space age. He bestrode the upper echelons of NASA with a "reputation for troubleshooting and no-nonsense management" and a forensic level of detail. With his perennial crewcut and general military bearing, Jim Martin was known as the "Prussian General" by some of those who worked under him. More than one *Viking* veteran says it was akin to having worked for an Old Testament prophet. Yet he also brought them out of the wilderness of technical and budgetary difficulties more than once.

Viking would test the engineers' ingenuity in realms approaching the impossible. To keep everything on track, Jim Martin developed a novel approach in the form of a top ten list of difficulties that needed the most urgent attention. Always near the top were the life detection experiments, which was hardly surprising, as they were arguably the most complicated instruments ever built, let alone sent into space. Within a box about the size of a gallon container, some forty thousand individual parts had to be crammed. Twenty thousand parts were transistors. There were some fifty electronic valves along with tiny ovens, bottled nutrients, and radioactive gases, plus a xenon lamp. To facilitate the flow of nutrients to the soil samples, flow pipes the width of human hairs were built out of a new form of rubber that was fashioned especially for the mission.

Small wonder that the total cost of the life experiments soared from $13.7 million to $59 million (roughly from $92 million to $400 million in 2018 dollars), a source of much embarrassment at the time. "Devising a biology instrument that held three experiments in a container less than 0.027 cubic meter in volume and weighing about 15.5 kilograms was more of a chore than even the most pessimistic persons had believed," the official NASA history of *Viking, On Mars*, records.

As late as the end of 1974, just months before they were due to fly, all three biology experiments were six months behind schedule. There was talk of abandoning them altogether. Martin was having none of it. His perseverance paid off as they were delivered on time and all worked flawlessly on the Martian surface.

The same could not be said of the *Viking* biologists, who did not mesh harmoniously. The biology team was prone to infighting that soured their working relationships and sometimes led to public squabbling. Arguments were assured when one experimenter publicly declared that his instrument was being compromised by the others. Lingering resentment resulted when one experiment team found that Antarctic samples declared sterile by another did indeed contain microorganisms. And during the actual mission itself, when one biologist said he could explain his results in terms of inorganic chemistry, an angry colleague retorted that he could explain his fellow researcher that way, too.

If that were not bad enough, many of the non-biologists in the scientific community felt that the life detection instruments were far too new, wouldn't work, and, because so little was known about Mars generally, any attempt to detect life was ultimately pointless. Sitting somewhere in the middle was Gerald Soffen, who, as project scientist for the mission, was the peacemaker.

Not only did the scientists argue among themselves, they bickered with the engineers, as is common on space missions. To the scientists, engineers were blinkered, by-the-book bureaucrats who possessed insufficient imagination to realize the grandeur of what they were trying to achieve. Scientists, to those who struggled to build hardware on time and on budget, were seen as airy-fairy ingrates, their heads perpetually in the clouds and forever wanting something at the last minute no matter how badly it would affect the overall mission. And, engineers sourly noted, when things went wrong, it was always described as an engineering failure, yet when the mission made the news bulletins, it was always a scientific success. All *Viking* veterans agree that Soffen managed to walk through these potential minefields with fairness and aplomb.

For their time, the three biology experiments on *Viking* were revolutionary. Taken altogether, they would look for signs of metabolic activity by indigenous organisms. The Pyrolytic Release (PR) experiment would detect the synthesis of organic material—that is, it would sense if any microbes had taken up any radioactive carbon from gases carried in a special chamber. The remaining pair of instruments assumed that life on Mars would be similar to terrestrial microbes that could have adapted to the Martian environment. The Labeled Release (LR) experiment, which grew out of the Gulliver instrument, was designed to look for decomposed organic material when microbes were "fed" on a nutrient. The Gas Exchange Experiment (GEX) would infer metabolic activity from changes to the atmosphere in the experiment chamber.

But that wasn't the whole story so far as biology was concerned. The lander samples would be analyzed by a mass spectrometer, which would check for the presence of organic molecules in the Martian soil. The biggest worry was that Mars might inadvertently become contaminated by Earthly microbes carried there by the *Viking* landers. A United Nations treaty signed in 1967 meant that any spacecraft sent to Mars would have to be sterilized to avoid just such a possibility. "All it would need was a fleck of paint from the lander to be analyzed, and our results would be out," says Gerry Soffen. "That was my nightmare." And with good reason: when the *Apollo 12* astronauts returned parts of the robotic *Surveyor 3* lander from the Moon in 1969, for example, engineers who examined the parts were astounded to find microbial contamination from launch.

The only way to ensure that *Viking* would not accidentally discover something similar was to sterilize the whole of the spacecraft. This in itself was no small endeavor, considering the delicate electronics that were being developed for the mission. Soffen says he fought hard "degree by degree" to keep both engineers and scientists happy in finding a compromise that would kill off any Earthly microbes but not do any damage to the spacecraft. The *Viking* landers were baked for over forty hours at temperatures reaching 112°C (234°F) to no lasting harm.

In tests, the *Viking* biology experiments found evidence for life on Earth in soil samples from all over the world (including those controversial ones from Antarctica, which had been declared sterile). That they could work anywhere on Earth was all very well, but that could not answer the nagging yet fundamental question of whether the biology instruments would work on the desiccated world next door. Clearly, there was only one way of finding out.

* * *

Both *Viking*s left for Mars in the autumn of 1975 and were bothered neither by serious glitches nor by any lingering vexations from the Great Galactic Ghoul once they arrived in orbit around Mars. *Viking 1* arrived first, in June 1976, with *Viking 2* hot on its tail and due to arrive in August.

On the evening of Saturday, June 19, *Viking 1* returned its first pictures of the Martian surface from orbit in staggering clarity. Whereas *Mariner 9* could see down to a scale of roughly 330 feet (100 meters) at ground level, Viking could make out features ten times smaller. Viking's better cameras revealed a far rockier landscape. The Martian atmosphere was clearer than it had been five years earlier when *Mariner 9* arrived in the wake of a global dust storm. What had been thought to be gently rolling terrain was cut by sharp lava flows, channels, and craters.

That meant there were probably far greater geological obstacles waiting to impede the *Viking* landers as they attempted to touch down on the surface. Yet, with the highest resolution on the ground limited at best to thirty-three feet (ten meters), even the *Viking* cameras could not discern rocks that could quite easily bring a lander's arrival to a premature end.

The first *Viking* landing had been planned for Independence Day to celebrate America's two-hundredth anniversary. Thanks to the new pictures, a landing on the Glorious Fourth was obviously now out of the question. "If one sets off as Columbus did to find a new world, he need not apologize for looking for a safe harbor," Jim Martin told a press conference, to a certain amount of grumbling from the reporters.

Over the days that followed, there were many sleepless nights, long meetings, and heated discussions over what exactly constituted a safe landing site. *Viking 1*'s orbit was changed five times over the next three weeks to allow eight new possible candidate sites to be examined in better detail. As many as four hundred people, from NASA higher-ups to undergraduate interns, were roped in to help in the anxious poring over new pictures and the counting of craters deemed necessary to declare a site "safe."

With the search continuing well into the second week of July, there was antipathy by some scientists at the continued delay in the landing. A possible traffic jam was looming. *Viking 2* would arrive in early August and would itself require care to get parked in the correct orbit. By then, the *Viking 1* lander needed to have been dropped down to the surface. Bleary-eyed geologists pored over the latest pictures throughout the night. To increase the likelihood of life, a potential landing site had to be "wet and warm" so far as the confines of the Martian atmosphere would allow. That meant as near to the equator as possible, and a low area with an appreciably denser amount of atmosphere. This extra warmth and atmospheric thickness might make it more likely microbes had congregated there.

On July 13, a place was found that seemed more promising. North of the equator, it appeared that an outflow channel had once fanned out over volcanic plains, smoothing out the surface in its wake. And if nothing else, its name beheld the lure of irresistible promise: Chryse Planitia, the Plain of Gold. At a latitude of 22.4°N and a longitude of 47.5°W, it represented a landing site on which biologists, geologists, and Jim Martin could all agree. Landing was set for a week later on Tuesday, July 20, 1976.

The biologists remained resolutely upbeat—in public, at least. While the scientist in charge of the Gas Exchange Experiment thought there might be a 50 percent chance of *Viking* finding life, privately most agreed with Gerry Soffen's assessment: one chance in fifty. And so the flight team prepared to land *Viking 1* on Mars, knowing that the time lag in communications—signals took nineteen minutes to arrive—meant the probe was on its own. By the time they found out what had happened to the Viking lander, it would have long since have been a fait accompli.

Against all odds, everything went according to plan. *Viking 1* separated from the orbiter without any problems, then descended through the atmosphere without any notable difficulties. At 5:12 a.m. Pasadena time on July 20, it successfully touched down on the Plain of Gold.

When the "made it" signal was received back at JPL, Mission Control erupted into sustained cheers and unrestrained joy. Jim Martin smiled the smile of one upon whom fate had bestowed great riches, as well he might. By a truly cosmic coincidence, *Viking 1* came to rest on the Red Planet seven years to the day after Neil Armstrong had walked onto the Moon.

Within minutes, the lander cameras started to return pictures, the first of which revealed a rock close by one of its footpads. "Oh!" cried Tim Mutch, the camera team leader. "Incredible! Incredible!" His incoherence was understandable. After eight years of planning, *Viking* was working perfectly on Mars. Normally taciturn project officials also weren't afraid to emote for the benefit of the press. "I can't think about the science right now," said one NASA top brass. "My heart's beating fast, and I had tears in my eyes this morning for the first time, I guess, since I got married."

The project team realized how exceptionally lucky they had been. *Viking 1* soon returned a panoramic view that showed it had come to rest perilously close to a boulder that could easily have smashed it to smithereens. What was christened Big Joe confirmed their worst fears. Nevertheless, after centuries of speculation, Mars became a familiar and aesthetically compelling landscape overnight. The Martian surface was revealed as a pleasant, rock-strewn vista with no obvious signs of life, very much as the biologists expected.

The next day, the first color picture revealed what was deemed quite a nice day on Mars. At first it was thought that the sky was blue, but when a camera scientist saw a wire sprouting from the lander body, he thought it didn't quite look right. Adjustments followed and the sky was found to be salmon pink, due to all the dust that was suspended in the atmosphere even when the conditions were comparatively clear.

That wasn't the only surprise. On its way down to the surface, the lander had measured the complete and unadulterated composition of the atmosphere for the first time ever. As expected, 95 percent of the air on Mars was composed of carbon dioxide, while just nudging 3 percent was nitrogen, finally confirmed by direct measurement as remote spectroscopy had never permitted. They had, one researcher involved said, learned more about the Martian atmosphere in *Viking*'s half-hour descent than in the previous half century of observations from Earth. Even the comparatively small amount of this elusive nitrogen was viewed as a possible fertilizer to enhance the prospects for biology.

The atmosphere may have been thin, but, comparatively speaking, it was dynamic so far as meteorologists were concerned. For the general public, there was also the novelty of the first weather forecast for another world after *Viking*'s first day on Mars. It was issued by Seymour Hess, who had designed the meteorological instruments that poked out of the body of the lander.

"Light winds from the east in the late afternoon, changing to light winds from the southeast after midnight," Hess announced with much relish. "Maximum winds were fifteen miles per hour. Temperature ranged from –83°C just after dawn to –33°C. . . . Pressure steady at 7.7 millibars."

In the months ahead, the weather became more variable as winter encroached. The lander instruments also detected weather patterns similar to frontal systems on Earth. There were no significant changes to the site itself, apart from variations in brightness and slight color changes among a handful of rocks. Even the onset of a global dust storm did not render the surface as opaque as some atmospheric scientists had feared.

Geologically, the landing site was fascinating. What appeared to be dunes in the distance were classified as drifts, meaning that they weren't transient like those seen in the Sahara. The surface soil was very finely grained and had physical properties similar to the sand on a beach on Earth where the sea has just retreated. When the lander arm dug out a small trench in the soil, it stayed put for many months afterwards. Given that Mars is completely dry, it was the electrostatic properties of the soil that gave it so much cohesion.

In the meantime, *Viking 2* had arrived. Its lander reached the surface successfully under even more trying circumstances six weeks later, on September 3, 1976. The geologists' sights had alighted on an area 4,000 miles (6,400 kilometers) northwest of where *Viking 1* had come to rest. It even had the more attractive and intriguing name of Utopia. If anything, though, there was much less consensus over the merits of safety. Some geologists were certain that the landing site would be replete with fields of soft dunes that would engulf the lander. In the end, Jim Martin's opinion that Utopia was much safer than Chryse prevailed.

Viking 2's descent was considerably more nerve-racking, as the orbiter was supposed to act in relaying its signals back to Earth rather like using a walkie-talkie link to a more powerful radio transmitter. This all-important relay failed due to a problem with the orbiter's communication system. Back at JPL, only a stream of engineering data would tell mission controllers that *Viking 2* had made it. On landing, direct transmissions would switch to a slightly higher data rate.

The relay failure meant the flight team was flying completely blind. It took twenty-one extra—and agonizing—seconds to confirm that *Viking 2* had made it safely down. The landscape at Utopia looked comfortingly similar to the first landing site, even though there were no dunes anywhere in evidence. The horizon seemed tilted, but that was because one of *Viking's* footpads came to rest on a rock. This made the horizon appear to tip at just over 8°.

So far as biology was concerned, there was nothing that looked remotely like Martian life-forms in the immediate vicinity. The cameras at both sites found no evidence whatsoever of either flora or fauna, for, as Carl Sagan later remarked, "So far, no rock has obviously got up and moved away."

Measurements by *Viking 2* showed that the soils in Utopia appeared very similar to those found by *Viking 1* in Chryse. Given that the two sites were geologically different, this suggested that the global wind patterns had evenly distributed the fine Martian dust around the planet.

Overall, there were more boulders than could be accounted for by excavation from the gouging out of nearby craters, which meant they must have come from volcanic activity or had been swept along by water or lava floods in ancient times. The rocks in Utopia were more pitted. This could have resulted from volcanism: pockets of air would have been formed as the rocks cooled, leading to an appearance similar to Swiss cheese.

The Viking scientists were agog at their success. "The feeling was there would be an even chance for the *Viking* landers," Gerry Soffen says.

"Our feeling was that one would fail and the other would be successful. We were astounded. Nobody was sure if we'd successfully land on Mars or not."

The *Viking* orbiters lasted far beyond the two-year warranty NASA demanded from project engineers to allow the return of useful information. *Viking 2* continued to transmit information from Utopia until April 1980, while the *Viking 1* lander was even pluckier and sent signals back to Earth until the end of 1982.

"The fact all four spacecraft worked for years was incredible," Soffen said. That it would take another twenty-one years before the next American spacecraft to reach Mars was something nobody could have predicted at the time.

* * *

The *Viking*s were sent to Mars for the express purpose of looking for life. On the evening of Friday, July 30, 1976, when the first results from the biology experiments were received, scientists gathered to see what *Viking 1* had found. "Oh, my God," said one as the data was printed off, "it's positive." Slowly but surely it seemed to become overwhelmingly clear there was evidence for Martian microbes. The laboratory's chief press officer went so far as to call the director at home: "They've found something."

The lights burned through the night in JPL's Building 264. Tired biologists excitedly scanned all the subsequent printouts, scarcely believing their graphs and readings. The numbers seemed to show that there was something very much alive in the soil. After centuries of speculation, years of preparation, and weeks of frustration in searching for a safe haven to land in, biologists had found their holy grail. "I could not believe my eyes," Gerry Soffen recalls. "It looked like there were signs of life from the data, and we sort of wandered around in a dream world."

One experiment team went so far as to order a crate of champagne and, with one eye firmly fixed on the historical record, signed their printouts. One day, they felt certain, they would have pride of place in museums like the Smithsonian Institution's National Air and Space Museum, opened just weeks earlier that bicentennial summer.

Not exactly. This unbridled optimism was premature, to say the least. By that last week of July 1976, the *Viking* biologists were almost as hermetically sealed as the experiments they had dispatched to the Red Planet. For many, the full glare of press interest was uncomfortable, and as one biologist, Norman Horowitz, was to remark, "if this were normal science, we wouldn't even be here—we'd be working in our laboratories for three more months. You wouldn't even know what was going on, and at the end of that time we would come out and tell you the answer."

None of them wanted to make mistakes in the full glare of publicity. What the press interpreted as equivocation was the normal conservatism of researchers; this was brought to bear when these first results were announced to the press that weekend. There was a great deal of activity in the soil, which publicly, one biologist was to declare, "may mimic—and let me emphasize that, may *mimic*—biological activity."

The surface of Mars was a truly alien environment, and the *Viking* biologists were uncertain of exactly what they were seeing. In the weeks that followed, as greater numbers of results were obtained and the experiments were run under different conditions, the pendulum of opinion swung back and forth over the possibility of life. Nevertheless, that first run-through had been staggering. After ten days on Mars, the initial findings from two biology experiments, which assumed terrestrial microbes could adapt to the harsh conditions of the Red Planet, found more or less what would be expected if such hardy life-forms were present in the soil.

The problem was the scale of what had been found. In both experiments, terrestrial life-forms would have taken time to grow in the biology chambers of the *Viking* lander. This would have started with gentle activity, becoming more intense over time. By contrast, the Martian results saw an immense burst of activity that, in Soffen's telling phrase, were regarded as "too much, too soon."

It was clear that the soils of Mars were far stranger than the biologists had anticipated. Most of the surface material was in the form of iron oxides, or "rusty dusts" as one scientist memorably called them. In fact, they contained peroxides and superoxides—industrial strength chemical agents—created by the unshielded ultraviolet radiation that streams down to the Martian surface. Much more reactive than the peroxides in bleaching agents found in domestic bathrooms, they released vast volumes of oxygen when humidified by the first two experiments at the end of July.

The Gas Exchange experiment (GEX) assumed that Martian microbes would be used to Earthlike conditions; that is, that they would be terrestrial-like in their behavior and metabolism but had adapted to the harsh conditions on Mars. An inorganic soup of nutrients was mixed with the soil sample and slowly humidified. The soil rapidly gave off oxygen, which at first glance would be expected if there were microbes present in the soil. Yet when the nutrient-rich soup was added, the oxygen tailed off. The spectacular oxygen peak showed there was something very active in the soil, but it was probably not biological in origin. Some biologists acknowledged that terrestrial microbes would neither release oxygen so quickly nor reproduce so rapidly, to generate such rapid activity.

More encouraging was the fact that the Labeled Release (LR) experiment produced a signal that definitely seemed biological. Its Geiger counter detected a very sharp rise of counts as the soil sample took up the radioactive Carbon-14. If there were microorganisms in the soil that had metabolized the nutrients, they would produce such radioactive results. The curve was so steep and formed so quickly that it seemed the sample was richer than any terrestrial soil. By comparison, preflight tests with freshwater samples from Lake Tahoe took 168 hours to reach the point that *Viking* got to in just forty hours. "The odds were overwhelming that nothing would happen at all," said Gilbert Levin, who had built the LR experiment, as he unveiled the graph to the press. "And when we saw that curve go up, we flipped."

The *Viking* biologists then had to wait a few days later for the findings from the Pyrolytic Release (PR) experiment. Norm Horowitz, the Caltech biologist in charge, had to incubate his samples first and then carry out his analysis in two separate stages, which took longer. Horowitz made sure not to hide the fact that he thought his experiment was far superior to the others. The instrument assumed Martian microbes would assimilate CO_2 and produce organic products as a result of being metabolized. As it wouldn't be affected by those puzzling oxidants in the soil, Horowitz had already declared, to the irritation of the others, that his "experiment is the only one that is designed to be purely Martian. . . . The others can say their results are ambiguous, but not mine."

The sample was illuminated by a xenon lamp in a chamber full of carbon dioxide that contained traces of the radioactive isotope Carbon-14. The gas was removed and the sample heated to drive off any of the radioactive carbon taken up. The initial measurements were six times greater than the anticipated value, suggesting that there was a great deal of activity—probably not biological—in the soil.

The *Viking* biology results continued to puzzle. In the weeks that followed, each of the experiments were run again—individually, and then together—under conditions that were different from the original run-through. The hope was that some clue to the strangeness of the soils would make itself apparent. So far as biology was concerned, that might provide a key insight.

The samples were fed greater amounts of nutrient, or were incubated for much longer. Sometimes, the amount of light and the temperature was changed. To avoid contamination, after each experiment run, the pipes, valves, and chambers were flushed with helium to clear them out. Nothing helped. The results got more difficult, rather than simpler, to understand.

Throughout the summer of 1976, the Gas Exchange experiment was run at different temperatures. The observed release of oxygen did not seem to be affected. It was clear there was a strong oxidant in the soil that was behaving in an odd way when the nutrient was added. Comparisons with the findings from *Viking 2*—as the soils of Utopia were appreciably colder and drier—showed that there was less oxygen released from the soil samples there. Reluctantly, the experiment team concluded that they were seeing a reaction between the water and oxidants in the soil.

The findings from the Labeled Release experiment were similarly perplexing. Although the "soup" fed to the soil sample was laced with Carbon-14, any microbes would give off telltale amounts of radioactive CO_2. Though this was indeed detected, the measured carbon dioxide ended abruptly before the soup was used up. When the experiment was run at different temperatures, the CO_2 always "plateaued out" at the same point on the scale. This would not occur with terrestrial microbes if they had been heated in the same way. Then again, Martian microbes might be hardier, as they had adapted to the harsh conditions on the Red Planet.

When the samples were sterilized before being tested, the peaks measured were not as high. That implied any microbes present had been killed off. Some of the scientists believed there was something in the soil that was producing unknown chemical reactions, which would explain the overall findings. Gil Levin was not convinced. As we will see in a later chapter, Levin believes that he did detect life and has largely been ignored. Even an official summary of the *Viking* findings points to the findings of the Labeled Release experiment as "the most controversial."

Norm Horowitz's Pyrolytic Release experiment was run nine times at both *Viking* landing sites. Sometimes the light in the experiment chamber was switched off, other times on. Sometimes water vapor was present, sometimes not. The results almost always produced a second peak, suggesting organic synthesis—in other words, nonbiological activity—was taking place. This could not have been photosynthesis, a key chemical reaction that often "feeds" life on Earth. As its name suggests, it needs the Sun to do that. On Mars, the results were the same whether they were run in light or darkness.

Water vapor is also needed for photosynthesis to occur. When water was added to the incubation process, no evidence for photosynthesis was ever detected. Heating the sample reduced the second peak; this finding was hard to reconcile with a purely biological interpretation of the take-up of radioactive carbon dioxide. Horowitz always felt that there was only a very small chance that there was life on Mars. Others were puzzled by his general skepticism. Horowitz was certain that his results showed some sort of inorganic catalyst that could synthesize small amounts of organic materials. He provided a pithy comment at the time. "One word was worth more than a thousand pictures," he said of his experiment results. "And the word was: no!"

By far the biggest stumbling block was the lack of organic materials found by the *Viking*s in the Martian soil, at least as far as instruments of the 1970s could tell. Life as we know it is based on the biochemistry of large molecular chains, which involve carbon as its structural backbone. At both landing sites, the *Viking* mass spectrometers found no organics down to a level of parts per billion. And though it could not determine exactly how carbon had arranged itself in its many molecular manifestations, the mass spectrometers could sense if carbon was indeed present. They did not find any.

In other words, if there were Martian microbes, where were their bodies? So far as Gerry Soffen was concerned, that meant game over. *No organics*, he thought, *no life on Mars*.

The answer may lie in the scavenging nature, chemically speaking, of the Martian soil. Unprotected from the continual bathing of the harshest of ultraviolet light, the soils would be bleached and oxidize anything in sight. Any organic molecules that happened to be around would be scavenged away.

The paradoxical results from the water-injected samples in the LR and GEX experiments were believed to be due to reactions caused by the peroxides in the soil that had been generated by that all-pervasive ultraviolet radiation. The best guess was a series of unexpected and unexplained chemical reactions between carbon monoxide molecules and certain rusty dusts on the surface. In other words, what was being seen were the products of super oxidation in the soils not biological activity.

The overarching impression that Mars was sterile came from the results from *Viking 2* in Utopia. Its findings were broadly similar, even allowing for the colder, drier environment farther north. "The fact they both gave the same sort of answers was important," Soffen says. "The answer was that there was no life on Mars."

Because the most aggressive oxidation reactions were found in the driest samples, it became a matter of faith to find somewhere "wetter" where the likelihood of life would increase. The more optimistic Viking biologists suggested there may be isolated oases for life where the water is. The soils surrounding the polar caps where permafrost prevails is seen as a likely place where extant life-forms may be ready and waiting.

Despite this, Gerry Soffen believed it highly unlikely Mars harbors life. In the immediate aftermath of *Viking*, he had come to the end of the road. After being immersed in the complexities of Martian biology for two decades, it was time to move on. In 1978, he took a sabbatical at Harvard University, and on his first day wrote "Goodbye Mars, Hello Earth" on the blackboard.

Soffen was not alone. A consensus similar to a morning-after feeling enveloped the *Viking* biologists. They hadn't found life on Mars for the very good reason that there wasn't any to find. "The failure to find life on Mars was a disappointment, but it was also a revelation," Norman

Horowitz wrote ten years after the landings in an excellent book about life in the solar system, *To Utopia and Back*. "We have awakened from a dream. We are alone, we and the other species, actually our relatives, with whom we share the Earth."

* * *

In time, all the *Viking* biologists moved on to other things, meeting on anniversaries of the landing, forever toasting their good fortune at being at the heart of this greatest of all scientific detective stories. Sometimes they would wistfully wonder at what might have been on that perplexing planet so near, and yet so very far from human experience.

"*Viking* was the greatest experience of my life," Soffen recalls. "In one sense it was the greatest joy, and I often wonder how do you pay back the taxpayer for letting us search for life on another world? But it was also the greatest tragedy, in the sense that we found no signs of life on Mars."

He moved on to other work, heading NASA's life sciences program for a while. He then moved to the agency's Goddard Space Flight Center, just outside the Washington Beltway, to become NASA's liaison officer with universities. He was content for a new generation of "astrobiologists," as they now call themselves, to ponder the question afresh, bolstered by work on Martian meteorites and the realization that life is hardier than it seemed in the 1970s.

Gerry Soffen was always around for what he wistfully termed "the gathering of the clans." He was eventually appointed to head NASA's own Astrobiology Institute, a loose affiliation of researchers around the United States who could meet and plan in cyberspace rather than take endless airplane flights, as they had done during *Viking*. He passed away, at the age of seventy-four, in 2000.

"Mars never quite leaves you," Soffen had said four years earlier. "I am just very grateful to have explored it and looked for life in my own lifetime."

By then, a new generation had come to the fore.

5

THE MEASURE OF MARS

For Dr. Michael Malin, there is nothing more exciting than the vicarious exploration of the Red Planet, roaming at will across the perplexing and utterly captivating surface of another world. Mars has thrilled him since he was a child and motivated him to become a scientist in the first place. "When I was young, I promised myself that I would never work at a job that wasn't fun," he says. "I have been very lucky to have had the opportunity to keep that promise."

Malin now has a ringside seat looking at other worlds. His cameras are sending pictures back from Jupiter, the Moon, and, over the last two years, the surface of an asteroid called Bennu. But the Red Planet is where he has made his name and, indeed, devoted most of his career. Unhindered by the geological jargon in which he must couch his professional work, Malin has made the magnificence of Mars comprehensible to anyone through his photography. Essentially, he followed the dream he first had as a young child, of making the unreal look real. It is no exaggeration to say that thanks to his efforts, we have gained the full measure of Mars.

"When I was a kid, I found pictures of stars and galaxies beautiful, but somehow detached and unreal," he says. "Pictures of the Moon and Mars were, for me, more real."

As part of that greater quest, Malin has made extensive studies of landforms on Earth that are similar to those on Mars. Over the last forty years, he has participated in geological field trips to places such as

Iceland, Alaska, Hawaii, Mount St. Helens, and, for someone interested in the Red Planet, somewhere that is de rigueur, Antarctica.

These places have all provided clues to what Mars is really like. Relating what is on the ground to what can be seen from above is an area of study that Malin has made all his own. "It became clear that to understand Mars, we needed observations that bridge the gap between the landers and the orbiters," he has said repeatedly in interviews. The Red Planet has come alive in front of his and now everybody else's eyes. It has changed beyond all recognition. "These instruments are transforming parts of Mars to places as familiar as our own backyard," Malin has said. "It is a great privilege to be able to do that."

Yet the Red Planet is more complicated than even Malin had ever thought possible. A steady stream of startling images has shown, from above and on the surface, that Mars was geologically and volcanically active until comparatively recently. The Martian surface is much more dynamic than had ever been thought possible. "Even in areas we pretty much thought we understood, we've been wrong," Malin says.

As well as triumphs, there have been frustrations. Malin has lost at least four cameras due to spacecraft failures. His achievement is all the more remarkable given a long and drawn-out battle he had thirty years ago with scientific bureaucrats. Strange as it may now seem, it nearly came to pass that America's return to the Red Planet after two decades might have been completely blind. There was a sizable body of opinion that didn't want any new cameras sent to Mars. Without them, our understanding of the Red Planet would have remained horribly incomplete.

* * *

When Neil Armstrong and Buzz Aldrin became the first human beings to walk on another world, in July 1969, Malin, then a nineteen-year-old undergraduate, was allowed into the inner sanctum of Mission Control in Houston to witness the events firsthand. It was a unique experience for anyone, let alone a Berkeley physics student. He had also been invited to witness the launch, and then a week later, the splashdown of *Apollo 11* in the Pacific Ocean aboard the recovery ship. The young Michael Malin was granted this extraordinary privilege because of his abiding interest in space and a certain naive chutzpah. "I was a true child of the Space Age," Malin reflects, "and I had bombarded NASA with letters to let me witness this astounding event in history."

Malin had no doubt about what he wanted to do. In the summer of 1969, the feeling was that after the Moon, the next big step in space would obviously be sending humans to Mars. "I wanted to be an astronaut," he says. "I wanted to be the first human being on Mars!"

Malin grew up in the San Fernando Valley in the late 1950s and early 1960s, when the space race was characterized by its infancy and intensity. At an early age, he began a scrapbook collection of newspaper clippings about space. He entered into correspondence with officials at the nearby Jet Propulsion Laboratory and, later, NASA Headquarters in Washington, DC. "I would write in with detailed questions," he says. "Today, I get the same kind of questions from kids. I realize how the system works."

In 1968, through his newfound correspondents, he requested to attend an Apollo launch. One letter landed on the desk of Thomas O. Paine, then a mid-level NASA official. That October, Paine invited Malin to attend the liftoff of *Apollo 7*, the first *Apollo* launch with astronauts aboard. Unfortunately, Malin had the misfortune to develop pneumonia. Not wanting to miss a true chance of a lifetime, Malin later wrote to Paine asking to attend the launch of the first attempted Moon landing.

By that time Paine had taken over as NASA administrator. Malin was amazed to find himself invited out to the launch of *Apollo 11* in July 1969. The Berkeley undergraduate gamely caught the cheapest flight he could find down to Florida. When he checked in to register at Cape Canaveral, Malin was momentarily scared. His name did not appear on the official invitees list.

To his and the protocol people's mutual astonishment, he was listed with the Very Very Important Personnel. The other names were Vice President Spiro T. Agnew and former president Lyndon Baines Johnson. "I'm sure that was Tom Paine's doing," Malin says today. "He had a wicked sense of humor." After liftoff, Malin managed to wangle an invite to Houston and the splashdown recovery. Only the backup astronauts ever had that amount of involvement in the flight from start to finish. "It was a once-in-a-lifetime opportunity, and I knew I was going to be involved in the space program."

After getting his physics degree from Berkeley, Malin returned to southern California to pursue a PhD in planetary science and geology at Caltech from 1971 to 1975. These were heady times for anyone interested in Mars. *Mariner 9* was returning a stream of observations from orbit, and the *Viking*s were being prepared for launch. As a student of Bruce Murray, then a Caltech professor, and Bob Sharp, a future collaborator and expert in geological morphology whose name is immortalized in the mountain at the center of Curiosity's landing site, Malin was in a unique position. When Murray became JPL director in April 1976, Malin moved across to the Lab with him. That same year, Malin was awarded his PhD; his thesis was entitled *Some Topics in Martian Geology*. He later coauthored, along with Murray and the late Ron Greeley, a standard geological textbook about the terrestrial planets.

Unfortunately, in the late 1970s, the bubble had burst for the first wave of planetary exploration. There would be no new US planetary missions from 1978 until 1989. Malin remained diligent in the doldrums. By the early 1980s he had moved to Phoenix, Arizona. Plans for an immediate follow-on to *Viking*, involving an ambitious roving vehicle, had soon been abandoned. More modest proposals were called for. As the coming man of Martian geology, Malin was elected to a committee that looked at what was then called a Mars geoscience mission. It was intended to answer basic questions about the surface chemistry of the planet.

At one meeting of the committee held at JPL, another group in the room next door was looking at climate questions and the role of water in Martian history. Lean times prompted what once would have been considered a drastic measure. "The two proposed missions had about 80 percent combined goals," Malin says. "It made sense to merge them."

In due course, the Mars Geoscience/Climatology Orbiter emerged, later renamed the Mars Observer. In this, its earliest genesis, there was no provision for a camera. Even today, there is always a residual antipathy to "imaging systems," as cameras are more formally known. The more conservative scientists feel that cameras are there for reasons of hype, glitz, and public relations, producing nice flashy images but nothing of particular scientific value. Those more attuned to the taxpaying public's needs realize that pictures are always a project's best foot forward.

To be fair, this prejudice is sometimes born out of the limitations cameras impose. They are heavy and take up valuable margins of mass and volume that would otherwise be filled by greater numbers of instruments. Cameras by their very nature need a lot of electrical power. They consume most of the capacity of already limited data channels to transmit their images back to Earth. There are also less lofty reasons, to be sure. Scientists are just as suspicious, jealous, and egotistical as the rest of us. The spokesperson for the camera team will find him- or herself on the nightly news bulletins and generally lionized by the press.

Characteristically, Mike Malin wouldn't let up. Almost grudgingly, meteorologists said that maybe a low-resolution camera might be useful in creating daily weather maps for the whole of the planet. Yet, attuned to the mood of the scientific community and the financial savings they could realize, NASA Headquarters did not include a camera when it first accepted a baseline design for the Mars Observer in 1984.

Malin remained undaunted. He proposed that it was better to have a camera as a backup than not at all. His first design was about the size of a kitchen trash can, with a mirror twenty inches (fifty centimeters) across. It would be capable of resolving features smaller than three feet (a meter) on the Martian surface, which would bring about a quantum leap in observing Mars. "I knew the resolution I needed to answer important questions from field geology on Earth," he says. "If you needed to know what was going on at the surface of Mars, you needed to go down to twenty-five-centimeter resolution."

A lot of his colleagues simply laughed at his proposal. So a more modest version capable of seeing down to a meter on the surface was proposed, taking advantage of new electronics then coming onto the market. In the end, Malin designed a camera that took up only 10 percent of the mass and power of the total spacecraft. "In everything except volume, I was small," he says with some considerable pride even today. The agency, however, would not commit itself.

What transformed NASA's institutional half-heartedness about images came from a simultaneous triumph and tragedy in space. In January 1986, *Voyager 2* reached Uranus. After a journey of nine years and five billion miles (8 billion kilometers), it had survived any number of technical mishaps to get there. It was deaf, arthritic, and slightly senile—which, as one bitter planetary researcher remarked, was a bit like the then incumbent in the White House.

The slowdown in planetary exploration at the start of the 1980s could be laid at the door of Ronald Reagan's presidency. One of the worst effects of Reaganomics was a desire to lop off the budget needed to keep *Voyager 2* "alive." Somehow it survived. Alas, the crew of the twenty-fifth Space Shuttle mission did not. They were launched into space at the same time as *Voyager 2* departed from Uranus. When *Challenger* exploded in the skies above Florida on January 28, 1986, so too did America's ability to send anything into space. The Shuttle fleet was grounded. NASA no longer had the wherewithal to send a backlog of Earth satellites aloft, let alone Mars Observer to the Red Planet.

Within weeks, the folly of relying totally on the Shuttle was brought home by a celestial apparition. In March 1986, the Europeans, Japanese, and Soviets sent probes to return information directly from Halley's Comet. America did not, foiled by Reagan's cutbacks to NASA's budget earlier in the decade.

It was a lesson not lost on NASA higher-ups, who suddenly demanded to know why there was no camera on Mars Observer. Malin needed little persuasion to revamp his earlier proposal. Two other camera designs were also considered. His own came out on top in NASA's estimation, Malin says, not just because of its technical excellence but because he took great care to look after the "production values." In his application, Malin used color and laser-printing—a novel and expensive technology at the time. In the 1980s, this made a deep impression. "It was probably the first laser-printed document anyone had seen at NASA Headquarters," he says. "Today it's a difficult concept to understand that such stuff was revolutionary."

And so, after innumerable trials and tribulations, the Mars Observer Camera was accepted "as everyone finally understood that the instrument would basically revolutionize our picture of Mars." Even so, Malin was aware that minimizing the development expenditure was going to be important. "If we built the camera ourselves, that's how we could keep the cost down," he says. And so, at Arizona State University, a team of engineers fabricated an instrument that was originally intended for launch in 1990 but was delayed until 1992 because of a backlog of Shut-tle-related problems.

And after Mars Observer was en route, disaster struck at the last min-ute. Just hours before it was due to enter orbit around Mars in August 1993, it disappeared without a trace. Never again that decade would NASA roll all its dice on one large mission—which cost $600 million. Any future spacecraft to Mars would have a budget ceiling of $150 mil-lion, or $300 million in today's dollars. This approach had serious conse-quences, not least for Michael Malin himself.

* * *

After spending most of the 1980s as a professor at Arizona State Uni-versity, Mike Malin was awarded a MacArthur Fellowship—known as a "genius grant"—in 1987 for his development work in many fields. These included geological studies of Antarctica, pioneering work in computer graphics, and the first-ever digital camera, designed to map the Martian surface in rich and exquisite detail.

The quarter million dollars bestowed by the fellowship "gave me some financial freedom," as he says. So Malin decided to return to his native California and set up his own company, Malin Space Science Sys-tems. His was the first private company involved in the direct scientific exploration of the Red Planet. And after the loss of Mars Observer, Malin could move fast. He turned to what he has called a team of "bright, young, gifted, and iconoclastic engineers."

Thankfully, spare parts for the Mars Observer Camera had been fashioned during development by those same engineers. These included spare mirrors and the graphite epoxy structure of the camera housing, which had been assembled but not tested. There were similar pieces of hardware for the spacecraft as a whole in laboratories all around the United States. Known as "flight spares," they represent an insurance pol-icy in the event anything goes wrong before launch. Rather than rebuild the whole mission, the most obvious way to meet the original scientific goals was to split the instruments up. They would all be flown separately, mixed and matched on whichever mission was next up to go.

The first follow-on was given the name Mars Global Surveyor. It could only carry half the payload of the original mission that had been lost. The heaviest hardware would be delayed for later flights to Mars. The delays also allowed engineers to take advantage of lightweight "composite" structures that had come into general use since the Mars Observer had originally been built.

With lighter structural components, the replacement craft could be boosted to Mars on a less powerful, and therefore cheaper, launch vehicle. Due to suspected problems with the original radio transmission and engine systems, a new propulsion system was built, as were more powerful amplifiers to allow constant transmission back to Earth. Instead of a tape recorder, which was susceptible to jamming, state-of-the-art solid-state recorders were used.

The Mars Global Surveyor became the first US orbiter sent to the Red Planet in twenty-one years, a curious hybrid of the new and the tried-and-tested. Pride of place was given to Mike Malin's camera. To save weight and reduce costs, the revamped spacecraft could not carry as much fuel. So, to brake itself into Martian orbit, it had to use the revolutionary aerobraking technique that, engineers fervently hope, will one day allow crewed missions to reach the Red Planet on limited fuel supplies.

When it arrived at its destination in September 1997, the Mars Global Surveyor had to repeatedly pass into the uppermost reaches of the Martian atmosphere to slough off speed. This ensured that a long looping "capture" orbit could be changed into a more or less circular one just 230 miles (370 kilometers) above the planet. "I designed the camera to work in that particular orbit," Malin says. "It would give us our best ever views of surface features."

* * *

In an increasingly digital age, Mars came under the scrutiny of the first truly digital camera for the best part of a decade. The Malin Camera, as it was usually called, was more formally known as the Mars Orbital Camera or MOC, giving it the same initials as the one flown on Mars Observer. The camera comprised three functions combined into one and worked, its designer says, like a glorified fax machine. As the spacecraft moved across the Martian surface, the camera swept up light and reflected it via a mirror onto the most sensitive image sensor ever sent to the Red Planet thus far.

It was all the more remarkable because it employed no moving parts. The camera focused by the action of a heater. With the same sort of sensors that a very fancy color photocopier might employ, the camera had a metallic mirror that changed shape by use of heat, which expanded and contracted the mirror in a way that could be carefully controlled. In this way, features below were brought into very sharp focus.

The Malin Camera employed a sixteen-inch- (forty-centimeter-) deep graphite epoxy barrel. The light from the planet below was split in three ways. This included the narrow angle field of view, which concentrated light down to a resolution of three feet (one meter) on the surface. In its wide-angle mode, the camera could observe down to 650 feet (200 meters), which made it ideal for looking at the whole of the planet. Light entering the wide-angle sensors was split through violet and red filters, effectively producing two cameras in one. By comparing wide angle images through the different filters, meteorologists got what they had originally asked for. They could distinguish fogs near the surface from high-altitude clouds.

The sensors at the back of the camera were similar to those you would find in a Hi-8 video camera of the era (and most security cameras today). Known as Charge Coupled Devices (CCDs), they were capable of far greater resolution than all earlier cameras sent to Mars. The relationship between their individual picture elements and the actual surface of Mars was easier to quantify. That meant sharper, more versatile images that lent themselves to greater enhancement.

Most digital cameras work by having an array of sensors that can take in a whole field of view. Malin's camera, however, was limited by both the electrical power available and the fact that it was traveling very fast—1.8 miles (three kilometers) per second—across the Martian terrain. Technically, the camera was known as a "push broom" instrument: it worked by sweeping up light as it headed over the surface of Mars. The actual area of the array was reduced so that the actual "push broom" itself was about two inches to a small fraction of an inch (five centimeters by a millimeter across)—revolutionary in 1986, when the camera was designed, but primitive by today's standards.

The CCD array was long and thin because, as with any camera system that is in motion, individual frames had to be "frozen" by shuttering. The elongated shape allowed for a much faster shuttering. That way, images of the surface did not appear as a blur. In fact, the Malin Camera had to shutter far faster than any spacecraft imaging system previously sent into space. A mechanical shutter would be out of the question. It could not blink fast enough. To capture the surface below, the camera had to repeatedly snap at shutter speeds as fast as 1/1,200th of a second.

The highest-resolution images required the pixels to be packed in very tightly when they were transmitted back to Earth. There was actually the equivalent of a digital logjam: the information recorded in each frame consisted of 40 million pixels. Given that 5 million pixels was the limit of how much data could be transmitted every second from Mars Global Surveyor, each picture had to be stored onboard. It was then read out in any number of sections. Also, the relative positions of Earth and Mars altered the amount of data that could be received. By 1999, for example, there was only about half as much data received on Earth because Mars was farther away from our world than when it had arrived two years earlier.

So, for the first time, a powerful chip in the camera stored the visual information. It then rationed it out to the spacecraft's data transmission system. "We can trickle it out at any rate you want," Malin explains. "So we take up the slack from the other instruments which use less data capacity. Whenever there's any spare capacity available, we will make use of it."

When the Mars Global Surveyor mission was completed in the mid-2000s, Malin reported, "The Big Picture is emerging from this camera. It is giving us a much more comprehensive view of Mars than we had ever thought."

* * *

The surface of Mars remains enigmatic, though. Where old riddles have been resolved, new ones have been thrown up in their place. The planet has emerged in a brand-new light, hinting that its past was more conducive to life than had originally been thought. "Every place we look, we see complexity rather than simplicity," Malin says. "And almost everywhere we look, we see dunes."

"Mars is really more Earthlike than we give it credit for," he says. And that, paradoxically, is difficult for someone who likes to understand the Red Planet in terms of the geological processes with which he is familiar on Earth.

Significantly, Malin's pictures have effectively extended the possibility for life in either direction of time. There is both old and new volcanism, hinting that Mars was geologically active for far longer than previously thought. Thanks to the largely unexpected persistence of this potential energy source, the possibility for microbial life has increased. The only disappointment, so far as the general public is concerned, is that his cameras cannot detect signs of life directly. "There's nothing I can do to distinguish microbes," Malin notes. "We already know that there aren't Martian giraffes or elephants. Unfortunately, nobody has proposed a hypothesis for Martian life which can be tested by my camera."

By far the biggest surprise was the identification of widespread deposits of sedimentary material, some with a thickness of tens of kilometers (or miles), on a scale that had not been anticipated. In some parts of the crust, it had been previously known there would be layers of sediments similar to those that had accumulated in lake beds (and subsequently seen close up by the Curiosity rover in Gale Crater).

From earlier missions, the most pronounced layering had been found in the canyons of the Valles Marineris, the vast chasm that stretches a fifth of the way around the planet's circumference. It had been suggested that the observed sediments were due to the accumulated buildup of water, a possibility that would have been good for the sustenance of life, as lakes would have repeatedly evaporated and been filled over time.

Malin's camera showed that this kind of layering is ubiquitous across the whole of the planet. It is especially notable with layers seen within the equatorial canyons, particularly an impressive one called Coprates Chasma, and also within crater rims. In other words, the layers can't just be because of sedimentary deposits. In the eastern extremities of Coprates Chasma, there is enough layering to have cut through 33,000 feet (10,000 meters) of rock. Those layers have been used to help decipher the history of the planet. "What we do see with the layering is, very early on in Martian history," Malin says, "when the geology was very much more dynamic."

Hitherto, it had been thought that many of the canyons would reveal bedrock fractured by the heavy bombardment that accompanied the birth of the solar system. Thereafter, they would have been covered by lava from subsequent volcanism. Yet Malin and his colleagues have found that the layered deposits span the entire length of the Valles Marineris more or less uninterrupted.

There are light and dark layers that follow the contour lines of the canyon. Because they are intact, they had to have been deposited *after* the end of the heavy bombardment. The structures and spectra—from other instruments—of these layers suggest that they resulted from lava flows. As a rough calculation, Malin estimates that there was ten times more lava than previously calculated during the first four billion years of the planet's history. That much molten material would cover the area of the continental United States to an equivalent depth of four miles.

"We conclude that volcanism on early Mars has probably been much more extensive than previously documented," wrote a team of authors, including Malin, led by Alfred McEwen of the University of Arizona, in *Nature* in February 1999. "And it must have affected the climate and near surface environments."

The Red Planet has also been volcanically active in more recent times. This means that there would have been more energy available for a greater time period during Martian evolution. This in turn means there could have been greater amounts of energy around to start biochemical reactions. The crust of Mars could have been a much more significant biological crucible than had ever been appreciated. To understand why, the role of craters has to be examined afresh.

* * *

To most people, one crater looks very much like any other. Five decades of planetary exploration have revealed that bodies across the solar system often exhibit telltale signs of these pockmarks to some degree or other. As such, craters provide a unique insight into what happened in the earliest years of geological history.

Cratering was ubiquitous. The birth of the planets was accompanied by heavy bombardment, when all the remaining flotsam and jetsam slammed into planetary surfaces just under four billion years ago. Thereafter, the bombardment rapidly tailed off. Yet there was still enough material orbiting the Sun to intermittently pepper craters across the surfaces of planets and moons.

Because of Earth's active geology, our home planet is an exception to this rule. Much of our cratering record has subsequently been erased, but a few remain. One of the most notable is Barringer Crater, near Flagstaff, Arizona, which has been pored over by geologists for many years. Significantly, the Apollo astronauts trained for their missions to the Moon in and around it. This helped them learn the geological techniques to help unravel the complex history of lunar evolution. They looked for the most interesting rocks. From these targeted samples astronauts collected in and around lunar craters near where they landed, geologists back on Earth have been able to piece together a coherent history of the Moon over geological time.

Craters act as chronometers thanks to that first, brief flowering of human exploration. The Moon rocks are the gifts that keep on giving. Because lunar samples can be dated from the radioactivity of their naturally occurring isotopes, it is possible to compare the actual ages of rocks picked up by the astronauts in and around the craters that ejected them. It's similar to the way that you can gauge, on a snowy day, whether the postal worker came before your spouse returned from work by the way their respective footprints have been covered by subsequent snowfall. Younger craters appear much fresher than older ones. The difference in age can be checked against the natural radioactivity of the rocks that were picked up by the astronauts on the Moon. Apollo enabled geologists to put actual numbers to their previously estimated ages.

In other words, the Moon calibrated crater counting, a technique that has become standard in planetary geology. It has allowed geologists to piece together histories of other planetary surfaces such as Mars. While it has not been possible to determine the age of features absolutely, the cratering record can at least show the order in which they formed.

Cratering on Mars has shown something surprising. Some small craters have been filled in completely, destroying any chance of dating that part of the Martian surface. This widespread infilling from accumulated debris and dust has meant bad news for geologists expecting to find bedrock within easy reach. At face value, this implies it will be much more difficult to find evidence for fossilized microbes on Mars. The bedrock, where the fossils would be located, is covered in extensive dust.

Small craters are significant; surface erosion will preferentially remove the smallest craters at the expense of larger ones. Counting smaller craters in and around the edges of volcanic plains on Mars has generated a more accurate chronology of when this more recent volcanism took place. The discovery of basalt—a fine-grained rock formed by volcanic activity—that crystallized 1.3 million years ago has been made from observations from the Malin Camera. "The crater statistics that we report here suggests that volcanism is continuing on Mars in current geologic time," reported another team in the same February 1999 issue of *Nature*. That is, volcanic activity has taken place as recently as roughly one million years ago, if not more recently.

* * *

Michael Malin's magical camera on the Mars Global Surveyor inaugurated a new era in scrutiny of the Red Planet. While some people would be satisfied with blazing the trail for others to follow, Malin was more enterprising in his thinking. He recognized that finally, after decades of neglect, the Red Planet was once more emerging center stage. At the start of the twenty-first century, NASA's Mars exploration program had set an ambitious target of dispatching spacecraft there nearly every two years, whenever Mars and Earth are at their closest in their respective orbits. Give or take a few delays, that is precisely what has happened. Almost every time an American spacecraft has headed to Mars, one of Malin's cameras has been aboard.

His next camera system after the Global Surveyor was THEMIS (Thermal Emission Imaging System), a device that extends its vision into the infrared. This instrument was developed, while Malin was still in Arizona, with his colleague Philip Christensen as a camera with additional infrared detection. Christensen tells a revealing story. In the late 1980s, there was a meeting at his house, when Malin complained of feeling unwell. Dr.

Christensen's wife, a leading obstetrician in Phoenix, ordered him to lie down. When he still complained, she took him to a nearby hospital, as she suspected he was undergoing a coronary, as indeed he was. She made sure he was hooked up to the right machines in time. "Except when Mike tells it, he always says we were ignoring him," Christensen says. "He had to drive himself and protests that nobody took any notice of him."

At times, their relationship has been equally fraught. Phil Christensen freely admits that he once didn't talk to Malin for nearly a year, even though they worked just down the corridor from each other. "I think we can both get a little stubborn," Christensen says. "But he is one of my closest friends."

Their instrument has been rather more harmonious in its working. It traveled aboard the NASA Mars Odyssey orbiter, which arrived at Mars in October 2001. Amazingly, both spacecraft and instrument continue to operate around the planet in good health nearly two decades later. Malin's camera and Christensen's spectrometer continue to map out the surface in tandem.

The spectrometer examines "thermal inertia," the tendency of an area to remain at a given temperature. In the near infrared, this reveals clues to the physical properties of the surface. Dunes, for example, will trap heat more effectively than rocks. So the "heat signature" of a certain area gives geologists a better sense of what lies below the surface. In particular, it has revealed how volcanic flows and ice under the soils have shaped many surface features.

The camera, which examines only in visible light, was built by Malin's company. It can resolve down to around a hundred feet or so, in conjunction with Christensen's spectrometer. The camera has been useful for peering into crater walls and searching out interesting material, which then shows up in the infrared. Together, their many discoveries include finding water ice at the Martian south pole, learning that some gullies were formed by melting snow, and even spotting weird gas jets bursting out from the Martian ice cap. "Mars Odyssey's enduring success has let THEMIS achieve a longer run of observations than any previous instrument at Mars," says Christensen. "It has provided the context for most recent Mars scientific research."

And context formed a key part of the next Malin camera.

The Mars Reconnaissance Orbiter arrived in March 2006. It too is expected to last until well into the 2020s. Behind its overall mission, explains the mission's project scientist, Dr. Richard Zurek at JPL, the aim was to follow the water. "Well, we knew there was water on Mars," he explains. "There's some in the atmosphere. There's ice at the polar caps. But, was there water elsewhere? The evidence that we are looking for is not only water today, but also ancient water on the planet."

The Mars Reconnaissance Orbiter features two Malin cameras, the Context Camera (CTx) and the Mars Color Imager or MARCI, which observes the state of the Martian surface every day. Malin designed the Context Camera to work in concert with two other instruments that have been described as taking a microscope to the Red Planet. These are HiRISE (a high-resolution camera built by Ball Aerospace), whose individual pixels are twenty-five centimeters (roughly ten inches) across, and a spectrometer called CRISM, described as a "chemical camera," which maps minerals associated with where liquid water has reacted with the surface rocks.

"If there were oceans and lakes," continues Zurek, regarding the water-based scrutiny, "they would have formed certain minerals that we should be able to detect. Finding them means we can look at that ancient history of Mars or earlier history of Mars."

Malin's instruments have provided a wider perspective on the surrounding areas mapped by the higher-resolution cameras. By March 2017, the Mars Reconnaissance Orbiter had completed more than 50,000 orbits of the Red Planet. Malin's Context Camera had taken more than 90,000 images, which meant that 99.1 percent of the planet has been photographed. Even allowing for dust, clouds, and camera snags (also called "bit hits," where information is sometimes corrupted), they have provided what Malin calls a baseline for observations, to discern any changes that take place between orbits, such as the fall of avalanches around the north pole that were observed in 2015.

By comparison, the HiRISE instrument has mapped only 3 percent of the surface. Such a zoom lens can see features as small as a desk. It has discovered details concerning fresh craters and fans of soil deposited by water in the ancient past, as well as more recently.

CRISM looks at the composition of rocks and where water may have pooled in the ancient past. Different minerals absorb infrared light at certain characteristic wavelengths. Telltale dips in the observed signals, known as absorptions, are similar to fingerprints in that they are unique to the chemistry that created them. "While rovers have the advantage of seeing the small detail of the rocks down to the sediments that form them," says Scott Perl, an investigator on the team, "they are limited to other observations over the distance the rover has driven."

CRISM can see the whole world view, which has another advantage. Both CRISM and the Malin Context Camera grab images in stereo for selected portions of the Martian surface. This is achieved by rolling the spacecraft as the images are electronically shuttered, which generates mosaics useful in unraveling the more complicated geology. Re-creating these images back on Earth allows geologists to see the surface in three dimensions, "with mineralogical information attached," in Perl's phrase. In other words, the effect of water, in terms of the minerals it has fostered, stands out.

Malin's other camera on the Mars Reconnaissance Orbiter has a visual acuity far better than a human eye: across five wavelengths in the visible and two in the ultraviolet. MARCI has generated Mars meteorology status reports for a decade and a half. The Malin website has published them virtually every week. Throughout 2018, for example, MARCI watched the Red Planet being engulfed and then, over the following year, recovering from the effects of a global dust storm.

For meteorologists interested in Mars, understanding global dust storms is a holy grail, especially the way in which they are triggered and how they then switch off. During this most recent storm, large columns of churning dust were spotted within the global murk by instruments onboard the Mars Reconnaissance Orbiter, including MARCI. Some of these dust columns took many weeks to dissipate and, it is believed, may well be involved in carrying greater amounts of water vapor into the upper atmosphere. There it will be broken down by the ultraviolet radiation streaming in from the Sun.

The weather-obsessed on Earth can now watch meteorological features on another world. "Weather patterns were fairly typical last week for this time of Mars year," the Malin website reported just before the Red Planet passed into conjunction (it appeared to pass behind the Sun as seen from Earth) in late August 2019. "Both the Curiosity Rover in

Gale Crater and the InSight lander on Elysium Planitia had storm-free afternoon skies all week."

By monitoring how the dust has been swirling around the landing site for the upcoming Perseverance mission, project scientists believe they won't encounter any problems during its landing in February 2021.

That all this visual information is streaming down from Mars is quite an achievement. Remember that only twenty years ago, there were not that many Martian spacecraft. Today, there is a small fleet of international spacecraft in orbit around the Red Planet. This network is delicate, as recent events show. At the start of 2018, NASA had two rovers working away on the surface; since then, Opportunity has been lost and Curiosity has had the mechanical problems noted earlier. Our measure of Mars is utterly dependent on mechanical whims and the vagaries of government funding. It is too easy to imagine a perfectly healthy Mars mission being canceled because of some other political priority. In March 2020, for example, the budget for the Mars Odyssey orbiter (on which THEMIS flies) was removed in the Trump administration's budget request for NASA for the next financial year. To have a constant stream of information from the Red Planet every day is no small achievement. It should never be taken for granted.

For now, though, the tsunami of data keeps on coming. As the Mars Reconnaissance Orbiter mission continues, all three instruments—and the Italian-built radar—have trained their eyes on the puzzling layered terrain that Malin's original camera revealed in such surprising detail. Many of the layers could have been caused by water-deposited sediments, or perhaps layers of volcanic lava, volcanic ash, or layers deposited by wind. It is still not possible to say with any certainty which of these scenarios was the most important.

For that, the full picture—the wide angles from the Malin Context Camera, the close-ups from HiRISE, and the mineralogy from CRISM—will be needed. All will be used in the years ahead to unravel these ongoing riddles about Mars and how it has evolved over geological time.

* * *

When telescopic observers in the nineteenth century drew the first maps of Mars, they discovered a bright, instantly recognizable feature that they soon adopted as their prime meridian. Greenwich on Mars was christened Sinus Meridiani, the Meridian Bay. Despite the fact it was later shown by orbiting spacecraft to be a completely desiccated feature, observations have indicated it may well have been an old ocean basin. The most recent Mars orbiters have discovered that its floor is covered in soft, easily eroded rocks. The layers spotted within them suggest that the area was likely flooded with water in the ancient past.

Within its first year of operation in 1998, the original Malin Camera also discovered a huge deposit of hematite in this region, which still stands as one of its most significant discoveries. Mars, it may be argued, is red because of hematite, an iron oxide formed by the interaction of water and rock. There is also a rarer gray "specular" form of hematite that has also been confirmed by Phil Christensen's infrared spectrometer. "Hematite is mineralogically fascinating," Christensen says. "It is the first, best indication of liquid water on Mars."

This particular hematite-rich region is very intriguing. On Earth, hematite tends to precipitate out in pools of water or hot springs. Hematite and iron share the same chemical formula, but hematite has larger crystals, akin to grains of sand. Coarsening up the crystals, so to speak, requires water—perhaps enough water to have also supported life. While it is not quite accurate to say that where you find hematite you'll find life, it certainly is very useful in indicating where microbial life might have congregated.

It was for this reason that the Opportunity rover was targeted to land in what has been renamed Meridiani Planum. When it did so in 2004, it found plenty of hematite (as did its twin, Spirit, on the other side of the planet). More recently, further observations from the Context Camera have been so promising that NASA has pegged this same region as a possible spot for humans to land. It is easy to understand why. "Even more intriguing is the possibility that the hematite may have initially precipitated from a large body of water," Christensen has said. "This is one of the best places to look for evidence of life on Mars."

Opportunity came to rest next to exposed outcrops of sedimentary rocks. The rover discovered telltale signs of an "ephemeral lake"—where mud was enriched in salts that had precipitated out as the water evaporated, leaving behind sandstones rich in sulfates. Similar salt flats populate parts of the United States, for example in New Mexico and Utah. The percolation of groundwater through these rocks generated hematite concentrations within them.

These findings buttress the notion that the water was around for quite a long period of time. Meridiani Planum is rich in history. A trained geologist could spend a lifetime exploring just one of its craters in the hope of finding ancient evidence for water and possibly ancient life. With the Context Camera, however, it has now been established that similar areas that may have been underwater in the distant past are located all over the Red Planet. The problem is not so much picking a spot to land but picking the best spot.

Opportunity and Spirit found plenty of hematite in and around their landing sites. Scientists anticipated much of the same thing where they wanted Curiosity to land. In 2010, all the cameras aboard the Mars Reconnaissance Orbiter revealed hematite within the proposed landing site in Gale Crater. As we have seen, Curiosity subsequently confirmed its presence at the base of its destination mountain, Mount Sharp, more formally known as Aeolis Mons. The mountain possesses an entire layer of hematite roughly 1,650 feet (half a kilometer) up the slopes. Indeed, one of Curiosity's more important findings has come from scrutinizing the layers within these hematite outcrops.

Measurements have shown that these layers were deposited in a low-energy environment. That is, the deposits that formed Mount Sharp came from their sitting in the middle of a large lake. As the water evaporated over time, the different layers were laid down. Further stratification, as the deposition of layers is formally known, arose from the wind (or perhaps even from transport of boulders and stones carried by currents in the water). "We've now found that this record of lakes is more or less continuous and has not stopped for a thousand feet or more," says project scientist Ashwin Vasavada at JPL.

In the fall of 2019, the rover continued to find telltale signs of water that had likely persisted for many tens of millions of years, if not longer. "We're still seeing it," Vasavada says of the evidence of prolonged water on the surface. "It has been expressed to varying degrees within that thousand feet. Sometimes it's these incredibly beautiful, two-millimeter-thick repetitive layers, like we saw at the very foot of Mount Sharp. Other times, there are little intervals where it looks like it may be drying up for a bit. We see more like sand dunes blowing through, interweaving, and then the lake-bed mud stones come back." That is "the bottom line," he says, in terms of finding water.

Curiosity has also been looking at hematite up close and personal—in particular, in the Vera Rubin ridge. According to Abby Fraeman, the rover planner and PhD geologist at JPL, the most interesting rocks are different colors. Another Malin camera—this time, on the mast of the Curiosity rover—has an unparalleled ability to discern subtle difference in color. The Mastcam, as it is known, has revealed that some of the rocks in the ridge are red, while other patches are gray.

"That was really unexpected, and we're still piecing together what that all means," Fraeman says. "It probably has some implications about waters moving iron around and oxidized or reduced iron when the ridge was forming."

This oxidation was common, since hematite is all over the crater. Figuring out how the environment deposited these sediments there is one of the more immediate goals for Curiosity. The geologists would like to know if the hematite was deposited there originally, or did it evolve over time? The chemistry of how this happened has serious implications for the way in which life might have been able to form in and around Gale Crater—and indeed, elsewhere on the Red Planet.

* * *

Mike Malin's company had a tantalizing taste of a landing on the Martian surface in 2004, when both Opportunity and Spirit made their nerve-racking descents. The aging Mars Global Surveyor spacecraft was still operating well and watched them land. The original Malin orbital camera confirmed their safe touchdown from above. While Malin's group did not build the camera hardware on either rover, their experience was invaluable. For about a week every month, operations people at Malin Space Science Systems helped with the commanding and sequencing of how both Opportunity and Spirit took pictures on the surface.

Malin's own personal experience of trying to re-create the thrill of landing—actually recording what happened—was one of the few occasions where the fates worked against him. When the Mars Polar Lander crashed near the south pole of Mars in late 1999, data had not been relayed from his camera during the descent. Unfortunately, communication had been lost before the camera was uncovered. (The first prototype of the weather camera, MARCI, had been lost three months earlier when the Mars Climate Orbiter crashed into the surface.)

Malin took it all in stride. Space exploration is like that. Even with the best will on this planet, some landers still do not make it all the way down to the one next door. In fact, it's a miracle that some do at all. Ten years later, the phoenix arose from the ashes. Designed as a replacement for the original Polar Lander, on May 5, 2008, the Mars *Phoenix* lander successfully alighted at its target, the incongruously named "Green Valley," a strange moniker for part of the north pole of Mars.

Phoenix sat in place, diligently taking measurements for fifteen months, well past its design lifetime. Only the tough Martian winter felled it. *Phoenix* also studied the weather, nearby dust devils, and the soil, in which it found calcium carbonate, which indicated the site was wet sometime in the ancient past. Unfortunately, Malin's revamped Mars Descent Imager also failed. This was because of something his company website elliptically refers to as "technical concerns on the spacecraft side of the interface."

Before launch, problems were found with the electronic circuitry that connected the camera to the *Phoenix* lander's main computer. The camera feed interfered with data generated by the gyroscopes that were crucial in keeping the lander stable as it descended to the surface. There was not enough time to swap out the balky electronics or perform a last-minute redesign. So Malin's Mars Descent Imager remained unused.

NASA eventually braved an attempt to turn this MARDI device on once the lander was safely down on the surface. For the first time, it was hoped, an instrument would be able to pick up sounds from around a spacecraft on the Red Planet. Alas, the instrument received nothing. So Malin could only silently watch as the other scientists reaped the benefits of their data.

As ever, his company was still very busy. Four years earlier, NASA had awarded Malin Space Science Systems an extremely lucrative contract to allow his unique expertise to be used in building the most powerful rover ever envisaged. A central element in the Curiosity rover's scientific detective work would come from the most advanced cameras ever sent to Mars. And finally, a descent toward the Martian surface would, it was hoped, be captured in exquisite detail by his latest version of the Mars Descent Imager.

It was a case of third time lucky. In August 2012, the last hundred seconds of the rover's "seven minutes of terror" were recorded in exquisite detail. Malin's descent camera took hundreds of color images at a rate of about four frames per second. This produced a high-resolution silent movie composed of some five hundred images. They have subsequently been assembled into a reconstruction that shows the drama of the landing in breathtaking detail. Nothing has ever been seen like it before.

The whole sequence starts with images taken "in flight," which show an area of 2.5 miles by 1.9 miles (four by three kilometers). Thereafter, the whole landing sequence plays right down to extreme close-ups just before touchdown, with detail as small as 1.5 millimeters per pixel. This sequence was crucial. It has allowed the Curiosity driving team to figure out exactly where the rover had come down on the crater floor, by matching the close-ups with images (in much less detail) from the Mars Reconnaissance Orbiter.

The descent camera also gave a brief survey of the surrounding area. That allowed the rover drivers back at JPL to pick out where Curiosity should first traverse. And it worked splendidly. Curiosity trundled off and found an ancient streambed just weeks after landing. It was the first of many pieces of evidence, as discussed in chapter 3, that showed that water had flowed extensively throughout Gale Crater. Malin's team continues to play a crucial role in the planning as the rover continues well into its eighth year of continuous operations on Mount Sharp.

As the months have stretched into years, two other Malin cameras on the Curiosity rover have also come to the forefront. One, the Mars Hand Lens Imager (MAHLI), is better known by the public for Curiosity's "selfies" (one of which appears on the cover of this book). Periodically, the rover stops peering at rocks and soil to take pictures of itself with the breathtaking landscape of Gale Crater in the background. As with his use of laser printing thirty years ago, Malin's aesthetic sense has served him, and by extension, us, very well indeed.

Martian selfies are an incredible public relations tool. They have generated many thousands of Twitter retweets and dozens of media stories. Yet there is also a scientific rationale: they show the condition of the rover in such a harsh environment, so that engineers on Earth can anticipate mechanical issues and deal with them. Curiosity has, as already noted, encountered wear and tear on its wheels far greater than expected. For example, without the camera, JPL controllers would have no idea whether Curiosity's wheels have suffered any further degradation.

But the main function of MAHLI is to take close-ups within Gale Crater in exquisite, microscopic detail. It can resolve features as small as 13.9 micrometers per pixel—quite something, given that the width of a human hair is about 75 micrometers. These close-ups build on the experience with the first-ever microscopic measurements made by the earlier Spirit and Opportunity rovers. Both showed sand-sized grains that the geologists had long suspected but couldn't confirm. MAHLI goes even further, discriminating between rocks, frost, and "fines" (features that are smaller than four millimeters, such as dust, silt, and sand) in extraordinary detail.

MAHLI takes natural color pictures using a type of detector called a Bayer Pattern Filter, named after an Eastman Kodak engineer. In it, an array of red, green, and blue filters sits directly above the light-sensitive detectors. They are arranged in the same way the human eye perceives color. In other words, by mimicking the way our retinas perceive light, they generate something incredibly close to what an astronaut would see on Mars. This process greatly shortens the time it takes to confirm that the color MAHLI sees is indeed true. Most of the earlier Mars surface cameras needed at least two or three images to achieve that same level of accuracy.

MAHLI can even take several images of the same feature at different focus positions. The camera cleverly combines these individual images so as not only to get a sharp picture of the target but to build a "range," as its focus can be adjusted. Such self-contained "stacks" of close-up images save time during Curiosity's busiest days. Less photo-snapping, more scientific return. Mission scientists can spend less time sequencing images and more on the analysis that produces the journal papers and conference proceedings they must generate to keep NASA funding going. And that is assured for quite some time. Curiosity should keep on rocking until its plutonium generator runs low, a scenario that is not anticipated for many years to come.

* * *

In February 2018, MAHLI acquired a set of images that had geologists scratching their heads. Dark, spoke-like objects—sticks in some estimations, bumps in others—about half an inch wide were observed. Some were in the shapes of stars, others described as "swallow tails." They are not crystals, as some had originally thought, but rather a kind of chemical covering in which crystals are embedded. They form on Earth when salts are concentrated in evaporating water, such as lakes that are beginning to dry up. In particular, they are known to surround crystals of gypsum salts.

Scientists are still puzzling out exactly how they formed. Perhaps the crystals came from a zone of wet sediment. Or perhaps the water bubbled up to the surface from underground and slowly evaporated to leave them behind. It's not even clear if these crystals originally formed in the area where they were observed. Perhaps, as the water dried up, another feature was there originally and then dissolved over time. What is left is a curious remnant of what had been.

Fairly swiftly, "life on Mars" conspiracy theorists started buzzing the "news" that the rover had found trace fossils of animal tracks. NASA quickly shot down such arrant nonsense. A variation on this perennial theme came in November 2019 when a retired entomologist claimed that there were fossilized insects and reptiles visible in some Curiosity images. Just because things vaguely look like something on Earth doesn't mean that they are the same on Mars.

Behind such claims, however, was something much more subtle, pointing at how conditions on the Red Planet could have been conducive to the formation of life many millennia ago. "So far on this mission, most of the evidence we've seen about ancient lakes in Gale Crater has been for relatively fresh, non-salty water," says Ashwin Vasavada. "If we start seeing lakes becoming saltier with time, that would help us understand how the environment changed. It's consistent with an overall pattern that water on Mars became more scarce over time."

While close-up features are important to Mars exploration, Mike Malin took care to include an instrument on the rover that would take in the wider context of the terrain that Curiosity covers. The Mast Camera (Mastcam) is a powerful still and video camera rolled into one: it takes images in color, multispectral color, stereo, and high-definition video. While MAHLI is obsessed with the fine-scale details, the Mastcam captures close-ups as well as the bigger picture: the landscapes, the sky, and the weather that surrounds the rover as it makes its way up Mount Sharp.

Geologists are interested in how dunes move across the Martian surface, and whether they are similar to those on Earth. At first glance, the two planets have surprisingly similar types of dunes, such as barchan or crescent-shaped dunes (named after the Kazakh word for their shape). They occur thanks to a prevailing wind. Dunes can thus indicate what kind of ancient climates there may have been on Mars. They show where the wind was flowing from when the dune was formed.

In 2015, Mastcam captured a sand dune in close-up so fine that it distinguished individual ripples. Observations like these are being combined with images from orbit to learn more about how a dunefield evolves over geological time. Their buildup and evolution will reveal if the dunes are indeed similar to or different from Earth. Ancient water, ancient wind, ancient climate—all are clues that planetary detectives need to piece together the story of how Mars evolved and what likelihood there may have been for life to have formed.

As NASA sets its sights on its most ambitious rover yet, Perseverance, Malin is once again along for the ride. His instruments will not only look for signs of where life could have been sustained in the surrounding terrain but also identify where to take promising "cached" samples for future retrieval and return to Earth.

While past Malin instruments scrutinized the geology, his latest contribution will also look out for biologically significant features within its field of view. In a very real sense, it is more like a microscope. It will zoom in on the telltale signs of organics, the carbon-based molecules that could indicate the presence of life in the ancient past.

Perseverance will have an improved "giraffe's eye," as Malin has termed his camera on the mast of the rover. Known as Mastcam-Z, it will take 3D pictures and video at astonishing speed. Looking through the eyes of Mastcam-Z will truly feel as though we are exploring the Martian surface right alongside the latest rover. And perhaps, before too long, the synthesis of all these enhanced techniques and technologies will produce something akin to an IMAX 3D presentation, a completely immersive view of the Red Planet that has never been seen before, let alone imagined back on Earth.

* * *

Once called the "cameraman on Mars," Mike Malin's instruments are playing a leading role in helping understand how and when life might have evolved on another world. Whether from orbit or on the surface, Malin's cameras are teasing out new details about the conditions where life might have been nurtured in the ancient waters of Mars. His cameras have generated fundamental discoveries about sedimentary details, the effects of water, and how sediments and lava harden into more resistant forms of rock.

The discovery of sedimentary rocks was not just a game changer, but the start of a whole new scientific ball game. The unraveling of sedimentary strata reveals the detailed history of how Mars has evolved. In the same way that a geologist could infer something similar on Earth by hiking through the Grand Canyon, the sedimentary layers on Mars speak to geologists of a more complicated past than they had once imagined.

Small wonder that Malin's peers have recognized his contributions to Martian studies, most recently with a prestigious award by the American Geophysical Union, which noted that his "foresight and conviction paved the way for many discoveries."

The next generation of Mars explorers will build upon Michael Malin's legacy. New cameras will doubtless discover important new details concerning possible environments that were conducive to life. From the materials involved in forming the rocks to the processes that led to their creation, Malin's cameras are helping identify where there might have been hydrothermal activity in the ancient past on Mars.

For Malin himself, the places he would love to explore directly are the weird and wonderful canyons at the heart of the Martian equator. "Valles Marineris is the most interesting place to go, because it gives you access to the deep crust of Mars," he says. "That's where I'd love to explore."

Now in his early seventies, he is unlikely to set foot on Mars in person. In the interim, until someone else does, he will keep building cameras until the money—or interest—runs out. Yet Malin remains upbeat, concluding with a sentence that distills the very essence of why we explore the next world out from the Sun: "The most excitement is going to come from the things we haven't discovered yet."

6

THE PATHFINDER

Proof of the potent interest in the Red Planet came on Independence Day 1997. A cut-price space probe the size of a VW Beetle came to rest on a floodplain and dispatched a tiny roving vehicle the size of a microwave oven across the rock-strewn surface. Mars *Pathfinder* lived up to its name and heralded America's return to the surface of the Red Planet after two decades. Pride of place was given to the rover known as Sojourner, which rolled into the history books when it started up and left its landing platform.

Over the next three months, Sojourner allowed geologists to reach out physically to the nearby rocks. The little prototype proved that rovers were conceptually valid. Until then, many engineers in the space business had tended to be sniffy whenever they were mentioned. The more traditional scientists treated them as hardly more than engineers' playthings, fine for when the press was there with cameras, but with little scientific value or technical merit. Without Sojourner's pioneering efforts, the world's space agencies would never have been able to plan or execute the complex missions they are now undertaking.

Later that same year, the Pathfinder team graced *Vanity Fair*'s "hall of fame"—as good a social barometer as any—in such illustrious company as James Cameron (then an Oscar shoo-in for *Titanic*), Ellen DeGeneres, and Princess Diana (who died only weeks after Sojourner first rolled onto the surface). The team's elevation into this pantheon was, the magazine claimed, "because they rocked the Red Planet," treating it as "their own private sandbox," the landscape "barren and unbecoming, like the opening vista of a Sergio Leone film witnessed through a bloodshot eye."

Foremost among the team, and achieving a certain notoriety, was an engineer who had often appeared on television in her bright red business suit. Donna Shirley, who was in charge of the rover, became synonymous with media coverage for the mission. Without her, a rover would never have flown to the Red Planet on the Pathfinder mission.

Shirley herself was no less a pioneer, who not only had to fight her corner but prove, as a female engineer, that she could manage and build flight-ready hardware. At times, both she and the Sojourner rover were largely thought of as unwanted, troublesome hitchhikers. Many older engineers at the Jet Propulsion Laboratory were openly skeptical and hostile to her work because of her gender.

Long before social media united the voices of ordinary women facing discrimination worldwide, Donna Shirley succeeded in a field that always liked to think it was breaking barriers. With a wry smile, Shirley sometimes referred to herself as Sojourner's mother, though others have christened her Dejah Thoris, the Princess of Helium from the old Edgar Rice Burroughs stories about Mars.

"The rover was the most fun job I ever had," she says, and in the final analysis, it wasn't a scientific journal that came anywhere near having the last word on her astounding perseverance. *Mirabella* magazine got it about right when it described her as "part proud parent, part car salesman, and part stand-up comic."

* * *

Donna Shirley had first dreamed of flying to Mars when she grew up in a small Oklahoma town. It may have taken her forty years to get there, but she got there in the end. Without her doggedness, the search for life on the Martian surface would not now be poised for its current renaissance. Sojourner proved how useful rovers were. Subsequently, ever more capable and sophisticated roving vehicles have made greater strides across the Martian surface to search out the most intriguing rocks and places of interest. First Spirit and Opportunity and now Curiosity have blazed a trail for just what can be achieved, with the prospect of returning samples from the Red Planet to Earth to be realized in the next decade. None of it would ever have been possible without Sojourner and Shirley.

Donna Shirley was different from many of her colleagues at the Jet Propulsion Laboratory, and not just because of her gender. She can describe things in plain, simple English. During the summer of 1997, you couldn't switch on a television set without coming across her, always dressed in her "lucky" red suit. When *Pathfinder* successfully made it down safely, she even shed a tear live on CNN. Her autobiography, *Managing Martians*, became a bestseller when it was published a year later.

She personified the change in NASA's approach in sending cheaper, more adaptable missions to the Red Planet. For a decade, she had been in charge of all of JPL's efforts in building rovers. By 1997, it was a task that had been made all the more difficult by dwindling budgets and a certain resistance to her—and not just because she was a woman, although she had encountered enough fully paid-up, card-carrying misogynists to impede her on the way. Though many of the engineers at the lab liked her personally, Shirley had never "delivered" flight hardware—that is, worked on a mission and actually built it ready for launch. This was the catch-22 of her career, as she saw it. Delivering a flight was an essential qualification in becoming an initiate into JPL's high priesthood, which, even by the mid-1990s, was still all male. "She's OK, but she'll never make a flight project manager," said middle-aged men who really should have known better.

Ultimately, Donna Shirley's vindication came on the afternoon of Saturday, July 5, 1997, the day after *Pathfinder* made its one-shot landing on Mars. Unfurled from the lander, the Sojourner rover slowly made its way onto the surface. Many of the same old hands who had criticized her throughout its development were now all smiles, high-fiving and full of mealymouthed congratulations. Not only had she had successfully delivered hardware, Shirley had outmaneuvered ("out womaneuvered," said one friend) others for a promotion to managing the whole of the Jet Propulsion Laboratory's Mars exploration efforts. It was a fitting culmination to her career, a complete turning of the karmic wheel, because the very reason she'd joined JPL in the first place was that she wanted to go to Mars.

* * *

Conventionality, compliance, and segregation were the social norms for women in the 1950s when Donna Shirley was growing up. In the face of conformity, she was inquisitive. She wanted to explore far beyond the flat horizons of the small town of Wynnewood, Oklahoma. Such ambitions were hard to reconcile with being the "carefully dressed daughter of the town doctor," as she described herself. Shirley loved English at school and wanted to learn to fly. "I read a lot of science fiction," she recalled, "and I built small model airplanes and hung them from my bedroom ceiling with twine." The optimistic projections of the future that regularly graced the pages of *Collier's* magazine made a great impact on her imagination. Arthur C. Clarke's *The Sands of Mars*, a story of human exploration, turned her mind toward the Red Planet.

When she was ten, Shirley attended an uncle's graduation, where she noticed some graduates were described as aeronautical engineers. "I asked my mother what that was and she said 'That's people who build airplanes,'" Shirley recalls. "And that's what I wanted to do." A decade later, on her first day attending class in the aeronautical engineering department at the University of Oklahoma, she was flatly told: "Girls can't be engineers."

So she proved her adviser wrong. After earning a degree in "professional writing" (with some journalism, which would prove useful), she had an all-too-common experience for women at the time. During one conversation about potential employment, she was asked to "discuss things" in the recruiter's hotel room. "We got up there, and it was clear he had other things in mind other than this job interview," she recalls.

Then fate took another turn. She was actually offered the job. Donna Shirley began working for McDonnell Douglas in St. Louis as the lowest of the low: a specification writer. She then took her degree in aeronautics, and by perseverance—what she terms dedication, creativity, flexibility, and more than a little moxie—she prevailed to become an aerodynamicist. That meant she worked on designs for the first missions considered for landing on Mars.

"I was a very junior engineer, but at least I was an engineer," she says. She applied for a job at NASA's Jet Propulsion Laboratory, who "were picky about who they hire," she recalls. To her astonishment, she was accepted. It was to be her professional home for the next thirty-two years.

* * *

To be a female aerospace engineer in the 1960s was never going to be easy. Despite the forward-looking nature of the space business, many male engineers had attitudes firmly rooted in the distant past. We will never hear all the stories of the many women who were belittled, sidelined, or discriminated against, but Shirley bravely stands out. She fought for the better part of two decades to become accepted in what she has termed "the boys' club" that was JPL. Partly, it involved picking her battles carefully. In the male-dominated hierarchy of the laboratory, that meant treading very cautiously indeed.

Unenlightened attitudes were the norm. In her first job in St. Louis, what she later called "her horny old boss, a married guy" at McDonnell Aircraft was chasing her around. She was able to outwit his attentions by subterfuge. Over dinner, she held his hand and said that if they ever did get together, their careers would be ruined. In this and less obvious ways, Shirley says she learned the passwords to join the boys' club. She then changed the passwords by stealth.

When she joined JPL, there were at least a handful of female scientists who would became preeminent in their fields. Marcia Macdonald was an expert in plasma waves who married Caltech's Gerry Neugebauer. As Marcia Neugebauer, she investigated the flow of solar wind on many subsequent space missions. ("Where Neugebauer studied waves," notes one witty chronicler, "Shirley made them.") A while later, Eleanor Helin came to prominence after working on the Caltech campus. She fought prejudice and academic snobbery to become the primary discoverer of Earth-crossing asteroids in the seventies and eighties.

These women were the exceptions to the rule, however. When Donna Shirley joined the laboratory, women were largely relegated to menial positions. Allowed to work on the switchboard or as secretaries, they were not seen as capable when it came to delicate instruments. One "authority" at the time suggested females would "become bored by the niggling work, get careless, and have to be fired."

In recent years, the singular contribution by what were known as the JPL "computers" has come to light. These were female engineers who were recruited to make the endless mechanical calculations—all by hand—for the best trajectories that could be flown to the planets.

Known more formally as Section 23, they reported to a supervisor who took care to hire only women to maintain social cohesion. Her mantra for new recruits: look like a girl, act like a lady, think like a man, and work like a dog.

Or perhaps that should have been "underdog," because that was what they were. Amazing as it seems now, the lab routinely still held beauty contests—Miss Guided Missile eventually became Miss Outer Space—until as late as 1970. Author Nathalia Holt, who has chronicled the work of the computers, has observed that, paradoxically, such anachronistic competitions had a positive influence on female careers. The women involved ended up becoming very close with their male colleagues.

These antiquated attitudes endured at least into the late 1960s. When the lab recruited Janez Lawson, a brilliant engineer, she was the first African American to be hired. "We sent people out to hire lady engineers, and they hired black lady engineers as well," remarked one JPL higher-up, revealing his own prejudices. But by the mid-sixties, greater social change had started to trickle through many staid and starchy institutions. When Donna Shirley joined JPL in 1966, she would become the first female engineer who worked on a flight project among the two thousand male engineers at the lab. In every sense of the word, she would be a pathfinder.

* * *

By 1966, plans for the Voyager mission to land a pair of large vehicles on the Red Planet were reaching fruition. Shirley was charged with designing an aerodynamically stable cone to ferry it down to the surface. "We were going to have to come in through the atmosphere," she says of what was then proposed. "And for that, we would need a heatshield. Nobody knew what the right shape was."

She experimented with the necessary shapes, eventually settling on a blunt cone with a spherical, conical nose with rounded shoulders. "So when it came to Pathfinder, we didn't have to reinvent that," she says. "We could use the same shape." When Voyager was canceled, it was a blessing in disguise. Not enough was known about the atmosphere and surface of Mars to make the mission remotely feasible. "If we had launched something to land in 1971," Donna Shirley says, "it would really have been silly."

Congress agreed and pulled the plug. As the most recent addition, Shirley was first to be laid off. So she joined a nearby company, where she was once more expected to be decorous. At one dinner, she was placed next to a potential client, and the boss "somehow expected me to be a hook for this guy." She refused, as she also did when he asked her to falsify some results, and quit. Fairly quickly, she returned to JPL to become a systems analyst.

Systems analysts were the unseen, unglamorous foot soldiers in the laboratory's hierarchy. They often worked on projects that had little to do with sending actual spacecraft into space. Indeed, thanks to a boss's friendship with someone who ran the Santa Clara County Forensics Laboratory, Shirley was tasked to look at how flight hardware could be adapted to automatically analyze drugs in crime labs.

A tidal wave of drug-related social problems had started to affect American society. Shirley spent a frustrating year only to find that the promise of the technology then available was far greater than its actual ability. As much as *CSI: Crime Scene Investigation* has, in more recent times, glamorized forensics, in the early 1970s the technology was still fairly primitive. (Intriguingly, the gas chromatograph "mass spec" instrument built for the Viking program is now commonplace in forensic laboratories and is much more accurate than infrared detectors.)

Nevertheless, Shirley's forensics lab experience was richly formative. It taught her an indelible lesson for the future. In technology, you have to back the right horse. It's not always necessary to ride the fastest thoroughbred. At times, as her Mars experience later showed, it was better to work with a foal out of sight of the most egotistical trainers and jockeys.

After impressing some of her bosses, Shirley got a job as an analyst on an actual mission, the first spacecraft ever sent to the planet Mercury. She walked into her new supervisor's office and asked, not unreasonably, "What does a mission analyst do?"

"It's customary to define your own job," he replied.

So she did. Somewhere near the top of the list was working out what the trajectory should be. As mission analyst for what became known as the *Mariner 10* mission, it was Shirley's task to sort out all the often-conflicting needs of scientists—"quite a circus," she says—and the pragmatic needs of the mission designers. Today, she recalls with glee that she made one abiding contribution to the project. "I picked the day we launched, November 3, 1973," she says with considerable pride.

After that, though, it was not easy sailing. Shirley was plunged into the basic irreducible verities of launching hardware into space. Everything that could possibly go wrong did—and usually at the worst time possible. On various occasions, *Mariner 10* ran out of attitude control gas, its instruments played up, and its automatic housekeeping functions went balky. "The spacecraft was broken the whole trip," she says. "But we worked around it."

The need to fix things on the hoof, when everything was literally falling apart, became a baptism by fire. It helped Shirley develop the necessary skills to prevail in sending a mission to Mars. A brief digression, studying how the lab could be involved in the energy business (her team was "a wild bunch" even by JPL standards for the mavericks they often employed), proved another dead end. As she wryly notes, "as soon as Reagan was elected, the energy crisis was declared over."

She then managed JPL's Mission Design Section before taking charge of the laboratory's contribution to the future space station in the early 1980s. Privately thinking the station a "four-headed monster," Shirley learned that there was another kind of culture at other NASA centers. "At JPL, you got up and just slugged it out in open meetings," Shirley notes, "[and] yelled and screamed and carried on and there was nothing done behind the scenes." So when, for example, she met the good ol' boys from the Johnson Space Center in Houston, they tended to clam up. "I'd always be the only woman in the room, and everybody else would be from the Deep South," Shirley recalls, "And so here's this pushy broad from this robotics center they don't like, sounding off." By the 1990s she had grown tired of pen-pushing and wanted a more responsible job. A JPL higher-up said they need someone to manage all the lab's robotics and automation work.

"I don't know anything about that," Shirley said.

"That's OK. The people who do know something about it don't want that job."

In her new role, she became interested in how rovers would work on

Mars and what technology would be needed. In the way of all space agencies, the Mars Rover program, which Shirley joined in 1987, grew very fast. Two years later, the Red Planet briefly flowered as a grand national goal after President George H. W. Bush made an ill-starred attempt to invoke Kennedyesque rhetoric. On July 20, 1989, the twentieth anniversary of *Apollo 11*'s landing on the Moon, Bush vowed to return astronauts there and also to Mars. Within two years, the initiative was dead, as were various options for supporting it by means of roving vehicles and robotic precursors.

The eventual cost of JPL's contribution alone was estimated at $6 billion (two billion for its orbiters, another couple for the rovers, and a further two for the sample return). That was chump change compared to the money needed to send humans to Mars. "It was about $400 billion," Shirley notes. "It was so expensive that NASA didn't want to tell Congress."

Before the plug was pulled, Shirley was asked to diagnose NASA's bureaucratic problems one summer in what she called "the no-wimps study." It looked at the state of management; characteristically, Shirley did not pull any punches. Once the National Aeronautics and Space Administration was formed, she said, "you've got to feed it." She termed the agency an entitlement program, for which the various NASA centers kept themselves in business by lying, cheating, and stealing. "And when you lie, cheat, and steal—surprise!—you can't deliver what you said you'd be able to do," she concluded in what was officially known as a handbook for Mars system engineering. When she presented her findings to the NASA higher-ups, they didn't want to know. After another hour-long yelling session, another woman came up to her in the auditorium at NASA Headquarters.

"How can you live with being yelled at by all these people?"

"I am from JPL," Donna Shirley said. "I've been yelled at by the experts."

* * *

Being yelled at by the boss soon became an occupational hazard for everybody else when Daniel Goldin was appointed NASA Administrator in 1992. His job, as he saw it, was to cut costs, bang heads, and set the space agency on a new course for the 1990s. Known for his no-holds-barred approach—"Captain Chaos" was one of his more printable nick-names—Goldin wanted missions that were "faster, better, cheaper" after a number of embarrassing and costly failures.

Goldin pointedly termed the Mars Observer, which was due for launch in September 1992, "the last ship out of port." At a cost of $600 million to build, half of which was needed for its launcher after repeated changes, it clearly represented the old way of doing things. When it was lost hours before arriving at the Red Planet, Goldin demanded a new, cheaper way. Though that was good news for Mars explorers, there was bad news for their paymasters.

Compared to the vast sums available in the past—in today's money, the Viking missions to Mars in the 1970s would cost some $3.6 bil-lion—most of the subsequent NASA Mars missions in the nineties were capped below $200 million, at a time when the space agency's budget was on a downward curve. The National Aeronautics and Space Administra-tion of the 1990s was frugal to the point of parsimony.

Technological advances helped. More nimble spacecraft employ-ing new computer chips would permanently reduce costs. Dan Goldin encouraged the use of small development teams who worked to a tight time line with off-the-shelf equipment. While the old JPL hands were cautious, for an engineer like Donna Shirley, it presented a golden oppor-tunity. She finally could be given a chance to deliver flight hardware. Nobody else wanted to take responsibility for managing such a risky enterprise. Shirley had nothing to lose by sticking her neck out, so she volunteered to manage a new generation of smaller rovers. As she recalls, that meant presiding over opposing factions within the laboratory.

Though JPL wanted to rove, its engineers argued endlessly about exactly how to do it. There was one group who wanted the rover to navigate autonomously by remote control. A mobility system invented by JPL engineer Don Bickler ("working in his garage") would allow any rover to scramble over obstacles in its way. Such a rover had no brain, was mounted on a small chassis, and formed the earliest prototype of what became known as the Rocky series of rovers.

Another group with a greater interest in driving the rover from Earth built a device called FANG (for Fully Autonomous Navigational Gizmo). A few months later, they miniaturized it into a ten-inch-long robot called Tooth. It could just about sense blindly for hockey pucks along a JPL corridor.

Reflecting on how to get to Mars, Shirley had a radical idea: why not put Tooth's brain on Rocky's body? They somehow scraped enough money together to build Rocky 4, a prototype that weighed all of fifteen pounds (seven kilograms). That provided the basis for an experimental flight program to develop a micro-rover that Shirley was asked to manage. As well she knew, this meant a Faustian bargain: she would finally get to deliver a flight project, but with very little money (and many thought it would fail anyway). In other words, it was an opportunity with a price, and a ridiculously low one at that.

Shirley's team had just under $25 million (roughly $50 million in today's dollars) to design, build, and test a rover that would one day travel to Mars. Only by scrimping and saving money that was grudgingly handed out by a number of different NASA offices could she keep the project on the rails. In 1993, there was an odds-on chance that Rocky 4 would have to be abandoned completely.

Still, Shirley retained some advantages. Her immediate team of a couple of dozen engineers was either too old or too young to have been part of JPL's glory days, when budgets were bountiful. Shirley determined to handpick individuals who were amenable to doing things in a new way. That meant older guys nearing retirement or youngsters with nothing much to lose. The older people at the lab had built fast, cheap, and cheerful missions in the sixties. At the time, JPL, if not NASA Headquarters, had accepted that it might lose some spacecraft due to inexperience. Shirley wanted to tap into that heritage. "They'd built spacecraft like that in the ancient past," she wryly comments. "And younger people who hadn't been on big missions didn't have that experience as a prejudice. They didn't know what we were doing was impossible."

There was not a lot she could do about bureaucracy or red tape, however, or the internal divisions that were inherent in the way the mission was actually funded. Pathfinder was paid for out of NASA's science budget, because it would return useful scientific information about the Red Planet. The rover, however, was, in NASA-speak, a "technology demonstrator," paid for by the space agency's technology research program.

Donna Shirley's hand-picked innovators also needed to be inoculated against the "not invented here syndrome"—the tendency to reject ideas originating from outside—that is often rampant in large technical organizations. While others around the lab made it clear they thought her project would not work, the engineering rationale for a rover was simple: before JPL engineers could ever contemplate roaming freely across the surface of Mars, they would have to learn how to crawl—literally. Their prototype rover was seen as a precursor to more ambitious vehicles. But nobody had ever imagined just how cut-price and penny-pinching its development would be. The vehicle later christened Sojourner became a triumph of last-minute expediency coupled with engineering skill and risky technological advances.

Shirley knew a lot was riding on the outcome. Without rovers, there would be no possibility of returning samples to Earth any time soon. (Today's most optimistic projections say a sample-return mission could happen in the late 2020s.) Worse, even before they had settled on a working prototype, Shirley knew that rovers had already developed an unenviable reputation as recalcitrance on wheels. They always seemed to break down at the worst possible moment. In tests, they behaved erratically and unexpectedly, usually when there were NASA higher-ups and film crews around. During a public demonstration of Rocky 4 in June 1992, a prototype rock chipper shorted out, hammering uncontrollably, until somebody put it out of its misery. If they couldn't get a prototype to work in a parking lot, how could they do it on Mars?

* * *

From the outset, Donna Shirley's team knew they didn't need that much intelligence on the rover. They could make clever compromises in the vexed business of its control, too. It was already abundantly clear that "we couldn't carry our own brain, unless it was very small," in Shirley's telling phrase. Once more, it was a case of backing the right technological horse.

"We knew we didn't have to go fast," Shirley says. "And we knew we didn't have to do too many things. The few things we'd do, we'd do well."

Crucially, they didn't need a live video feed. There would be few hazards crawling across the surface at a top speed of 0.0037 miles per hour, equivalent to a centimeter a second. So they could use a simple still camera that weighed all of an ounce. They could fashion that from off-the-shelf digital storage devices (Charged Couple Devices) that would use the Martian cold to "store" the images.

Computing power was also catching up with expectations: in 1996 and 1997, IBM's ridiculously hyped "Deep Blue" narrowly beat chess master Garry Kasparov in several rounds. The central computer used on the *Pathfinder* craft was a variation on this same machine that had been suitably adapted to the task.

Another perennial problem was just how autonomous the rover should be from *Pathfinder*. Make the rover smart, and it would be too expensive; dumb it down, and operators in Pasadena would spend all their time uploading commands. They would waste precious time as the rover would be idle on the surface, waiting for the commands to arrive.

They decided on an even teenier brain for the rover that used an antiquated chip by most standards, even at the time. It could store only 8-bit-long words—shown to comedic effect in the movie *The Martian*, where Matt Damon's stranded astronaut uses Sojourner to contact Earth—so it was hardly efficient for programming. Shirley's team relished the opportunity to do something well-nigh impossible. To simplify things, a new way of prompting the rover's motion, called "behavior control," was developed. It took as its cue the motion of insects. They don't have to know what the obstacles are, just how to avoid them. Insect-like autonomous control algorithms allowed the rover to move with as little intervention as possible from the ground.

"Autonomy is all very well if you know where you are on the Martian surface," says Shirley. "There are always going to be errors with our aiming. The only intelligent thing which Sojourner does is stay out of trouble." As a result, T'ai Chi was the preferred mantra invoked by some project personnel for the speed of the rover's motions.

Slowly but surely, the vehicle started to take shape. Topped by a large, flat array of blue solar panels, Sojourner looked like a golf cart with a psychedelic sunroof. As Shirley recalls it, the reaction from most people at JPL on seeing the rover for the first time was an entirely predictable "Pah! What a pesky little thing!" The rest of the time she had to convince the general public that it wasn't actually a scale model.

To some extent, inbuilt resistance to the rover was understandable. After the loss of Mars Observer, the original goal for the first mission that followed it was simply to get *Pathfinder*, then seen as the forerunner for a series of automatic weather stations to be established on Mars, to the Red Planet. There were enough difficulties in getting this curious, petal-folding machine ready, without adding another layer of complexity. To critics, the rover was a tiny machine that would divert funds and attention and, as many older JPL engineers confidently predicted, would most likely not work at all.

Many of the people working on the *Pathfinder* spacecraft itself were openly skeptical of and bitterly opposed to a rover's inclusion. Yet Donna Shirley knew how to get their attention. She could give scientists interested in Mars a brand-new instrument to play with. As well as the still TV camera needed for navigation, Sojourner would carry a device that could nuzzle up to rocks and determine their chemical composition.

The Alpha Particle X-Ray Spectrometer (APXS)—which had to be replaced with its backup the day before the eventual launch—worked by bombarding whatever it was looking at with a cocktail of exotic particles. The way in which they interacted with the Martian surface characterized its chemistry.

"The secret history of Mars," Donna Shirley explains, "is in the rocks, not the dirt, which has all been weathered and chemically changed. So they were really interested in having this spectrometer go up to a rock."

* * *

There was, nevertheless, still a great deal of residual mumbling about how exactly scientists could use this pesky interloper. A vociferous minority pointed out that they'd be perfectly happy with a sample arm. As Shirley patiently showed them, they would then only be able to poke at rocks that just happened to be within easy reach. A simple statistical analysis revealed that there was only a 60 percent chance that a static lander's arm could ever reach any of the particular rock types that they might want to explore. A rover could pick and choose at will.

The greater challenges were technological, not least in overcoming the extreme conditions of cold on the Red Planet. There was then, as now, no simple answer nor anything like a miracle solution for keeping the rover warm and working. The diurnal range of temperatures runs from the freezing point of water down to −110°C. The rover's onboard electronics could not work below −40°C, so a special "warm electronics box" was built to contain all the electronics and batteries. Running the electronics and heaters during the Martian day warmed this box, but they needed help from what were termed radioisotope heater units. These were essentially three pieces of weapons-grade plutonium, each the size of a flashlight battery, that decayed and provided extra thermal energy.

They were also highly controversial. In the environmentally friendly 1990s, space launches were sometimes boycotted by those who thought they represented a hidden danger, the sending of dangerous nuclear material into space. Whereas *Pathfinder* used plutonium as an additional heat source, other space missions—such as *Cassini*, a complex mission to Saturn, on which Donna Shirley had briefly worked—needed it for all its energy needs in the chilly reaches of the outer solar system. There, the Sun is but a bright star, too feeble to provide sufficient energy for solar panels.

The *Cassini* project, which was launched four months after the *Pathfinder* landing, consumed some $25 million ($50 million in today's dollars) in the filling out of forms and due diligence to allow the passage of plutonium into space. The price of the paperwork alone was equivalent to the total cost of the Sojourner rover. What Shirley calls the "A-Team" worked on *Cassini*, "and we were kind of the B-team," such was their standing in the laboratory's hierarchy.

During the Martian day, the rover would be powered by the Sun, although there was a backup provided by lithium thionyl chloride batteries. Sojourner's solar cells were efficient enough to convert the weak sunlight into power. If they broke down, the batteries would provide a few days' worth of electrical power. To save weight, they were not rechargeable.

The incredible cold on the Martian surface meant that there could be no lubrication. Indeed, Shirley's colleagues had to remove all lubricants in the wheel motors, as they would stick. The whole body of the rover was effectively and efficiently insulated by lightweight silica aerogel. Weight was always at a premium, so featherweight aluminum was used for the chassis of the vehicle.

Eventually, a six-wheel framework was chosen over a four-wheel design because that meant greater climbing capability. The wheels were fashioned from stainless steel and were just five inches in diameter. Each wheel could move up or down and move independently of each other. Motion sensors stopped Sojourner from tipping over, although one side could tilt up to 60° before it would get into trouble. The wheel joints bent along with the shape of the surface underneath them. They could scale small rocks up to ten inches (twenty-five centimeters) high.

To do all this, the rover was given a revolutionary system of suspension that involved neither axles nor springs. It was known as a rocker bogie. A bogie is a device in which the wheels are connected to levers that swivel on the ends of a pivot to move up and down. "Since they rocked back and forth and were on bogies, we christened it the 'rocker-bogie' system," Shirley says. "So that's why I christened the prototype Rocky."

Shirley often wondered whether the rover would actually triumph. At times, the handicaps in its way were so great as to appear insurmountable. Tony Spear, *Pathfinder's* project manager, had made it abundantly clear that he didn't even want a rover onboard. He viewed it as little more than a technical parasite that was sucking up money that he needed. Worse, it could screw up the whole mission. He seemed to regard Shirley as "a real amateur, not having flight hardware experience, and so on, not to mention the fact I am a woman." This led to many arguments, culminating in such a noisy confrontation between him and Shirley in the middle of 1993 that the engineer in the next office walked in.

"Can you keep it down a little?" he asked. "I'm trying to meditate."

* * *

In 1994, Donna Shirley was promoted upstairs. She was made, to no little fanfare, the manager for the whole of JPL's Mars Exploration Program. Her role was to come up with a coherent plan for the next ten years of missions, to be laid out logically and systematically. Launches would occur more frequently and employ constant, steady funding. Her appointment felt a lot better, she wryly noted, than being crowned Miss Wynnewood thirty-three years earlier.

Over the next three years, she still kept an eye on *Pathfinder*. There was no guarantee it would work. There was literally no backup because "you could do these things cheaply when you only had one." The lander would head straight down to the surface of Mars, taking just five and a half minutes to descend. In December 1996, *Pathfinder* left Earth on time and on budget, and all the omens seemed good. Once it reached its destination, however, there was very little margin for error. It would come in faster, certainly compared to Viking, and so "they had to redesign the heat shield to recognize [that]," in Donna Shirley's words.

Even worse were the airbags, chosen to cushion the lander's fall onto the surface. In repeated tests out in the Mojave Desert, they failed not just once, but many times. They snagged on sharp rocks and sometimes tore open. The airbags, in Shirley's recollection, "were just seen as absolutely silly and everybody was laughing at them and nobody believed they would work."

During their protracted development, they would have to be made thicker and more resilient. There were four layers of abrasion-resistant material, which were tested in a large vacuum chamber on an extended bungee cord "with the nastiest rocks you ever saw," in one estimation. In the end, the bags took up a quarter of the weight of the whole spacecraft.

In the event, *Pathfinder* came in on the dark side of the planet, shadowed from the Sun in the early morning of Independence Day, 1997. As it barreled down toward a region called Ares Vallis, the lander came in on a parachute. As with the more recent Curiosity lander, once the radar locked onto the surface, the lander descended along a bridle like a fireman dropping down a pole. When *Pathfinder* was fifty feet above the surface, the rocket motors inside a backshell fired to further slow the craft down.

Then the all-important airbags inflated, the bridle was cut, and the motors kept firing to pull the parachute and bridle away. At 3:00 a.m. local time on Mars, *Pathfinder* hit the surface exactly as planned at fifty miles per hour. With its airbags fully inflated, *Pathfinder* bounced, rolled, and tumbled across the surface. It bounced at least fifteen times along a distance later estimated as half a mile before coming to rest.

When it came to a complete halt, the airbags were retracted. Each had its own "rip patch" attached to a motor that would bring them in. *Pathfinder* came to rest right side up, and bang on schedule. Just before noon in Pasadena came the words from Mission Control that even the mission manager himself couldn't quite seem to articulate.

"EDL Comm is reporting a signal is barely visible." And then for the benefit of the listening world, he added: "That's a very good sign, everybody."

They had done it. "Hallelujah! Back on Mars!" exulted the *Viking* biologist Gerry Soffen, articulating the mood among the older generation of Mars researchers. "I've been holding my breath for twenty-one years. Now I can breathe again."

And so began the media circus. The worldwide response to the *Pathfinder* mission was nothing short of astonishing. Its online coverage scored the greatest number of hits ever recorded on the internet at that time. Four days after the landing, there were 47 million hits on NASA's website, and 150 million by the end of the week. So, in the same way it is said that President Roosevelt's "Day of Infamy" speech "made" the medium of radio in 1941 and the first Moon landing did the same for television, *Pathfinder* lived up to its name in online coverage.

"Nineteen ninety-seven doesn't sound like very long ago," Donna Shirley says, "but the internet was still relatively new. So it was the defining moment for the internet."

* * *

Pathfinder came down in an area that was, at least geologically, the most interesting they could find. Ares Vallis formed the mouth of an ancient valley that fanned out onto an otherwise unspectacular northern plain. To geologists it was a "grab bag" site, a floodplain where as diverse a collection of rocks as possible would be waiting for them to examine. Most had been washed onto the volcanic plains from the southern highlands by catastrophic flooding in the earliest epochs of Martian history.

Pathfinder did not disappoint. Its cameras revealed rocks of all shapes, sizes, and textures as well as rounded boulders, which had probably rolled in from the ancient highlands and were deposited by the floods. There were darker, angular rocks that had been excavated by craters as well as rocks that jutted out of the bedrock below. They spoke of an intriguing history of the planet.

In close proximity to the lander were a group of jagged rocks, all tilted toward the northeast, suggesting that they were lined up as water flowed by. High-resolution images showed a variety of rock textures, some of which seemed to be layered, others of which were sedimentary. Farther away there were what appeared similar to gullies, drift deposits, and terraces that had been cut by repeated flooding. On the horizon were two streamlined lips of a crater that almost inevitably came to be known as "Twin Peaks."

It was yet another geological wonderland. Perhaps as recently as three billion years ago, Ares Vallis was awash with water caused by volcanic heat that had melted the subsurface permafrost. Catastrophic flooding had deposited the big rocks during the first rush of water. Then, as its volume decreased, dust and smaller particles settled around them.

Yet, paradoxically, *Pathfinder*'s immediate surroundings seem to have been unchanged in the two billion years since. As the waters in the ancient past dissipated away, wind took over as the main agent of geological change. The wind steadily eroded away the variety of rocks where *Pathfinder* had come to rest. What it all implied, in the words of the JPL project scientist for *Pathfinder*, was that Mars really would be a good place to look for evidence of life.

"One major requirement for life as we know it is water," says Dr. Matt Golombek. "*Pathfinder* showed clear evidence that Mars was wet and maybe not all that different from the Earth early on at a time when life got going here. That's why a lot of us find Mars such an interesting place to study."

* * *

On its second day in Ares Vallis, Sojourner exited the ramp, turned right, and then, give or take a few minor hiccups, spent the next fifty-six days roaming around the landing site. Roving on Mars turned out to be a piece of cake. "Sojourner worked way beyond our expectations," Donna Shirley says. "We have learned that we can drive around the surface of another world."

Any initial problems were minor. After touchdown, the rover and lander didn't talk to each other for a while because of a temperature difference with their modems. It was a common enough occurrence on Earth at the time for it not to be seen as a major design flaw.

The choice of modems had been yet another compromise. For the rover to communicate with the lander, they needed to be specially qualified to handle the extremes of cold on Mars. Shirley's people wanted to use off-the-shelf modems to save money. The manufacturer said their modems would be out of warranty if they flew into space. Only by prevailing upon the company were they able to get them specially adapted for use in space.

Once the modems talked, the most common errors thereafter involved software. The rover used the lander to stay in contact with Earth. "You can't plan for every contingency on the surface of another planet," Donna Shirley says. "You have to make decisions once you have had a look at the terrain."

A few days after Sojourner unramped, Shirley's team had their first traffic accident when they attempted to scuff the soil to see how much resistance the rover encountered within the soils and overshot a rock, so that Sojourner's back wheels were raised off the ground. Then the rover's compass started to drift, and the vehicle wouldn't go where it was supposed to. So they learned to work out where they were heading from simply counting up the number of revolutions of the wheels.

After all the hassles, the controllers had some fun. In the days after the landing, the geologists tried to outdo one another by coming up with ever sillier names for the more prominent rocks in the vicinity of the lander. This was nothing new: twenty-one years before, *Viking* geologists had done exactly the same. Donna Shirley found it all hilariously liberating.

"It's like having seventy kids in a toy store," was how one scientist described what they all got up to. One young geologist was appointed as the "nominator," with whom others had to check before they stuck their labels on a photographic panorama of the surface with the increasingly ridiculous names. Just about the only criterion was that the names be straightforward and humorous.

Nearest to the lander was "Barnacle Bill," so named because it looked like an old weathered rock, and others were christened "Yogi," "Flat Top," "Hippo," "Chimp," "Shark Wedge," "Scooby-Doo," and "Squid." There was even the "Rock Garden," wherein a curious variety of chipped rocks could be found. For Matt Golombek, who had chosen the landing site with such variety in mind, it was justification for his claiming "I deliver" to his colleagues. And indeed, when the first results were returned from Barnacle Bill, it produced a scientific shock.

*   *   *

After spending ten hours snuggled up to Barnacle Bill, Sojourner's X-ray instrument found that it was strangely familiar. It contained far more silica than most rocks examined anywhere else in the solar system, let alone Mars. Barnacle Bill could almost have come from anywhere on Earth.

High-silica rocks here come from volcanic eruptions that are fueled by the sinking of rock into the interior. "That implied it had been differentiated quite a number of times," says Matt Golombek today. In other words, a great deal of chemical cooking has taken place. "The way that happens on Earth is by plate tectonics," he adds.

On Earth, this process is smoothed along by the oceans. It had hitherto always been thought to be uniquely terrestrial and not possible on other worlds. At face value, this finding suggested the surface has been heated and changed many times. Their discovery implied Mars had been much more geologically active than had previously been thought. "The high silica in that rock was way out there," says Matt Golombek. "It threw us at the time. But these days, it doesn't feel as strange because Spirit, Opportunity, and Curiosity have all found the same thing."

A few days later in July 1997, Sojourner snuggled up to Yogi—described as being "like a '57 Chevy with its two-tone colors." It had a similar composition. Indeed, nearly all the rocks examined by *Pathfinder* were found to be similarly high in silica. So, too, are many other rocks observed by subsequent rovers since then.

The best explanation is that weathering, especially from accumulated moisture over geological time, is a global phenomenon on Mars. Over the eons, as the climate changed, there would have been periods where water could interact with surface rocks. That has changed the composition of material immediately beneath the rock surface. "That would be enough to just begin to do some weathering of the most susceptible minerals," says Golombek. "The material that's left over is the least soluble."

So the levels of silica are artificially enhanced compared to compounds that readily dissolve in water. This has been confirmed by subsequent rover missions, which have abraded the rock surfaces to peer underneath. "At various depths as you dig into the rock," Golombek adds, "we clearly see the silica increases the deeper you go."

Sojourner then journeyed over to Scooby-Doo and was halfway through the "Rock Garden" when contact was lost with the lander, fifty-eight days into the mission. At the end of September 1997, *Pathfinder*'s own rechargeable batteries failed. JPL lost all contact with both lander and rover.

The flight team was optimistic they would hear from Sojourner again. In the end, they did not. Signals were returned a couple of times thereafter in early October, but that was it. With no lander available to act as a relay for its own signals, Sojourner was lost on the surface of another world. At the end of October 1997, they were still trying to determine why the lander had gone into a hibernation mode. "For all we know, the rover is still working on the surface of Mars," Donna Shirley said at the time.

Contact was never made again. In any event, *Pathfinder* lasted three times longer than expected. Sojourner lasted twelve times as long. The rover explored an area greater than two thousand square feet (two hundred square meters) and performed many millions of individual measurements of the rocks and dust within that perimeter. More than that, Sojourner's lasting legacy showed not only the importance of roving on the Martian surface but also the potential of what cheaper and ultimately more ambitious missions could accomplish. "Without *Pathfinder*," Matt Golombek concludes today, "there would have been no subsequent Mars exploration program."

* * *

After her media appearances in the summer of 1997, Donna Shirley became a highly sought-after public speaker. Nevertheless, she often said that she found the management of the engineers back at JPL more fun. That was the diplomatic version, for as she has later observed, her work in managing the whole of JPL's subsequent greater Mars missions was disappointing. "I did not pull that off," she says today.

Although there was a logical plan for the next steps, the recalcitrance of the human element came into play once more. All the project managers beneath her didn't like each other, wouldn't communicate, and, worse, the funding available for all their projects was spread too thinly. Despite wanting to avoid what she calls the earlier "orgies of exploration," it didn't quite work out like that. "You learn something," Shirley explains of what she wanted to achieve, "and then you design the next mission based on what you learned."

Political pressure for spectacular results meant there was not enough time between missions for any new technology to be developed. "You pretty much have to go with what you have got," she said. NASA Administrator Goldin was pushing for a sample-return mission as early as 2005, which Shirley still thought, as she did in 1966 with the original Voyager, was ridiculously ambitious and way too soon.

At least after *Pathfinder*'s success, people no longer yelled at her at NASA Headquarters. Yet nobody in Washington, DC, wanted to contradict the boss, Administrator Goldin. Worse, what she terms engineering hubris took over. Shirley diagnosed this as: "I'm smart. I can figure out how to make this work." Fairly soon, she realized that the follow-on missions to *Pathfinder* patently would not work. What NASA Headquarters wanted to do was simply not feasible.

"The problem was that the HQ managers were all scientists," she said. "They did not understand engineering." When she told them that they were biting off more than they could reasonably chew, it was perceived as whining. "They were just feeling a tremendous amount of pressure from the administrator and weren't going to take him on."

Something had to give. The Mars Climate Orbiter, launched in December 1998, employed a propulsion system designed for English measurements. Flight controllers at JPL assumed they were metric. "They had far too many things to do and just didn't check," Donna Shirley notes. When it fired its engines to enter orbit around Mars, the orbiter came in too shallowly and burned up. Two months later, contact was lost with the Mars Polar Lander before it came in to land. As Shirley noted, "Nobody really knows why the lander failed."

Both were doomed, she says, because they were too ambitious, technologically speaking. The project engineers "worked their tails off," but it didn't matter. A later inquiry noted that, exactly as she had warned, both had been given 70 percent of the budget they actually needed to get the job done properly. While Donna Shirley herself had broken through a glass ceiling, "Faster, Better, Cheaper" had crashed through a floor. She took early retirement in 1998, knowing full well what would happen.

"I was kind of sad to do it," she says. "I didn't want to be associated with the failures."

Eventually, she moved back to her native Oklahoma, where she became assistant dean of engineering at her alma mater. Today, she remains in demand as a public speaker, even though she is now in her late seventies. One perennial question she is asked is what could she ever do for an encore. "I haven't found anything as intriguing than landing on Mars for the first time," she freely admits.

For Donna Shirley personally, her greater achievement was being as much a pioneer as the spacecraft she helped send to Mars along with its rover. "I think just being able to accomplish as much as I was able to accomplish with a lot of strikes against me," she says, "mainly being female—the aerospace industry is a cold warrior kind of industry, and it's not easy for a woman to do well in it."

* * *

During the worst times, Donna Shirley had to endure a ridiculously petty argument over the naming of the rover. She organized a nationwide competition that led to a NASA higher-up bawling her out over the phone because she had not followed "proper procedures." The eventual choice of the name Sojourner Truth had less to do with political correctness than a desire by Shirley to get young people involved. The fact that it was named after an African American reformer from the time of the Civil War who traveled the country advocating the right of people to be free was an incidental benefit.

Some of the older people at JPL disagreed. "I'd had enough of all this affirmative action crap about blacks and minorities," said one veteran JPL flight engineer (in the hearing of the older author) who was taking early retirement in the early 1990s.

Today, it is doubtful that those who have followed in his footsteps would think such a thing, let alone articulate it. Where once Donna Shirley had been the only female engineer on staff at JPL, the situation for women has markedly improved. By the year 2000, 405 out of 2,548 employees were female. That figure now remains steady at around 20 percent overall. Yet the most recent figures show that in 2016, only 15 percent of the greater teams for JPL missions to the planets were female (compared to 5 percent at the turn of the century). During the most recent landing on Mars, of InSight in November 2018, the core team was spread equally between men and women.

Many women, Donna Shirley believes, steer clear of engineering. The competitive, confrontational approach appeals more to boys. "You're expected to be tough to make it through," she says.

In recent years, the last great taboo has also been openly discussed right across society: harassment and sexual misconduct, which for many years were simply ignored or swept under the carpet. In the aerospace industry, as elsewhere in society, careers are at stake. Often, young researchers are the most reluctant to bring allegations of harassment forward. They depend on more senior colleagues to help further their careers. Privacy concerns also shield harassment investigations from public view, so it is hard to say how common these claims are.

Even today, in the era of #MeToo, we will never know how many women have been affected. As noted earlier, Donna Shirley later recorded some of the indignities she had to face. When asked about the situation in 2001, she stated that she didn't think things were as bad as they had been for her, "but it used to be there wasn't any sense in reporting it because nobody would do anything about it, so it didn't make any difference."

Yet attitudes have changed for the better. Nathalia Holt's superb book on the JPL "computers," *Rise of the Rocket Girls*, has redressed the balance about the crucial role they played in dispatching spacecraft to the planets. NASA's treatment of African American women during the 1960s has also been highlighted in Margot Lee Shetterly's equally excellent book *Hidden Figures* (and the film that followed). It showed how these women—highly qualified mathematicians and physicists among them—were forced to take low-paying jobs, use "colored" washrooms far from their offices, and even eat and drink separately from their white male colleagues.

NASA has taken care to signal its greater virtue in employing a diversity of women. It hosts a "Women@NASA" website and regularly releases profiles about women working at the agency. Even a cursory glance at the firing room for launches or the "trenches" during landings on Mars reveals that women are more involved in space missions today than even a decade ago. The same is true for people of color.

The astronaut corps, which contains NASA's most famous women and some of the easiest records to access, reveals just how far the agency has progressed. The first female astronauts were finally selected in 1978, nearly two decades after the first male Mercury pilots. The story is now well known about the "Mercury 13," a cadre of female pilots, including the remarkable Wally Funk, who were selected at the same time as crew-cut male test pilots in 1961 but never allowed to fly.

Among the Shuttle astronauts chosen in 1978, Sally Ride was the first American woman to fly in space in 1983. (However, Ride, a recognizable public figure until her death in 2012, never disclosed that she was involved in a longtime relationship with a woman, Tam O'Shaughnessy.) Female astronauts later went on to do spacewalks, command the Space Shuttle, and take charge of the International Space Station, but always after men did it first.

Greater numbers of men have flown in space than women and hold the record for having attained more cumulative time in space. For individual astronauts, men continue to hold the prominent astronaut records, such as amount of time in space across several missions, amount of time in space on a single mission, and number of missions in space, to name a few.

NASA had planned to run the first all-women spacewalk in March 2019 from the International Space Station. But it had to be put on hold because the "off the peg" spacesuits were uncomfortable for females and had to be altered from their male baseline design. Seven months later, Christina Koch and Jessica Meir participated in the first all-female spacewalk in October 2019.

Slowly but surely, many of the barriers that Donna Shirley had to face are being removed. In talks, her advice to young girls is simple: "Just do what your heart tells you to do," she says. "One of the things that I teach in my course is to follow your passion." Despite the hurdles, she is pleased that the situation has changed for girls who might become the engineers of the future. "What I have ended up being is a role model to everybody else," she says, even though she did not have similar role models herself.

Perhaps today a young girl who wants to fly will be able to explore Mars directly for herself. When she does, she should raise a glass to the true pathfinder in this story: Donna Shirley. And one day, in the not too distant future, an astronaut—hopefully female and/or a person of color—will land in Ares Vallis and come across the little rover that did: Sojourner. As an epitaph, Donna Shirley's own words are apt: "It wouldn't have been there if it wasn't for me."

7

WATERWORLD

For life to have occurred on the Red Planet, water would have been an important prerequisite. The emerging portrait is that liquid water was around on the Martian surface for sufficient periods of time to allow for life's emergence. Today researchers are using a variety of different forensic techniques to determine what happened to it. Exactly when that water flowed reveals clues as to how the Red Planet has evolved in the billions of years since. In teasing out the details, the riddle of Martian biology may finally be solved.

The basic building blocks for life may have been available in the Martian waters. If conditions were clement enough, they could have evolved into simple life-forms that could well have self-replicated. Finding evidence for both ancient or extant organisms, likely incubated by that primordial water, is one of the most compelling yet immensely difficult tasks ahead.

Perhaps there was even a planet-wide ocean, although many debate its existence and a number of experts think it was only ever transitory. While nobody has the complete picture, in the last few years, there has been a change in scientific attitudes. The northern plains of Mars contain some of the smoothest, most featureless expanses seen anywhere in the solar system. Some believe that is evidence for features caused by the lapping of this ancient sea.

The Mars ocean hypothesis started as a minority viewpoint. Now some geologists estimate that the volume of water on the surface would have been sufficient to cover the planet with a layer a third of a mile (a half kilometer) deep; a large enough basin would be the obvious place for it to have pooled. From this "What if?" scenario, various lines of geological evidence converged to suggest there may have been an ancient ocean on Mars. Though controversial, it represents yet another clear link for the possibilities that life may have existed on the Red Planet in the ancient past.

* * *

Earth's own oceans are, and always have been, a cradle for life. They are a crucible for biological activity: from the simplest cellular plants to the largest mammals such as whales, all of which have been able to adapt to a wide range of conditions. Whether in the warm, near-surface seas of the tropics or the icy depths close to the poles, life permeates the whole of the ocean. Indeed, the ocean floors have any number of advantages for the evolution of life. They would have been largely protected from the violent activity that was taking place above in the ancient past. The occasional asteroid impact and any volcanic activity would have had little effect on the stable environmental niches below.

The most remarkable species are those in the deepest reaches of our oceans that have acclimatized to the immense pressure and dark. Hydrothermal vents that appear as bizarre smokestacks—rich in thick black smoke that emanates from visibly boiling lava—have opened a whole new vista for the possibilities of life elsewhere in space. Within them may be found tiny wormlike creatures that derive their energy from some very fancy chemistry, not photosynthesis. At these abyssal depths, there is no sunlight available in the all-pervading murk. The enormous significance of these creatures is that they have evolved totally separately from much of the life found on the rest of our planet.

So far as our own species is concerned, three hundred million years ago fishes evolved into amphibians, which then moved onto the swamps and coastlines of Earth. Our future, give or take a few environmental accidents, was more or less assured. If life ever began on Mars, it is already evident that it never got this far along the evolutionary journey.

Nevertheless, the most recent spacecraft have shown that the Red

Planet had an early Earthlike period that lasted for geologically significant periods of time—perhaps millions rather than billions of years. There was a window of opportunity for life to have formed. "That window closed pretty early, and any remnants of that life are probably dead," says NASA biologist Chris McKay. "But the significance is that it could have opened."

The key to unlocking that window is water.

* * *

The geological evidence for water on Mars is persuasive. Innumerable channels crisscross the Martian surface and appear, to the untrained eye, like dried-up riverbeds. Careful analyses of their shape and size reveal distinct differences and subtle changes compared to terrestrial riverbeds. They offer important clues about the way in which water probably bubbled up and carved its way across whole swaths of the Martian surface.

Geologists have classed the channels by size, highlighting the distinct and different ways they were formed. Immediately after the *Viking* landings, it was thought that Mars had had a denser atmosphere early in its history, which would have allowed rain to fall. But more detailed analyses suggest that rainfall cannot explain the range of features that have subsequently been observed.

These include large outflow channels, seen in the mainly equatorial regions, that are more recent than some of the features seen on older terrain. These large outflows have few tributaries and leave behind tear-shaped islands many hundreds of miles long in their wake. Some geologists have concluded that repeated floodings from these outflow channels filled transient lakes that may well have been deep enough to form an ocean whose outlines are suggested across the northern plains. There is evidence that deltas formed where water flowed across the surface and accumulated into standing bodies of water.

South of the equator, there are medium-sized channels joined by tributaries. The final type of channel is the most controversial and the source of many abiding mysteries. Known as the valley networks, they are the most Earthlike in terms of their drainage patterns. When they were first observed, the most obvious explanation was that they were created by rainfall in a warmer climate. They are, however, the most troublesome to explain. Opinion has shifted because, as one participant in the debate notes, "those who have tried to model the atmosphere to allow for rain don't like a warm, wet Mars for very long."

* * *

There is no doubt that the amounts of water that flowed across the surface of Mars were of truly biblical proportions. "Humans have never experienced events like this," says Dr. Matt Golombek, the JPL geologist who came to prominence with the *Pathfinder* mission. The findings from Ares Vallis, where *Pathfinder* landed in 1997, revealed the evidence for water beyond all doubt.

The largest channels are found in the equatorial regions of Mars, where they fan out onto the northern plains upon which *Pathfinder* landed (as had *Viking 1* at a more northerly location). Most run for many thousands of miles, and some stretch many tens of miles across. Though they appear similar to dried-up riverbeds on Earth, they likely originated when catastrophic floods of water erupted and covered the surface.

The water that flowed through these channels was anywhere from a hundred to ten thousand times the amount currently discharging from the Amazon basin. Even conservative estimates suggest that corresponds to a fifth of a cubic mile (a cubic kilometer) of water being released *every second*. The broad, gentle ridge where *Pathfinder* came down most likely resulted when an elongated tail of debris was carried along by one such episode of flooding.

Speed was truly of the essence. Features close by the *Pathfinder* lander in the "Rock Garden" appear inclined and stacked, as if they were placed there by remarkably rapid flows. The lips of the "Twin Peaks" crater toward the horizon also show streamlining characteristic of fast water flow. It has been estimated that the water surged along at speeds of up to roughly 170 miles (275 kilometers) per hour. That was equivalent to all the water in the Great Lakes fanning out into the Gulf of Mexico in a short period of time. "The catastrophic flood at the *Pathfinder* landing site probably lasted a couple of weeks," Matt Golombek says. "There are places on Earth where things like this have also happened."

Significantly, *Pathfinder* spotted what is known as a conglomerate, a smaller rock that resulted from where pebbles and soil had joined together when they had been repeatedly rolled around by the flow of water. This lonely rock was itself a revelation. "The rock wasn't all that rounded," Matt Golombek says today; it likely formed elsewhere and was carried in by the floodwaters. It had clearly not been "streamlined" for any length of time.

Taken in a wider context, these great outflows of water coursing across the landscape potentially formed transient seas.

* * *

Martian geology provides a clue as to where all this water might have come from. When the large equatorial volcanoes emerged in Tharsis early on in Martian history, they likely cracked open the crust at the western end of the Valles Marineris. This would have generated heat so quickly that subsurface ice would have melted (though other causes such as impacts could have produced much the same result). The energy released could have prompted flooding on an unprecedented scale farther east. What geologists have called "chaotic terrain" on Mars likely saw the sudden release of water and the subsequent collapse of the surface.

In the southern hemisphere, which has an older appearance because the cratering record has not been erased, there are medium-sized channels joined by tributaries. Here, spring sapping likely occurred. Springs sprang up and flowed, effectively eating into the channel margins and reappearing later upstream, a process that repeated itself time and again. Some of these channels have a "fretted" appearance. A number of geologists suggest this is characteristic of glacial activity gouging flat-bottomed channels, aided by meltwater that came from the advance and retreat of glaciers.

The controversial valley networks, with their many treelike branches, comprise narrow, V-shaped valleys that stretch for a few tens of miles. They appear similar to terrestrial rivers in their drainage patterns, but, as one geologist says of their origins, "whether you need rain is a very different matter altogether." More recent imagery has shown that the heads of the valleys are different from those seen on Earth. On Mars, they appear much shorter and more open in their drainage patterns. Many of these valley networks are restricted to the more heavily cratered southern hemisphere, which suggests that they too are ancient surface features. They have not been removed by any more recent glacial activity.

Opinion is divided as to how they formed. The first detailed images merely fanned the controversy, for in those images it no longer appears as though they were formed by running water. Many of the valley networks now seem to owe their patterns and their relief to subsidence and collapse. In some places, they could have been produced by groundwater or glacial flows.

The problem is that the computer models that have tried to reproduce the early climate on Mars cannot reproduce such drainage patterns under a warm climate or certainly as a result of rain. It now appears that Mars was much colder than many people first thought after the *Viking* missions in the 1970s. Though there have been warmer periods, nobody really knows how long they lasted, and the same goes for exactly how transient any water on the surface was.

* * *

One leading expert on the question of water on Mars is Dr. Michael Carr, a veteran of nearly all of America's first wave of planetary missions. Now retired, he worked for many years for the US Geological Survey in Menlo Park, south of San Francisco. He was chief of the Branch of Astrogeological Studies and has written two standard reference books on the Red Planet, entitled simply *Water on Mars* and *The Surface of Mars*.

Born in Leeds, Carr joined the brain drain from the United Kingdom in the 1960s and was leader of the camera team for the *Viking* orbiters. As a result, he has had a unique perspective on all the subsequent debates. His Yorkshire accent long modified by living in California, he talks quickly and excitedly about the implications of water. "It is a convoluted story. It has taken us a long time to understand Mars. Everybody's struggling on the question of water." Almost in jest, he adds, "I think we'll get definite evidence for life before we can work out the global inventory of water on Mars."

It was Michael Carr who calculated that the Red Planet may have had sufficient reserves of water to have covered the planet with an ocean

a third of a mile (half a kilometer) deep. But he does not believe that the
water was ever in the form of an ocean. Rather, it was trapped underground as ice, which formed a planet-wide water table. His reading of
Martian geology is that most of the water flows occurred in the first half
billion years of the planet's history. Even then, he believes they were very
sporadic.

Carr believes that the Martian climate was never much warmer than
it is today. Perhaps there was always an extensive layer of ice immediately
below the surface. In the early history of Mars, there was a greater flow
of heat from the interior—perhaps five times more than that estimated
today. Certainly, in those first epochs the Martian core was more energetic. This greater warmth naturally melted the ice deeper underground.

That means water existed as a liquid much nearer to the surface than
it does today. There would, however, have been a layer of ice directly
below the surface, effectively insulating the substantial subsurface layer of
water underneath. The ice layer would have been thinner than it is under
today's climate conditions.

Groundwater would thus have been trapped under the near-surface
frozen layer, known as the cryosphere. Meanwhile, extra pressure built up
underneath due to the extra heat flow. When that pressure became too
great, this natural dam burst. It punctured the cryosphere and allowed staggering amounts of water to escape from below the surface. Carr's feeling
that Mars always had a cold climate in the past is significant in this scenario. "The surface was frozen," he says. "It was that which kept the water
from getting out, as there was no other way to release it. Occasionally, an
impact or a fault erupted and that would have released great floods."

Because gravity forces water to migrate downward, the water tended
to move from high ground to lower regions in the earliest epochs of the
planet's history. That would have limited anything approaching a water
cycle on Mars. The release of this water would have been a one-shot trick
in the ancient past. After it had burst out of the highlands, it probably
couldn't percolate up through the water table to reach the mountains
again. As evidence, Carr points to the fact that the rugged highlands of
Mars are crisscrossed by the narrow valleys that appear to have drained
into lower-lying regions.

To some geologists, it appears that two of the largest and oldest

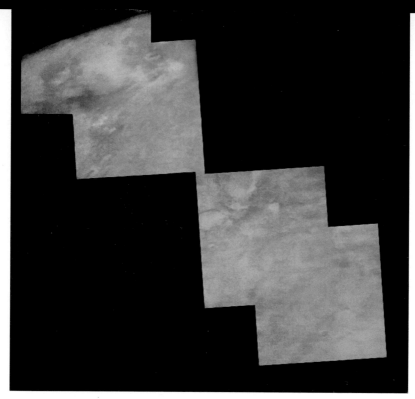

The first close-ups from *Mariner 4* in July 1965 as it tracked across the southern hemisphere of Mars, revealing unsuspected craters on the surface.

The first successful orbiter, *Mariner 9*, arrived in late 1971 when the Red Planet was covered in a thick blanket of dust. The three dark spots are volcanoes, whose existence changed perceptions of Mars overnight.

NASA astrobiologist Christopher McKay, a leading light in our current scientific search: "To understand how life would evolve on Mars, you have to go to Antarctica."

The Dry Valleys in Antarctica, close by McMurdo Sound, have the nearest equivalent on Earth to conditions on Mars. They are unexpected in their beauty and hosting of unusual microbial life-forms.

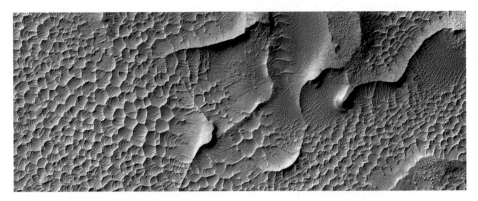

Frozen in place: Dunes observed in January 2020 by the Mars Reconnaissance Orbiter. Its HiRise camera can see down to 10 inches (25 centimeters) in scale. Within a lifetime, our quest to understand the Red Planet has returned photographs of Mars with greater resolution than those for Google Earth.

A glacier on Mars—seen here in the midlatitudes—suggests that episodes of glaciation have occurred in fits and starts throughout its history and more recently than originally thought.

The first selfie from the InSight lander in Homestead Hollow, as it prepared to place "the wok" (at the center) on the Martian surface. The wok serves to protect the sensitive seismometers from the local environment.

Bruce Banerdt (fourth from left), celebrating the InSight landing in November 2018. He has led the first dedicated effort to peer inside the Red Planet using new techniques.

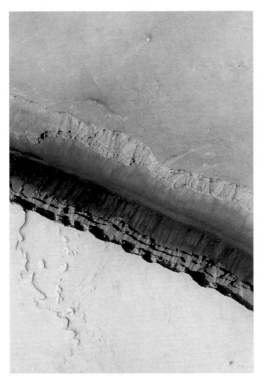

Recurring slope lineae, seasonal features that were thought to suggest fresh water may have flowed across the surface today.

Cerberus Fossae, a series of faults in the crust that are the source of the first marsquakes.

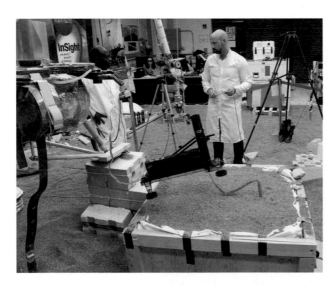

Mole problems: Troy Hudson working in the test bed to understand why the heat flow experiment has not been able to drill.

A panorama of the Martian surface, containing 1.8 billion pixels and composed of more than 1,000 images taken by Curiosity between November 24 and December 1, 2019. The Gale Crater rim is visible in the distance ahead. On the far right (facing page) is Mount Sharp, formally known as Aeolis Mons.

Adam Steltzner immediately after the landing and deployment of the Curiosity rover in August 2012. An unlikely rocket scientist and failed rock star, he says he simply got curious and followed that instinct, which led him to Mars.

Yellowknife Bay, a landscape unlike anything ever seen on Mars before, explored by the Curiosity rover as it made its ascent through the foothills of Mount Sharp.

Wear and tear, one of the greater worries for the Curiosity mission—the cumulative damage to the rover wheels, forcing the operators to be careful in their driving.

Abigail Fraeman, planetary geologist and rover-driving planner, who delights in the surprises that lie in store.

The *Mariner 6/7* infrared spectrometer, whose initial results caused a gasp when its readings seemed to suggest there was methane frost on Mars (later retracted).

Project scientist Gerry Soffen and the Viking lander. Arguably the most complicated spacecraft built at the time, the Vikings looked for life on the surface.

The first image of the surface of Mars, taken by *Viking 1* within minutes of landing, shows small rocks and dust that are ubiquitous on the surface.

The man in charge: Jim Martin pouring the champagne, flanked by two NASA administrators to his immediate left, celebrates the first successful landing on Mars on July 20, 1976.

Two trailblazers: Donna Shirley, who fought for the first rover, Sojourner (shown here to scale), to be included on the Pathfinder mission in 1997. "The most fun job I ever had."

The day after landing, *Pathfinder*'s air bags (shown here, left) worked, and the Sojourner rover was ready to explore the geological paradise of an ancient flood plain.

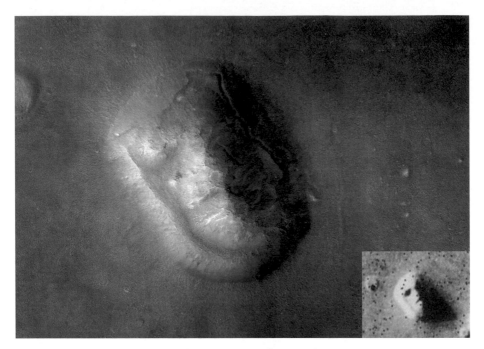

Face to face: a windswept mesa in a region called Cydonia, shown in glorious detail.
It is not an alien artifact, as some interpreted the earlier images (inset).

Dr. Michael Malin, whose work has transformed Mars into a familiar world, shown in a test shot for one of his cameras.

The Occidental tourist, Percival Lowell, in South Korea. His time in East Asia was formative for the Boston Brahmin, who became utterly convinced there had to be life on Mars.

Lowell's Mars, a world covered in canals that only he and a handful of others mapped, using the most powerful telescopes of the day.

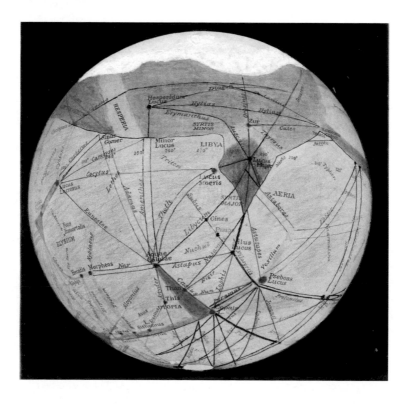

The Pearl Harbor moment in 1996, when a NASA team announced evidence
for fossilized remains inside a meteorite from Mars, a claim that soon
split the scientific community and is still controversial today.

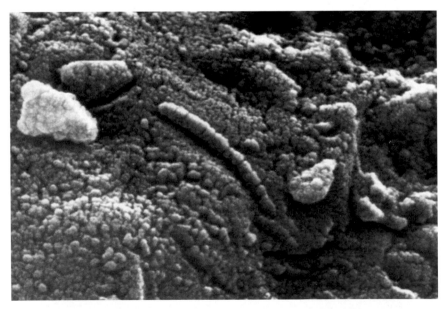

"The worm": a fossilized remnant of ancient life or an optical illusion?
A microscopic segment of a tiny feature in one of the highly magnified views
of that same rock that fell to Earth.

For the first time in forty years, NASA will be looking for signs of possible biologically significant material with its Perseverance rover.

In the test bed at the Jet Propulsion Laboratory, engineers put the new, improved wheels through their paces for their next mission.

Named after Rosalind Franklin, the pioneer of DNA, Europe's first-ever Mars rover will dig deeper into the surface, which some believe will provide the evidence life existed.

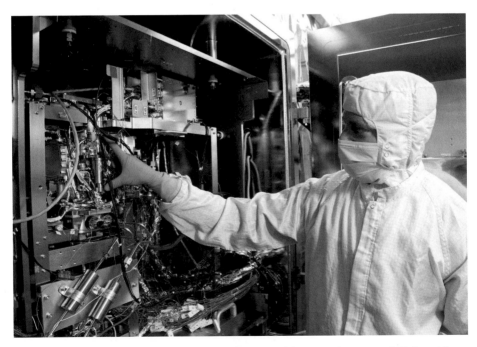

An improved version of the self-contained chemistry laboratory known as SAM used by Curiosity is being flown on the Franklin rover, with additional capabilities to look for life.

craters on Mars, Hellas and Argyre, both in the ancient terrain of the southern hemisphere, contain sediments. That indicates that vast amounts of water were pooled within them, perhaps collecting there after flowing down from the nearby mountain ranges. Most importantly, says Carr, the Martian climate changed quite suddenly around 3.5 billion years ago.

"The erosion rates on Mars are extremely slow," Carr suggests. "We can determine that from crater counts, and see how they have been wiped out. The obvious thing is that they haven't." In other words, nothing much has happened, geologically speaking, in billions of years. Erosion on Mars occurs at only one ten-thousandth of the rate seen on Earth. "There was a dramatic change in erosion at that time," Carr says. "The climate must have been very different."

The valley networks provide an important clue. Their origin remains controversial because of the climate conditions needed to allow them to form. If they are similar to river valleys on Earth, a warmer climate would be required for clouds to develop and allow rain to fall. Indeed, the earliest attempts to explain their formation implied that Mars was "wild, wet, and windy" in its ancient past. The only problem is that suggesting the Red Planet was hotter goes against all astronomical knowledge.

All stars tend to be dim and cool in their earliest epochs. Four billion years ago, the Sun was roughly 70 percent as bright as it is today. Both Mars and Earth would have been close to the freezing point of water. What is known as the "faint young Sun paradox" is a puzzle to climatologists, because on Earth, our primordial oceans were obviously not frozen over. Perhaps episodic changes in greenhouse warming helped form life on Earth—and, thus, perhaps something similar could have happened on Mars.

This hypothesis requires the Martian atmosphere to have contained sufficient carbon dioxide to trigger a greenhouse effect. As we will see, that generates further riddles.

Michael Carr says that if conditions were warmer on Mars, then they did not last for much longer than a few millennia. The extent of the erosion that resulted from water flowing across the surface is limited and restricted locally. Many ancient craters remain almost intact, and most show few signs of degradation. From the evidence returned by the *Viking* missions, the valley patterns suggest that they were formed during a short period of time.

If Mars was ever remotely "wild, wet, and windy," it was akin to an accelerated stage of puberty in the young planet's history. But how could the Martian climate have changed so dramatically 3.5 billion years ago? Certainly, by that stage, the greater bombardment—from asteroid and meteorite impacts—would have already gouged out the larger craters in the southern hemisphere.

Indeed, one of the greater puzzles in Martian geology is the north-south divide, sometimes known as the "hemispheric dichotomy problem." In simple terms, the smooth plains of the north provide a stark contrast to the heavily cratered terrain to their south. What could have smoothed out such vast tracts that stretch across the northern hemisphere of the Red Planet?

* * *

The obvious answer is that there was a large ocean covering the northern hemisphere at one time or another. Thirty years ago, Timothy Parker of NASA's Jet Propulsion Laboratory and other colleagues argued that there are features similar to shorelines seen on Earth in the high-resolution images taken by the *Viking* orbiters. Thus started the notion that there may have been large bodies of water, whose size and longevity depended on the climate of the time. There may also have been spillways between the various watery expanses, spits of land, and even terraces surrounding a number of larger basins in lower-lying areas.

The idea was taken one step further with the proposition that there was a large standing body or "pool" of water four times the size of the Arctic Ocean in the ancient past. This hypothesis was first set forth by a team led by Victor R. Baker, then head of the hydrology department at the University of Arizona. They called the body of water the Oceanus Borealis. "I see Mars as a much more dynamic planet throughout its history," Baker says. "The Martian ocean would be a short-lived feature that persisted for many thousands of years."

Throughout the history of Mars, there was sufficient heat to generate vast outflows of liquid water across the surface and, crucially, to replenish the aquifers and permafrost below. Large amounts of water and carbon dioxide dissolved within the waters were repeatedly released, creating a warmer atmosphere. This allowed an ocean to form. These transient greenhouse conditions were aided by other gases like methane, which trap heat much more efficiently.

The ocean would have also allowed a water cycle to occur. Water would effectively be recycled by evaporation, which would then cause precipitation in the colder uplands of the southern hemisphere. The water would either trickle back across the northern plains, or else percolate through the crust to top up the water table.

Evidence that large basins of water reformed over and over again came from Viking pictures of Martian features that are similar to "tidemarks" seen around Lake Bonneville on Earth. It is known that the Utah salt lakes were much more extensive in the past. The geology of the surrounding area has preserved concentric tidemarks as the lakes retreated.

However, more recent missions have altered this idea of what an ocean, or oceans, might have been like. The laser altimeter aboard the Mars Global Surveyor, which mapped the polar ice caps in such detail for nearly a decade starting in 1997, was used to create the first detailed topographic maps of the northern plains. While they did appear very smooth—"We don't know of anywhere smoother," said one of the altimetry team in the late 1990s—the cumulative results showed a number of difficulties with the theory.

Though many of the shorelines did indeed follow the edges of what was shown to be a flat surface, the smoothness itself varied by a few kilometers (or miles) in elevation. In other words, if it had been caused by an ocean lapping across the local surface, the average sea level would have risen and fallen alarmingly. The same shorelines do not form a straight line when superimposed over higher-resolution images that have revealed surface landforms in greater detail.

Yet it might well be that the lie of the land on Mars today on a global scale hasn't always been the same in the past. What are known as "crustal deformation models" show that if you took away the effect of the Tharsis bulge, the weight of that cumulative plug of volcanic rock in the crust removes some of this shoreline elevation. The rise of the large volcanoes has distorted what was a more uniform sea level. But critics contend that even taking such factors into consideration, not all the shorelines fit.

In any case, others suggest that Mars has changed so much over geological time, it is very difficult to understand what the shorelines are telling us about an ocean. "It's still alive as an idea and something people are working on," notes one participant in the debates. "The laser altimetry doesn't throw doubt on the ocean theory as a whole, but it does pretty much discredit the shorelines as they have been drawn up in the past."

* * *

Though the existence of any sort of ocean on Mars is contentious, it remains a very appealing idea that certain geologists would like to be true in some sense. There is enough evidence available for it to be viable. All it will take is one really good piece of evidence to prove it beyond all doubt. As a consequence, there are variations on the theme.

Some believe there was a single ocean that formed under warmer conditions and then evaporated over time; others think that there were many separate transient seas that formed and reformed as the climate changed; and to account for that climatic deterioration, some believe that the original ocean froze over and warmed underneath, which would fit the climate models.

"It all started off with these two polarized views," notes one wry observer of the debate, "but they are coming together where there is a middle ground—namely, that Mars was always colder than Earth, but there were stages of a warmer climate."

For now, the debate really centers on what the geologists believe they are seeing and what the climate scientists believe their models are telling them. The geologists are trying to get everything to fit together. There would have been a certain amount of water, but whatever estimate they come up with, it cannot be made to fit into climate models. Conditions were neither wet enough nor warm enough.

However, the scientists are getting to a place where they are meeting in the middle. Those who support the northern ocean are suggesting it didn't have to be there for millions—and certainly not billions—of years and probably was transient. But, as more than one authority has pointed out to us, there is no clear meaning of just how transient they might be.

If water existed on the Martian surface for a sufficient length of time, a significant hydrological cycle might have developed. In some ways, standing pools of water that are replenished are a much better bet for biology. "Hundreds of little pools are better than one big ocean," NASA's Christopher McKay notes. "The reason is that you get greater productivity and more photosynthesis."

More recently, there has been an intriguing finding that speaks of what might well have happened to all the water that may have pooled in one or more large expanses. Hypanis Valles is an ancient river flow that opens out onto a delta running across the equator, one of the finer exemplars of the north-south divide on Mars. At one point, it had even been considered as a possible landing site for the Curiosity rover, as it shows signs of sediments that opened out across a delta. In recent years, the region has been mapped in detail by the THEMIS spectrometer and the high-resolution camera, HiRise, aboard the Mars Reconnaissance Orbiter.

In the summer of 2018, a team of international researchers suggested the delta could have been the "planetary plughole" into which all the water from an ocean sank. Detailed observations implied there was a water cycle—"active about 3.7 billion years ago," said one participant—and that it started to shut down thereafter. It is not definitive proof that an ocean ever flowed away, but another of the team noted that "these geological features are very hard to explain without one."

Others remain unconvinced. What can be said is that a great deal of water would certainly have percolated through the soil to form a layer of permafrost. The ice sheets detected by the various radars in orbit around Mars are proof that water is there in large amounts. But quite a lot more was likely lost to space over geological timescales. As we've noted, with a gravitational pull only 40 percent that of Earth, Mars could not hold on to much of its original water.

* * *

Much of the original atmosphere of Mars has been blasted into space thanks to the accumulated impact of large asteroid-sized bodies. Not only did these impacts excavate giant holes in the surface like the basins of Hellas and Argyre, they would also have expelled significant amounts of the original Martian air, which the weak gravitational influence of the planet couldn't "hold back." Similarly, the bombardment that accompanied the birth of the planets is thought to have removed most of the original atmosphere of Earth.

Volcanic activity, in the form of fissures of magma or gases escaping out of the crust, would have formed replacement atmospheres on both worlds. Well before life could have formed, there would have been the first appearance of the physical phenomenon known as the greenhouse effect, now much maligned because of its association with human-caused global warming. On Earth, water vapor would have been a major player. Into this volcanically active cauldron, steam would have condensed to form heavy banks of clouds and coalesced, as temperatures cooled down even further, to form oceans.

When that happened, some three billion years ago, Earth would have become a largely ocean-covered world, constantly refilled from underground vents and the condensation of vapor. This process might also have been aided by the delivery of water from comets. After that, it is hard to determine what happened next. "When we look at the Earth's history," cautions Bruce Banerdt, the InSight principal investigator, "we run into a brick wall, because the Earth is so active, the evidence for all these processes has gotten basically erased."

On Mars, something similar could have taken place, but the aftereffects would have been very different. The atmosphere that was left behind and subsequently evolved due to volcanic outgassing was a paler imitation of the planet's original gaseous envelope.

There is a way of trying to work out cumulative atmospheric loss by studying how energetic particles from the Sun interact with the Martian atmosphere now. In a very real sense, our robotic envoys have arrived in the nick of time. Spacecraft such as NASA's Mars Atmosphere and Volatile EvolutioN (MAVEN) mission and Europe's Mars Express have shown that the solar wind is stripping out the uppermost layers of the atmosphere. The latest findings show that the death of the Martian atmosphere has been a long, protracted affair over millions of years.

"We have been able to measure this loss in more detail," says Professor Bruce Jakosky, lead scientist for the MAVEN mission. "We've looked at the processes which drive the ion production and acceleration and escape. What we've determined is the Martian atmosphere is being lost at a rate of two or three kilograms per second." As a result, they have been able to extrapolate back much farther in time. "What that allows us to do is get the integrated total loss over last four billion years," Jakosky says.

It would take a few hundred million years to completely remove the present-day Martian atmosphere. Interestingly, this would not be across the board, so to speak. The gaseous constituents of the atmosphere are being selectively pulled apart. Hydrogen is the lightest element and will escape from a planetary atmosphere at the drop of a hat. It may also be lost to space by these same energetic interactions in its chemically bound-up form with oxygen, as water vapor. (Earlier Mars missions have shown that mainly ionized oxygen was being removed by as much as ninety tons a day.)

There is evidence that in the ancient past the Sun was emitting much greater amounts of ultraviolet radiation that were accompanied by more intense solar winds. "The loss rate would have been much greater earlier in Martian history," Jakosky says. "The conclusion is the atmosphere of Mars was stripped of a couple of times as much atmosphere as the Earth has."

This latest work by Jakosky, a professor at the University of Colorado, Boulder, and others has focused on the isotopes of argon. These are chemically inert and do not react with other atmospheric molecules. Argon isotopes can only be removed by a process called "sputtering"—in effect, a form of interplanetary billiards. Electrically charged ions within the solar wind hit the atmosphere of the Red Planet at very high speeds. They physically knock other atmospheric molecules into space. The observed ratios of argon isotopes show just how extensive this has been.

Using argon as their guide, the MAVEN scientists have been able to work out which other atoms and molecules are likely to have been lost, including, crucially, carbon dioxide (CO_2). The cumulative loss of gases to space has been the single most important factor in how the atmosphere of Mars has degraded over time. "Though carbon dioxide has gone into the crust and there are carbonates and frozen CO_2 in the polar ice caps," Jakosky says, "the loss to space has been by far the greatest."

* * *

Glaciers also provide geologists with a cop-out, a kind of halfway house between the notions that Mars was either always freezing or that it was very much warmer in the past. As a process, glaciation has undoubtedly occurred on the Red Planet, whether there were primordial oceans or not. Earth's glacial flows, as on Mars, have ridges marked by boulders and sediments. There are also deposits that are characteristic of terrestrial features, which have been formed by streams running under glacial ice.

On Earth, glaciers will only gouge out surface features if the glaciers are large enough. Invariably, these glaciers were not completely frozen, so they must have flowed under warmer conditions. On Mars, more extensive midlatitude glaciations have been discovered from radar measurements of subsurface ice, which bear upon both the climate and the geology. One school of thought suggests that the Martian ice sheets were very thick and sublimated away—that is, changed into water vapor without first forming liquid water.

Others think that they were more like alpine glaciers: they grew and flowed across the surface, retreated, and then, when conditions were colder, grew again. Which of these processes was more important remains elusive. For example, one might try to determine the ages of subsurface ice from crater counting, but there needs to be a wide enough area to be statistically significant.

"My hunch is that the midlatitude ice goes all the way back to the onset of this glacial period," says Professor Jack Holt of the University of Arizona. Long-suspected cycles of climate change are supported by the layered terrain around the poles of the planet. It seems clear that most of the midlatitude ice was gone by three billion years ago. "There was some kind of transition period," Holt says, "but we don't know a lot about it."

The more recent warm periods would probably not have been stable beyond a few million years. Before that, there would have been distinct periods of glaciations that would have filled any oceans with ice sheets. Each successive outflow of water would also "top up" the atmosphere. Conditions would be cold enough for glaciers to exist for only a few thousands of years, until the climate got warmer again.

If you sprinkle dust on the polar cap, that will lower the amount of sunlight reflected back to space. That in turn raises the temperature and lessens the amount of carbon dioxide that gets frozen onto the poles. It is, however, hard to keep volatiles like carbon dioxide down on the ground.

Glaciologists have estimated that Earth and Mars have polar ice deposits of similar thickness, although the terrestrial ones cover a greater area. On Mars, the northern permanent cap is about ten times larger than that at the south. Water vapor is much less volatile than carbon dioxide. Because there is limited cycling of water vapor on the Red Planet today, it would tend to stay put. The most obvious current source for water on Mars is the north polar cap: during the summer, it is exposed and mixes in with the atmosphere. Any available water could have evaporated by the next winter season. Sublimation of water ice occurs during the spring and summer, but nobody knows how long this has been taking place, nor how effective the evaporation has been over time.

It also appears that the southern polar cap could be a "sink" into which CO_2 has also deposited. It has been calculated that somewhere between ten to twelve inches (twenty-five to thirty centimeters) of ice freezes out on top of the permanent cap on Mars today, just over the course of one year. Working back in time, there would have been a remarkable cumulative effect. What are termed "sequestrated CO_2 deposits"—the accumulation of dry ice over geological time—show that the Red Planet would had to have been very different in the ancient past. "When they were part of the atmosphere," Jack Holt says of these deposits, "it basically doubles the amount of CO_2 available for atmospheric pressure."

The situation is complicated because climate, water, and carbon dioxide all behaved in unknown ways on the Red Planet in its earliest history. If a scientific model seems to work for one aspect—say, the geological features observed—it doesn't really do so for the others. Much more work is required to understand what the exact role of glaciation was in the history of Mars and its climate.

* * *

To truly place Mars in its correct physical perspective, we have to look at the next world from ours toward the Sun. Venus, named after the goddess of beauty, is nearer to Dante's vision of Hell. The planet is smothered by dense clouds, some of which contain sulfuric acid. The surface below chokes on impossibly high levels of carbon dioxide, with temperatures high enough to melt lead. Venus is also subject to violent storms, from which thunderclaps have reverberated around the planet for up to fifteen minutes at a time.

Thanks to a handful of dedicated space missions, the surface of Venus has been unraveled by radar instruments that have penetrated the thick, featureless clouds. The great surprise is that while Mars hasn't changed much in the last billion years or so, Venus appears relatively "fresh," geologically speaking. It is covered by telltale signs of extensive lava flows, which do not seem to be older than a hundred million years or so. Indeed, today's geology on Venus is completely dominated by volcanism. Some geologists believe volcanoes may still be active, although the evidence is questionable.

One thing seems certain. Venus's extensive volcanism stopped short of further evolution into plate tectonics. That Venus fell short of this stage of geological evolution probably had to do with how much heat was available to drive it from within the planet's interior. Venus and Earth are roughly the same size. They probably contained similar amounts of radioactive elements within their cores when they formed. So why are they so different?

All the recent geology on Venus seems "dry." Yet geologists believe that water played a significant role during the early history of the planet. Water affects how the crust moves and lowers the temperature at which the rocks melt. On Earth, rocks buried below the surface, thanks to subduction caused by tectonic activity, liquefy at far lower temperatures than they normally would, due to the incursion of seawater.

Radioactive dating of the Venusian atmosphere has hinted that there were vast reserves of water in the planet's dim and distant past. It is very hard to "make" Venus without water. So perhaps the young planet was like the early Earth, covered by a primordial ocean that served to trap heat and maintain clement conditions. Yet at some point, the two planets diverged, and the water on Venus boiled off. The culprit is another potent ingredient in the chemical crucible of planetary evolution: carbon dioxide.

Along with water, carbon dioxide provides a common thread between the trio of somewhat similar planets in the inner solar system, Venus,

Earth, and Mars. Carbon dioxide prompted a greenhouse effect on all of them. What happened to Venus ultimately sheds light on how the Red Planet may have developed. All three planets were hewn out of the same sort of dust and gas some 4.5 billion years ago. The starting points for the chemical composition of their crusts, and hence their surface rocks, were broadly similar. Their atmospheres coalesced out of the same kind of gases vented by volcanism or released by the larger impacts. After the Sun became stable, surface temperatures were roughly similar on all three worlds. Each of the planets evolved along separate and distinct lines, and the root of their diversity comes from the way in which carbon dioxide acted in its role in underpinning the greenhouse effect.

Without carbon dioxide, conditions on Earth would not have been clement enough for life to have evolved. The amount of carbon dioxide in our atmosphere was regulated as it dissolved in rainwater. This would have produced carbonic acid, which eats away at rocks by the process of weathering. This process created carbonates, which are flushed out into the oceans. On Earth, they are usually recycled by the action of marine biology. Carbonates collect on the ocean floor due to the death of aquatic life-forms. After that, they are incorporated into the rocks by the chemical conveyor belt that is plate tectonics. As they are then subjected to higher temperatures and pressures, the carbon dioxide is released back into the atmosphere via volcanic eruptions on Earth—but on Mars, those conditions are unlikely.

* * *

Climatically speaking, carbon dioxide acts like an atmospheric thermostat. It traps sunlight by simply not allowing heat to escape back into space in the form of infrared radiation. On Earth, the thermostat setting compensates for itself: if the climate becomes cooler, greater amounts of carbon dioxide are released than are removed by natural geological processes. This then causes the climate to warm up, but not so much that it gets out of hand. The carbon dioxide is broken down by the action of sunlight and escapes into space. The net result is that Earth's greenhouse setting is reduced back to the normal, "regular habitable planet" setting on the thermostat dial.

On Venus, there was no compensation for this heating. The green-house effect galloped off without restraint. Mars, on the other hand, could not release enough carbon dioxide into the atmosphere in time to keep temperatures at a stable, or warm enough, setting. The terrestrial thermo-stat is precisely—some might say precariously—balanced. That stability is just about right. If all terrestrial carbon dioxide is not replaced, then it would be completely gone within ten thousand years. Our biosphere helps fine-tune this recycling. Plants, for example, assimilate the carbon dioxide, which is then incorporated into geological sediments when they die.

On Venus, measurements made by innumerable Soviet Venera space-craft chronicled the chemical ratios of hydrogen and deuterium, its rarer isotope. Deuterium is slightly heavier than hydrogen, and that extra weight is enough to keep it bound to the planet by gravity. What this means is that hydrogen will preferentially be lost compared to its heavier chemical cousin, deuterium.

It seems reasonable to assume that the original ratio was about the same as Earth's. Today, though, there is a surplus of deuterium on Venus compared to the lighter hydrogen. This suggests that there may have been oceans on Venus, but the lighter hydrogen, bound up chemically as water, evaporated. Since it was nearer to the Sun, Venus was hotter. Its predominantly carbon dioxide atmosphere probably began to thicken sooner, helped along by the water vapor that was released by the enhanced evaporation of the oceans. In other words, the primordial seas of Venus simply boiled away.

Solar radiation helped by breaking the chemical bonds within the water on Venus, which would otherwise keep its constituent hydrogen and oxygen bound together. The hydrogen escaped to space as the oxygen reacted with the molten surface rocks. In the case of Mars, however, the planet was too small to hold on to its original inventory of liquid water. The Red Planet did not develop a strong enough internal heat source to continue churning the crust and go on to form plate tectonics. When the heat ran out, active geology across the whole of the planet probably terminated. That would have stopped any recycling of rocks dead in its tracks.

At the heart of this discussion is yet another curious paradox, a considerable riddle that confronts scientists as they come to terms with the

Red Planet today. If there had ever been sizable amounts of water in a denser or warmer atmosphere of carbon dioxide, there should be telltale evidence in the form of carbonates on the surface. Repeated removal of carbon dioxide in the presence of water would have generated untold quantities of it. Under warmer conditions, carbon dioxide would be ever more likely to dissolve into any surface water that was lying around. So where are the carbonates on Mars?

Though NASA's Spirit rover identified carbonate-rich outcrops in the walls of Gusev Crater where it landed, such finds have remained elusive for the Opportunity rover, and, so far, for Curiosity. The next NASA rover, Perseverance, is about to land in one of the few places where carbonates have been detected from orbit. Why the rest are missing is still an open question.

Certainly, carbonates do not appear in the quantities that may be expected from vast swaths of the surface having been covered in oceans of water. If the water in the ocean was very cold or very acidic, that would have stopped them from forming. It has also been suggested that they may be stored out of sight, deep within the crust of the planet and, most importantly of all, locked up in the polar ice caps.

Recently, a team lead by Bruce Jakosky at the University of Colorado, Boulder, has suggested that even the poles of Mars have little carbonate available to warm the surface of Mars. There might be another explanation for this missing carbonate at the poles. "We might expect the poles to have solid carbon dioxide ice, not carbonate," explains Professor Jakosky. "Carbonates are a mineral form of CO_2 in which it has combined with another element such as calcium."

Other researchers have concluded that from the way that heat would be conducted under a thick polar cap, the CO_2 would be sublimated out—that is, changed from an ice to a gas without ever being a liquid— because of the curious ways in which the minuscule solar heating would be conducted underneath the poles. Professor Jakosky also points out that there is very little carbon dioxide in the polar regions, even allowing for the seasonal deposition observed today.

"There's none to speak of at the north pole," he says, "and only enough at the south that, if it were put into the atmosphere, it would only double the atmospheric pressure."

* * *

One of the dominant processes during the early evolution on Mars was volcanism. With the Red Planet volcanically active in the ancient past, its surface would have been sprayed not just with lava but with a pervasive, acidic fog. Volcanoes would have belched out clouds of gas that tended to become acidic, thanks to the action of unfiltered ultraviolet radiation upon them. When these poisonous clouds descended to the surface, they would leach any carbonates away. Another important question is how long the water was actually active on the surface. How long did it have a chance to interact with the rocks?

"What you are faced with is an acidic or oxidizing environment by virtue of the volcanic gases that came out, which then get oxidized by the weird photochemistry of the Martian atmosphere," says David Des Marais, a biochemist at NASA Ames. "This would tend to destroy any carbonates."

Biologists like Des Marais are trying to assess the chemical effect that water has had on the surface. How have the rocks been chemically altered? In this process, not all water behaves uniformly. Its temperature, ice content, sulfate content, and other chemical properties such as acidity are important factors to consider. A deeper understanding of the water chemistry thus commends itself. "After all, life is a chemical phenomenon," Des Marais explains. "A lot of what we talk about in terms of habitability of an environment deals with the chemical aspect."

Carbonates themselves are weakly acidic. That means they, too, would tend to get caught up in the curious chemical cocktail that was the early Martian environment. Perhaps it is no wonder, then, that orbital spectrometers have yet to see them. These instruments don't penetrate very deeply into the surface—a few microns of dust on the surface can block spectral readings completely. The tiniest layers of dust will cover everything. "The acidic fog would have been easily attainable and zap any of the carbonates right at the surface," says Des Marais. "The volcanic environment would have made life miserable for any corresponding carbonates at the surface."

Besides carbonates, the Spirit rover discovered soils and rocks rich in silicon oxides, most likely produced by hydrothermal activity. Such rocks require an interaction between the water and the crust of the nascent planet. Water plus volcanic crustal material produces the kind of rocks observed at various landing sites across Mars. "That makes it more consistent with what a habitable planet would get up to," David Des Marais notes.

* * *

Hydrothermal vents may have been key to the creation of any life-forms on ancient Mars. Work pioneered largely by Bruce Jakosky in Colorado suggests that early in Martian history, there was sufficient volcanic activity to create complex biochemistry out of the heat and water involved. Given that lava flows as recent as thirty million years ago have been detected on the flanks of the Tharsis volcanoes, the hydrothermal activity may not have been restricted to the ancient past.

The valley networks in the ancient terrain could also have been prompted by repeated hydrothermal activity on a local scale, which could have been persistent into the recent past. "There is growing evidence that hydrothermal activity was important in Martian history," Jakosky notes. "If we find the surface expression of hydrothermal vents, we might find evidence for prebiotic chemistry, which is even more compelling."

It was for this exact reason that the Spirit rover was targeted to land in Gusev Crater to the east of Elysium Planitia in January 2004. A well-preserved impact feature one hundred miles (160 kilometers) wide, Gusev was chosen because, seen from orbit, there had obviously been water flowing through it. Over Spirit's five years of operations on the surface, it amassed enough evidence for carbonate and hematite, both associated with aqueous environments.

"We came looking for carbonates," said Dr. Phil Christensen, who has also searched for them from orbit with his THEMIS instrument. A miniature version on the Spirit lander showed that they were present in the distant crater walls. "We're going to chase them," he said a few weeks after the landing. And when they were found, the variety of rocks showed that Spirit's landing site had been "a hot, violent place with volcanic explosions and impacts," in the words of Professor Steven Squyres, the project scientist for both Spirit and Opportunity.

The discovery of hot springs—what another observer calls "explosive hydrothermal vents"—was a major milestone in the mission. "Water was around, perhaps localized hot springs in some cases and trace amounts of water in other cases," Squyres added. Taken together, the findings were consistent with conditions being right for life to have started on the longer road to evolution, thanks to water and energy being available in this site on the Red Planet.

* * *

In August 2012, the Curiosity rover landed in Gale Crater. It was chosen for its sedimentary layers and the presence of minerals that all speak of ancient water. Within weeks, the rover had hit the geological jackpot. First it found an ancient streambed, and then, as it traveled across the plain toward Mount Sharp (Aeolis Mons), it saw patterns in the sand, rocks that were formed in the presence of water in the form of rivers and lakes.

"The lakes weren't just an ephemeral, temporary thing," says Ashwin Vasavada, the Curiosity project scientist. "Our best estimate is that it's at least tens of millions of years of lakes—and potentially much longer, like a hundred million years."

What the Curiosity rover has found while driving up toward Mount Sharp buttresses many of these findings. The rover's instruments have shown that organic molecules, nitrates, and sulfate-rich rocks are painting "a picture of a site with water," as Ashwin Vasavada has written, "the raw materials of biology, and the energy sources required for microbial metabolism."

Its cameras have observed rounded pebbles, which indicate they had come from streambeds. Then, other interesting features popped up: beds of sandstone that, as at *Pathfinder*'s site, were inclined in such a way as to suggest they were lined up thanks to the flow of vast quantities of water over a short period of time.

As ever, the Curiosity scientists have had to follow Sherlock Holmes: eliminate other explanations to find, among what was left, the probable cause. So they went through several hypotheses: perhaps they resulted from dry sand dunes or other stream deposits. But after sampling the sandstone beds, the best explanation is that they were small river deltas.

"Where a river empties out into a lake, you get a near-shore deposit that is very sandy," Vasavada explains, "because the water suddenly slows in speed, and the sand-sized particles drop out, but the silty stuff keeps going into the lake."

What he calls "Geology 101"—"how the pebbles define the river environment"—explains how these sandy deposits create a distinct delineation between the river and the lake. At that point, Vasavada explains, "you'd expect there to be very fine-grained mudstone where the lake actually was, because that fine-grained silt settles out in the lake and then turns into rock there."

What happened next to the water was probably more important. Further evidence has shown that Gale Crater was once a rich cornucopia of rivers, deltas, and lakes. The water flowed through and didn't just well up in isolated patches. As Curiosity has come up toward the central mountain, it captured more imagery of sandstone deltas that were pointing toward the middle of the crater—directly toward the feature NASA informally calls Mount Sharp and lists more formally in maps as Aeolis Mons.

"That was weird, because normally on Earth you'd see this progression when you're staring at a giant basin," Vasavada explains. "If you're entering Death Valley or something, you'd say, 'Okay, well, at the bottom of the basin is where the water went, and that makes sense. That's where the lakes were.' But here, we're staring at a mountain."

That was unheard of and, as a result, ushered in a giant scientific leap of the imagination. Nobody thought such sediments could create a feature that stands more than three miles (five kilometers) high. The layered-deposits explanation, discussed in chapter 3, also fits the progression of sandstone that Curiosity has been observing.

"We got to Mount Sharp finally, two years after we landed," Vasavada says. "And lo and behold, the rocks at the base of the mountain that we encountered were mudstone." That is, the kind of mud that—on Earth at least—accumulates at the bottom of lakes, most often with each rainy season. Layers of mud show what has been happening over geological time.

The very odd thing is that the layers keep going higher up the mountain. "We saw one- to two-millimeter-thick layers of mudstone, layers within mudstone, that formed the base of the mountain," Vasavada explains. "Fast forward now, and we've now climbed over nearly four hundred meters [1,300 feet], and those mudstone layers haven't stopped yet."

In other words, it was wetter for longer.

These observations, like those from earlier rovers, have expanded the time period when Mars may have had conditions conducive to life. Curiosity first examined fine-grained sediments as it made its way across Gale Crater, then came across a succession of fluvial deposits, and then greater amounts of really extensive mudflats, which, at face value, lasted for millennia.

"The lake beds in Gale Crater are exciting for their biological prospects," Vasavada noted in a summary for the journal *Physics Today* published in March 2017. "But their presence and layering at the height of volcanic activity 3.7 billion years ago is distinctly odd. It challenges current understanding. The great outflows of water are difficult to reconcile with models."

* * *

If, in its earliest epochs, Mars was wild, wet, and windy, carbon dioxide could have persisted for a billion years if it was replenished efficiently enough. To maintain such a carbon cycle would require a Martian atmosphere that had five times the atmospheric pressure seen on Earth today. Such an atmosphere would produce greenhouse warming to heat the surface of Mars well above the freezing point of water.

One caveat might have created a curious side effect. Very-high-altitude crystals of carbon dioxide ice would form in such a dense atmosphere on a smaller planet like Mars. These cirrus-like clouds would then reflect sunlight back into space, cooling the surface to below the freezing point of water. Although other greenhouse gases could warm the surface—ammonia and methane have been suggested—they cannot be recycled as efficiently as carbon dioxide. They would act as "one-shot" greenhouse gases.

Sunlight would break down the gases, chemically combining them into other compounds from which they could not be released. Over millennia, their contribution to keeping the atmosphere warm would be negligible. More recently, others have shown that other greenhouse gases, such as ozone and sulfur dioxide, spewed out of volcanic eruptions, could have selectively warmed the upper atmosphere. These gases would have dispelled any icy clouds that might reflect sunlight back to space.

Caution is required. We should not be seduced by the implications of computer models. All theoretical studies are limited not just by the information available but also by the assumptions built into them. "The failure to produce the needed climate forcing (the ratio between sunlight absorbed by a planet and what is radiated back to space) with sophisticated climate models now stands in stark contrast to the strong geological evidence for persistent lakes in Gale Crater," noted Ashwin Vasavada in *Physics Today*, "where there also is no evidence of ice cover and where vigorous surface runoff and a supply of abundant sediment were likely."

The farther Curiosity goes up Mount Sharp, the more puzzling its findings are. The sedimentary layers that it keeps on confirming could result from many of the individual processes discussed in this chapter. All could have kept the young Mars above the freezing point of water. Volcanism would have played a persistent and key role in maintaining moist conditions throughout Martian history. Volcanoes on the Red Planet have given out untold quantities of carbon dioxide. The heat that also poured out may well have recycled carbon dioxide, which would otherwise have been locked up in the form of carbonate rocks, while sustaining temperatures. Water also was a very potent contributor to greenhouse warming, either in the form of vapor, or as lakes or seas on the surface. All helped trap sunlight more efficiently.

Geological evidence suggests that volcanism may have been episodic. If so, carbon dioxide could have been released to periodically "top up" the warming, perhaps for as long as a billion years or so. As Martian volcanism died down, carbon dioxide removal would have been greater than its replacement. The planet would have been warmer only at its equator and in the lowest regions where the atmosphere was densest. Whatever water was around would have effectively dissipated along with the atmosphere. But this loss may also have been gradual and could have taken place in distinct cycles of climate change. That seems to be the implication of the layered terrain at the Martian poles.

All the available evidence suggests that the evolution of the Martian climate has been a curiously protracted affair. Like the bouncing of a rubber ball, these episodes of warming would have been intense at the start, accompanied by dramatic floods of water and extreme changes in temperature. Eventually, they would have died out and become less frequent in time, like the diminishing bounce of a ball dropped from a height.

There is no consensus as to how long these periods would have been, nor how many bounces of the ball resulted. There were periods of a lot of activity on the surface, with a great deal of erosion, then periods when not very much else happened. "Mars is a completely different planet a billion years ago compared to three billion years ago," notes Dr. Zach Dickeson at the National History Museum in London. All that can be said with any degree of certainty is that conditions have deteriorated from earlier states into the freezing tundra-like world we see today. The exact details of how it changed so dramatically remain unknown.

* * *

Over the last decade and a half, the trio of cameras aboard the Mars Reconnaissance Orbiter have revealed something very surprising. Between orbits, fresh water has flowed across the Martian surface. It is one of the most controversial discoveries ever made about the Red Planet.

In 2011, the HiRise camera revealed a number of streaks in gullies and craters a few meters wide in the spring, which seemed to lengthen during the summer. Over the winter season they disappeared as temperatures cooled, then reappeared the following year. The obvious conclusion was that water had flowed across the surface. Such "salty tears," as the influential journal *Science* called them, were intriguing.

These features have been christened "recurring slope lineae," from the Latin word for "thread." "When we first saw them," says Richard Zurek, the Mars Reconnaissance Orbiter project scientist, "they were very enigmatic because you needed high resolution to see them." Gesturing toward the desk in front of him in his office at the Jet Propulsion Laboratory, Zurek says the lineae "are only the width of this desk," on the order of a yard or so across. "If it were happening on Earth, you would say, 'That is water being released and wetting the soil material, and it's flowing down the hill and it's only doing it during the warm season, so that means, hey, that's probably when you would've had water.'"

That certainly seemed to be the case. In September 2015, the HiRise camera team revealed that many hundreds of these streak-like features had been discovered. They seem, at face value, plausibly to be caused by liquid. These dark, narrow streaks were observed on the sides of steep, rocky exposures, such as the walls of craters and furrows that mark the surface.

As the Mars Reconnaissance Orbiter has repeatedly passed over these sites, it has monitored their changes over the seasons. The CRISM instrument—the "chemical mapper"—uncovered hydrated minerals within these curious features. At first, project scientists interpreted these findings as the result of briny water flowing on the surface in recent times—"days, something of that order," said one. Their spectral signature appeared stronger during the Martian summer and weaker during the Martian winter. The seasonal difference in temperature would be enough, it was claimed, to let any liquid flow. So where did the water come from? That was, and remains, a considerable mystery.

Two years after announcing these findings with a news release that "Confirms Evidence That Liquid Water Flows on Today's Mars," NASA quietly inserted a correction. More recent observations suggest bone-dry dust slides skidded down the slopes to form such features observed by the Mars Reconnaissance Orbiter.

Most of the slopes where the lineae had been observed were steeper than 27°. The angle is important because water easily flows on much smaller inclines. So why were these features not ubiquitous on all levels of slope? Sand, for example, tends to be more sticky and cohesive, especially in the frigid conditions on Mars. Sand flows have trouble making it down a slope unless it is quite steep. And while, as CRISM suggested, they showed some signs of hydration, further observations essentially rule out the hopes for running water.

"They turned out, at least on the lower slopes, to be very close to what's called the angle of repose," says Rich Zurek, "resulting in dry avalanching, just granular particles moving down a slope, falling down a bit, basically." If there were some that had been seen on shallower inclinations, the case for water would be stronger.

The lineae remain, in a word, weird. The spectral mapping by CRISM suggests the presence of water was likely involved in the formation of the lineae. Perhaps they are caused by water pulled from the atmosphere (because there are no supporting measurements of humidity, it is difficult to know). Or perhaps it is some unknown process. The water in the streaks might be different in different regions. "If it is too salty, they would be flowing all year round," said one researcher in 2015. "We might be in that Goldilocks zone."

For the features to be liquid water, they would need to contain a great deal of salt. On Mars, greater salt content would keep the water from evaporating: the saltier the water, the greater chance the water would have of remaining on the surface.

So far as biology and the prospect for life are concerned, caution has prevailed. "The short answer for habitability is it means nothing," says NASA's Christopher McKay.

8

CLAIMS

Percival Lowell was the sort of polymath who populated the nineteenth century with languid nonchalance. His was an illustrious Boston dynasty: his brother became president of Harvard University, and his sister was the imagist poet Amy. The Lowells were so blue-blooded that, according to one Massachusetts wit, they spoke only to their fellow Brahmins, who spoke directly to God. In his youth, Percival was given a small telescope with which, as his brother and biographer later noted, he saw for himself "the white snow caps on the pole of Mars crowning a globe spread with blue-green patches on an orange ground." The Red Planet made a lasting impression.

Percival Lowell studied at Harvard, where one of his math teachers "spoke of him as one of the most brilliant scholars ever under his observation." After graduating in 1876, he made the grand tour of Europe with a classmate. Subsequently, he spent his twenties attending to the family business, managing trust funds, running a cotton mill, and acquiring the wealth that allowed him independence.

Travel became a greater preoccupation during his early thirties. In 1883, he made the first of many trips to a country that, as his biographer notes, enthralled him. Percival Lowell "was fascinated by Japan, its people, their customs, their tea houses, gardens and their art." Contemporary reports say he learned to speak basic Japanese in three weeks—"faster than I ever saw any man learn a language," his traveling companion marveled.

After traveling around the country, Lowell was asked to accompany a Special Mission—the first-ever American diplomatic one—to Korea. While shuttling between Seoul and Tokyo, he later helped write the Japanese constitution. For centuries, Japan had been closed to the outside world by imperial decree, its feudal society insulated from foreign influence, with Nagasaki as the sole port designated for international trade. Two years before Percival Lowell was born, Commodore Matthew Perry's ships came up through Tokyo Bay on July 8, 1853. The nation that had for so long been insulated from the rest of the world would be forced to embrace it.

Japan's assimilation with external societies would soon come to preoccupy Percival Lowell. It may explain why he became so utterly convinced about life on Mars. He was one of the few people of his age who had ever experienced such a profound culture shock. Its effects, it may be argued, inspired him to consider life on other alien worlds. Ironically, he provided words that could apply equally to Japan or the Canals of Mars. "We know now what was long unknown," Lowell wrote to a friend from Japan, "that true seeing is done with the mind from the comparatively meagre material suggested by the eyes."

* * *

Always fascinated by astronomy, Percival Lowell had taken a six-inch telescope with him around the Far East, where he continued to make observations, notably of Saturn. Even before he had left for Japan, something happened that would have far-reaching consequences. As the annals of his own observatory later noted: "In the summer of 1877 occurred an event which was to mark a new departure in astronomy—the detection by Giovanni Schiaparelli of the so-called canals of the planet Mars."

That autumn, the Red Planet was at its brightest for a century and, at its nearest, was just some thirty-five million miles (fifty-six million kilometers) distant. Two curious discoveries soon became headline news around the world. The first, and less sensational, occurred at the United States Naval Observatory, when its director, Asaph Hall, discovered two tiny moons around Mars. Hall named them after the acolytes of the god of war, Phobos and Deimos.

The observations by Schiaparelli were more controversial. From the Brera Observatory in Milan, he saw something unusual on Mars itself. He observed dark lines bisecting the Martian surface with such uniformity that, in his own words, he thought they could almost have been "laid down by rule and compass." These fine delineations seemed to separate what he believed, as was common at the time, were darker continents. He suggested the tracts were channels of water.

Schiaparelli was careful not to ascribe their creation to the handiwork of sentient beings. He simply reported what he saw. Sometimes these lines "were vague and shadowy," at other times "clear and concise." Such regular features were controversial because, until 1886, nobody else had reported seeing them. Thereafter, only a handful of astronomers ever did.

In his native Italian, Schiaparelli referred to the lines as *canali*, meaning "channels," which almost inevitably became anglicized as "canals." In an era when canals were vitally important commercial arteries in the industrialized world, it is hardly surprising that the notion of similar artifacts appearing across the astronomical gulf became popular. "This was the beginning of a controversy that has continued since," an obituary of Schiaparelli later remarked, "and in the capable hands of Professor Lowell has been made to assume an importance it really does not possess."

Lowell devoted the rest of his life to Mars. By the time of the next close opposition of the Red Planet in 1894, he was ready. He enlisted the help of a prominent astronomer, fellow Bostonian William H. Pickering, to supervise the design and construction of an eighteen-inch refractor in the Southwest over the next year. Lowell realized he needed to decamp to a less cloudy location than Massachusetts, which was also hampered by the mists rolling in off the Atlantic. Flagstaff, Arizona, at an elevation of 1.25 miles (nearly two kilometers) above sea level, with endless skies and fortifying air, was ideal.

The town was best known as a railhead that, in the words of a popular song, opened up the golden gate to California. Flagstaff's inhabitants warmed to Percival Lowell. They deeded him a hill, on which he established the observatory that today still bears his name, and helped lay a road from the railhead. Lowell built a mansion in the grounds of the observatory, where his garden became famous for its produce (his squashes and pumpkins were prized). Visitors got used to the observatory's cow, called, oddly, Venus.

From May 24, 1894, until August 3, 1895—"almost every night," as his brother recalled—Percival Lowell made observations of Mars, which ultimately totaled more than nine hundred. What he saw changed everything, elevating Mars into the public consciousness in a way that has never been equaled. To Lowell's eyes, the Red Planet was a world covered by a vast network of canals, the construct of a remarkable and ancient civilization, obviously more advanced than ours, which used them to effectively irrigate the planet by the seasonal melting of ice from the polar caps. He believed that meltwater was carried from the ice caps in the spring season to produce a dark blue "collar" that moved toward the equator. Lowell wrote: "Meteorological conditions carry [water] to deposit at one pole, then liberate it and convey it to imprisonment at the other, and this pendulum-like swing of water is all in the way of moisture that the planet knows."

It seemed to make sense, because otherwise Mars appeared completely dry. There was no evidence from spectroscopy for water anywhere on the Red Planet. The regions that earlier nineteenth-century astronomers had characterized as seas could not be wet: they didn't reflect sunlight. Lowell believed that the darker regions were vegetation because of their blue-green appearance, which seemed to change with the seasons.

Each spring, what Lowell termed a "verdure wave" swept along the canals from the poles to the equator, in the opposite direction to that which occurs on Earth. This so-called wave, which moved as a perceptible darkening across the surface, gave the impression there was something very definitely alive on Mars. There seemed no reason it would happen naturally, unless it had been engineered by an advanced civilization.

In 1896, Lowell published his first book on the subject, called *Mars*. It was eloquent in both its simplicity and the compelling portrait of a venerable civilization. To coincide with publication, he gave a newspaper interview in which he said, "I have no doubt that there is life and intelligence on Mars. No creatures resembling us are there. Local conditions, such as the thinness of the atmosphere, forbid it, but there are creatures of intelligence."

Within two months of beginning his observations, Lowell had

become convinced of what he was seeing. Now it became his greater task to persuade the rest of the world about the canals. "Gradually more and more observers began to see the finer markings and the canals on Mars," his biographer wrote, "while finally the question of their existence was set at rest, when it became possible to photograph them."

* * *

In reality, the scientific community was aghast at Lowell's claims. Reactions ran the gamut from polite professional interest to incredulous indignation. Many scientists considered Percival Lowell a fantasist. Other observers, such as E. E. Barnard, who peered at Mars from the Lick Observatory in 1894, saw only a ruddy surface "broken by canyon and slope and ridge." Later on, the prominent Franco-Greek astronomer E. M. Antoniadi suggested that Lowell had exaggerated the lines, which he believed were "a maze of knotted, irregular, chequered streaks and spots" at the edge of perception.

Yet observing Mars was something for which Percival Lowell was particularly well-qualified: he possessed the keenest pair of eyes ever examined by his optician in Boston. Lowell could see with his naked eye newly erected telegraph wires in the Arizona desert several miles distant. His visual acuity was so great that he could see lines where nobody else could.

At the time, however, the controversy became too much for the man himself. By 1897, Lowell, now in his early forties, suffered what he termed "a complete breaking down of the machine." Those endless hours at the telescope eyepiece had taken their toll. Although, as his brother notes, Percival could subsist on little sleep "while stimulated by the quest, the long strain proved too much."

During the four years from 1897 to 1901, he traveled all over the world. In London, he met Sir William Huggins, a pioneer of spectroscopy, who had already noted that "there is no conclusive proof of the presence of aqueous vapour in the atmosphere of Mars."

The problem was that the explanation proffered by "respectable" scientists to counter Lowell's arguments were even more preposterous. Neither chains of craters from volcanic eruptions nor rows of impact craters peppering the crust could explain the regularity of the perceived canals.

And so, in two subsequent books, *Mars and Its Canals* (1906) and *Mars as the Abode of Life* (1910), Lowell warmed to his theme. In the latter, he explained: "In our exposition of what we have gleaned about Mars, we have been careful to indulge in no speculation. The laws of physics and the present knowledge of geology and biology, affected by what astronomy has to say of the former subject, have conducted us, starting from the observations, to the recognition of other intelligent life."

Nobody else saw anything like the regular, linear features that Lowell was convinced he had been seeing, but the canal controversy still raged. In May 1903, Lowell observed a "projection" for seventy minutes, which he thought was an enormous cloud on the Martian surface (he believed he had seen its reflection). The press viewed it as a signal from the Martians. In particular, the *Daily Mail* noted that when something similar had been described in *The War of the Worlds*, the H. G. Wells novel published in 1898, "just such a projection indicated the commencement of that terrifying invasion."

This sort of speculation seemed eminently believable. In 1902, for example, a French widow had offered 100,000 francs to the first person to make contact with alien life. Mars was excluded on the grounds it would be too easy. Others later suggested the construction of vast mirrors to signal messages to the Martians. In the same way aliens always talk English in the movies, the plan was to use Morse code, with which Martians were supposed to be conversant. More esoteric still was the idea of building the proof of a Pythagorean theorem in the Siberian steppes, a triangulation with which Martian mathematicians would be entirely familiar.

* * *

Lowell's canals spread into popular culture. In 1899, one of Carl Jung's patients went into a trance, traveled to Mars, and beheld a landscape of canals. That same year, Nikola Tesla, a pioneer of electricity, detected radio signals he thought were from Mars: today, it is widely believed he had accidentally come across Guglielmo Marconi's first radio transmissions across the English Channel. In 1921, Marconi himself picked up signals on his yacht that he, too, felt had originated on the Red Planet.

The more Lowell saw canals, the louder the naysayers became.

Some objected that the canals were either too thin or too thick. Lowell explained their slimness by the way that the human eye can resolve linear features far below the theoretical limitations of visual acuity. That they seemed too thick was because, like estuaries on Earth, vegetation would grow either side of them and be more visible.

With hindsight, it is pretty clear what was happening. The simplest explanation was by far the most convincing. It came in 1903, when the British astronomer Walter Maunder asked a group of schoolboys to copy a drawing of Mars from which the canals had been removed. Many of the boys produced "canals" themselves. Maunder concluded that "the integration of the eye of minute details too small to be separately and distinctly defined" led to the scrawling of lines where there weren't any.

Yet the lure of Lowell's vision was much too powerful. In May 1905, Lowell announced, in the twenty-first bulletin of his observatory, that he had been able to photograph the canals. Many newspapers around the world reported the finding and carried the photograph. Two years later, Lowell's assistant, Earl Slipher, started to photograph Mars in earnest. He devoted the rest of his life to the enterprise. By the early 1960s, Slipher had amassed 126,000 photographs of Mars and published them in 1962. Only a handful ever showed what appeared to be lines on the surface. Even then, they could be interpreted in other ways. Nevertheless, in 1964, the year of Slipher's death, his canal-filled map of Mars was officially adopted by US Air Force cartographers.

In 1907, as the historian William Graves Hoyt notes, "the long-simmering canal controversy turned to a full boil" when others weighed in. Most important among them was Alfred Russel Wallace (who had independently confirmed Charles Darwin's findings on evolution). Now in his eighties, Wallace penned a whole book to refute Lowell's findings.

Wallace pointed out that the canals were only possible if the Red Planet was completely flat. If that were the case, however, the water would flow and irrigate the land of its own accord. The canals would not have been necessary. Others pointed out that the Martian ice caps were flimsy—"snow soufflé," Wallace called them—some of which would evaporate "the excessively scanty amount of water thus obtained would render any schemes of worldwide distribution of it hopelessly unworkable."

That same year of 1907, when the *Wall Street Journal* did its roundup of the year's news, it concluded that Slipher's first photographs of Mars were of greater impact than any of the fiscal losses from the "Bankers' Panic" that had halved the value of the stock exchange. As Hoyt remarks, the photos did not reveal canals, and "the pot boiled over, so to speak, and all but put out the fire."

In time, things got even more bizarre. Lowell's colleague, William H. Pickering, suggested that giant mirrors be set up on the plains of Texas to reflect sunlight and flash messages to the Martians. That they would understand Morse code was also assumed by a writer who came up with a variation on Pickering's theme—that giant black cloths should be placed across the whiter plains so that signals could be sent to our Martian cousins. By 1912, Edgar Rice Burroughs adopted Lowell's ideas in his first novel, *The Princess of Mars*. There, he records, the natives "had found it necessary to follow the receding waters until necessity had forced upon them their ultimate salvation, the so-called Martian canals."

By the end of the first decade of the new century, Percival Lowell's health had deteriorated. He retired to spend more time in the balmier climes of the Mediterranean. He died in November 1916, thanks to "an attack of apoplexy," in his biographer's phrase; it was likely a stroke. "He lies buried in a mausoleum built by his widow close to the dome where his work was done."

Though the Lowell Observatory begat useful studies, notably the discovery of Pluto in 1930 and details of the expansion of the universe, its founder's work completely polarized the scientific community. Planetary astronomy effectively died in the United States in subsequent years. In the words of one later researcher, "after Percival Lowell wrote all these crackpot books about Mars, planetary science had no reputation and nobody wanted to touch the field with a ten-foot pole."

As late as 1964, some of Lowell's surviving staff at the Lowell Observatory were still claiming that there were canals on Mars. In his book from 1962, for example, Earl Slipher still beat the drum for his old mentor: "Since the theory of life on the planet was first enunciated some fifty years ago, every new fact discovered has been found to be accordant with it. Not a single thing has been detected which it does not explain."

Alas, no. In 1965, eleven photographs from *Mariner 4* showed what had been suspected all along: the canals of Mars were a figment of Lowell's, and others', imaginations.

If only Percival Lowell could have known it, he had produced his own fitting epitaph for the canals. In his first book on the Red Planet, published in 1896, he had written: "If astronomy teaches anything, it teaches that man is but a detail in the evolution of the universe and that resemblant though diverse details are inevitably to be expected in the host of orbs around him. He learns that, though he will probably never find his double anywhere, he is destined to discover any number of cousins scattered through space."

A century later, that statement still rings true, but it has been shown beyond doubt that sentient, canal-building beings never existed on Mars. Lowell's canals were a product of his peculiar visual acuity and wishful thinking and were fostered by the compelling eloquence of his writing.

Though the canals may be gone, the notion of a civilization capable of erecting vast structures remains. A hundred years after Percival Lowell started to build his observatory in Arizona, something very odd happened in the vicinity of the Red Planet—so much so, it was claimed in some quarters that an alien civilization was behind it.

* * *

In the early evening of Saturday, August 21, 1993, an air of quiet expectancy hung over the Jet Propulsion Laboratory. Two hundred million miles (three hundred and twenty million kilometers) away, the Mars Observer spacecraft was about to enter orbit to begin the most intense scrutiny of the Martian surface ever attempted.

Despite their upbeat mood, there was a palpable sense of disquiet. The JPL engineers knew that they had paid a terrible price in getting their spacecraft to the planet. Mars Observer could fail at the last hurdle, because it had flown with a problem in its fuel valves that couldn't be fixed. This had left officials stuck "with a bit of a nightmare," in the phrase of one participant. It wasn't a secret but was hardly a matter to which they had drawn attention in press briefings.

The spacecraft had been cobbled together from parts of an Earth-orbiting weather satellite, along with hardware developed for a military space communications project. It had been sold to Congress as the first in a production line of spacecraft that would be customized for wherever they were to be sent in the solar system. That never happened: it turned out that actually adapting each spacecraft for a new destination would cost more than if it had been built from scratch.

A terrestrial weather satellite is protected by Earth's magnetic field and doesn't need to hibernate for months until it reaches its destination. The Mars Observer had been waiting for years to launch, particularly after the delays following the *Challenger* accident in 1986. Because of fiscal belt-tightening by NASA, its original 1990 launch opportunity was lost and it had to stay put on Earth for another twenty-three months, until the next launch window opened in 1992. For that, it was switched to a "traditional" launch vehicle, a Titan IV booster.

Ironically, the Titan IV malfunctioned on launch in September 1992. Minutes after liftoff, the radio transmitter onboard the Titan's third stage did not turn on. So JPL controllers had no idea what was happening out beyond the skies of Florida. Thankfully, exactly on cue, the Mars Observer spacecraft itself made radio contact, assuaging fears that all was lost.

Eleven months later, everything was going to plan as Mars Observer prepared to enter orbit around Mars. The previous week, Mike Malin's camera had taken a trial picture, which showed the Red Planet in much greater detail compared to *Viking* pictures taken from a similar distance away. Officials wanted as much data as soon as possible. A month after it arrived, the Red Planet would be in conjunction with the Sun. From Earth's perspective, Mars would pass behind our daytime star. So the first continuous data from Mars Observer, and its target itself, would not emerge until the start of 1994.

That Saturday night in August 1993, Mars Observer was three days away from the planned firing of its engines. JPL engineers had started to prepare the spacecraft's propulsion system for the all-important burn to slow Mars Observer's speed and enter orbit. In the weightlessness of space, however, propellant will not just flow around, as in your automobile fuel tank. The propellants have to be kept pressurized so that they will be forced through fuel lines by a small tank of helium.

Helium was chosen because it will not react with the propellants or corrode the fuel lines, nor is it flammable. To ensure sufficient pressure can be attained, small valves are needed to control the flow of the fuel by controlling the precise amount of helium pressurization. That was where the valve problem comes in.

During development, the valves had never been tested for the duration of the flight to Mars. In Earth orbit, the helium tanks would be emptied within seventy-two hours of launch, but a journey to Mars was a different proposition altogether. It later came to light that the longer the valves were left closed, the greater the chance they might not work, and a journey of eleven months was, to quote one engineer, "pushing the envelope."

If the valves were opened early en route to Mars to see how well they worked, the cumulative loss of helium could result in insufficient pressurization at the journey's end. With nothing to force the fuel into the engine, there would be little chance for a successful insertion into Mars orbit. However, if the valves were opened later, as had been originally planned, they might have passed their shelf life and be more susceptible to mechanical failure.

The decision had been made to open the valves as planned. As a matter of course, when this was done in the early evening of August 21, the spacecraft's radio transmitter had to be switched off. There was a danger the transmitter's tubes would shatter when JPL commanded the firing of pyrotechnic squibs—small, controlled explosions—to force open the fuel lines. So the radio system was switched off. It would switch back on again minutes later, when the fuel tanks were fully pressurized. At least, that was the plan.

All seemed to be proceeding well. Glenn Cunningham, the deputy project manager and a veteran of many JPL flight projects, was in his office eating Chinese takeout, anxiously awaiting confirmation that the signal from Mars Observer had been reacquired after pressurization. In other offices around the laboratory, engineers listened in earnest to the usual comforting tones of mission controllers. As minutes passed, Cunningham started to wonder whether another tracking station would be asked to search for the signal expected from the craft. As Earth turns, one station of the Deep Space Network would hand over to the next to ensure continuous monitoring of the signals expected from space.

Anxious minutes ticked by, which drew out into hours. Nothing was heard from the spacecraft that Saturday, or ever again. "Mars Observer just disappeared," Cunningham explained, his voice still emotional and disbelieving a few months later. "We never had any indication that there was a problem. It wouldn't have been so bad had it not disappeared. We were devastated."

In 1989, the Russian *Phobos 2* spacecraft had disappeared—due to a computer failure, it later transpired. JPL had been called in to help in the fruitless task of searching for it. Unlike Russian spacecraft of the era, all American probes had the capacity to return signals back to Earth after a few days passed. When the Mars Observer disappeared, Cunningham, like many others at JPL, felt that, with Mars orbit insertion coming up, the craft would automatically regain contact.

"It took me three months to accept that we had lost the mission," he says. "Until November, I was hoping that it was an electronic glitch that we'd work around."

What went wrong?

The exact cause of the accident can never be fully known, because the craft was out of radio contact. But the results of an official investigation identified one likely cause. The Mars Observer motor was fueled by nitrogen tetroxide and a form of hydrazine known as UDMH. When they came into contact, there would have been a controlled explosion, and the thrust needed to fire the spacecraft into orbit would have been achieved. Instead, it is believed that vapors of the tetroxide had got into the pressure system and, when the helium tank was pressurizing up, a fuel line simply burst. Mars Observer would have started to spin ever more uncontrollably. There was no chance it could have ever entered Mars orbit.

The loss was awful enough without a series of bizarre accusations that NASA had somehow scuppered the mission on purpose. For some, Mars Observer's disappearance was proof that there was a conspiracy to stop "the truth" from getting out—the "truth" being that the space agency had discovered evidence for an alien civilization on Mars and was keeping it hidden.

Perhaps we shouldn't be too surprised. So far as the Red Planet is concerned, many people on the next planet in toward the Sun are emotionally involved with it in unexpected ways. By Monday evening, two days after the failure to reacquire signals from Mars Observer, a small group of protestors had already gathered outside JPL's main gate. They carried placards claiming that NASA was lying about the loss. The notion persisted for years afterward that Mars Observer was an example of a sinister government cover-up.

* * *

The whole business had its origins on the evening of Sunday July 25, 1976, when the *Viking 1* orbiter swept over Cydonia in the northern hemisphere of Mars. Even by the standards of the Red Planet, the geological forces that sculpt the surface in this region are highly unusual. In the thin, freezing atmosphere, wind-blown dust carves out strange features that sometimes appear even stranger when they catch the light of the early morning or at sunset.

Viking 1 trained its cameras on an area that had been thought of as a possible landing site. It was about six p.m. local time, so the Sun was low on the horizon, and the shadows were quite extended, particularly around the strange features that stood out from the surrounding terrain. In one *Viking* image, slightly off-center, appeared something odd that drew attention to itself because it vaguely looked like a human face.

When the frame was analyzed by scientists at JPL, it caused some wry amusement. The next day, Gerry Soffen, the *Viking* project scientist, elaborated for the benefit of reporters gathered at the Jet Propulsion Laboratory: "Isn't it peculiar what tricks of lighting and shadow can do? When we took a picture a few hours later it all went away; it was just a trick, just the way the light fell on it."

Summer is traditionally the "silly season" for newspapers, so it was no wonder that the next day a few dutifully reproduced the photograph. The Face on Mars, as it became known, took its place alongside some of the other strange stories emanating from around the Viking project.

An Italian writer named Renuccio Boscolo had already claimed that a vast underground city had been discovered by NASA, but the evidence had been suppressed. How he came by this information was not explained at a press conference he held in Milan. Rumors had also surfaced of the supposed discovery of methane, which can be a biologically produced gas. It was an even neater touch, in the immediate aftermath of Watergate, to suggest that the Central Intelligence Agency had somehow prevented the finding from being disclosed.

The notion of a Face on Mars persists today, even though geologists always knew that the feature was nothing more than a naturally occurring, windswept mesa crumbling on a freezing northerly plain. The Face looked vaguely facial because, in our everyday experience, we are used to faces. Five years after the demonstrators congregated at JPL, the Mars Global Surveyor—built from flight spares of the Mars Observer—got an even better look. There was hardly a conspiracy to stop the pictures from getting out.

The Face on Mars is hardly a new phenomenon. It is yet another facet of the irresistible lure of the Red Planet manifesting itself in popular culture, exactly like Percival Lowell's "discoveries" a century before. Alien artifacts are much more exciting and easier to appreciate than discussions of surface chemistry. Conspiracies sell. And for a conspiracy theory to work, it must have any number of elements: the evidence must be ambiguous, perhaps suffused with a sense of cover-up; there must be a villain; and above all else, it has to be extraordinary. With NASA and the Red Planet, all these criteria have matched up only too well.

"These claims are absolutely absurd," counters Glenn Cunningham. "I'd like to find out as much as the next person about the Face on Mars. But we're not going to throw away a decade's work and $600 million as part of a cover-up."

The supposed scuppering of Mars Observer hardly makes sense when you consider such a find's most obvious beneficiary. "That is the really odd part of these claims," says Michael Malin, who has seen many cameras fly to Mars in the past three decades. "When people look for conspiracies, somebody has to suffer. If NASA found evidence for life on other planets, it's going to benefit."

Malin should know. Shortly after midnight on Sunday, April 5, 1998, his camera aboard the Mars Global Surveyor finally got a good look at the Cydonia feature and its surrounding landscape. "We did what we normally would do," Malin says. "Video resolution images were made available on the web once we had the full data and we had been able to analyze it. We knew we'd be damned if we did and damned if we didn't. We wanted to get the most interesting part out to the taxpayers as soon as we could."

The image was then passed electronically to JPL and simultaneously released on Malin's and the NASA Mars websites. Within minutes, the image had been downloaded on computers all around the world, to be analyzed and pored over. As expected, the image showed a pile of rocks, a badly weathered mesa that had been buffeted and cracked by the extremes of Martian climate. There was no Face. Subsequent observations by other cameras sent to Mars show exactly the same rocks.

If "the Face" was gone, the controversy remained, just as Malin and others always knew it would. "Most of the proponents of conspiracy theories take issue with just about everything I have ever raised on this issue," Malin says. "I have tried to get the Face people to propose something I could test. If the pictures could answer a question once and for all, that would be great. The problem is that I can't get them to define what they need for proof. They are never going to be convinced."

* * *

Gilbert Levin claims he discovered life on Mars in the 1970s and cannot be dismissed out of hand. Levin was principal investigator on the Labeled Release experiment on the *Viking* landers that caused so much excitement on its first run at Chryse Planitia. Levin believes life was found in the summer of 1976, but his conclusions have fallen on deaf ears because he dared question the scientific consensus.

His story begins on the very same beaches where the television series *Baywatch* was filmed. In the 1950s, Gilbert Levin was a public health engineer for the State of California charged with investigating pollution limits for Santa Monica Bay. Each morning, Levin would drive a jeep and judiciously pick up water samples. It would take three days before a series of classical microbiological analyses would reveal exactly how badly polluted the water was. It would take a further couple of days before limits could be calculated.

In other words, a week had passed before the quarantine limits were ready, by which time the ebb and flow of pollutants would have changed. The same problem occurred when Levin moved to Washington, DC, as a public health engineer, where he was charged with assuring drinking water quality. It would take a week of tests before the results were known. Levin then had the idea of using radioisotopes. These would be sensitive enough—and much quicker—to detect the metabolism of any bacteria in the water sample.

This was the unlikely birth of the experiment known as Labeled Release or "LR" on the *Viking* mission. In the 1950s, Levin would simply take a sample of water and feed it to a "broth" of radioactive lactose and measure the amount of carbon dioxide produced by the bacteria. It took only thirty minutes and could be performed in a test tube by use of a Geiger counter. The greater the number of bacteria, the bigger the signal. How large the signal was revealed how much contamination was present.

In 1952, he published a paper to this effect, expecting the idea to catch on. In the end, only the state of Illinois and the City of London adopted it, and even then, only for emergency use. "The reason was very simple," Levin says today. "Radioisotopes were anathema following the first atomic bomb explosions." Though the radioactivity contained within each sample was minute—ten times that amount could be drunk safely by a human being each day—many authorities shuddered at the thought of coming into contact with even low-level radioactivity.

Nevertheless, Levin continued his work at Georgetown Medical School, refining the technique at a new company he set up called Biospherics Inc. Had he not attended a cocktail party organized by the bureau chief of *Newsweek*, he might never have met T. Keith Glennan, the recently appointed head of America's newly minted National Aeronautics and Space Administration, in 1958.

Over martinis, Levin discussed NASA's future with Glennan. Somewhat ingenuously, the engineer asked, "Are you thinking of looking for life on other planets?" Glennan replied that he had just recently appointed a physician to look into this and other issues concerning life sciences. Foremost among his assignments was the more pressing, politically expedient goal of how astronauts might survive in space. But, yes, life on Mars was an obvious possibility, Glennan added somewhat casually. In due course, the fledgling space agency considered Levin's proposal for a life detection instrument. Unlike most biologists who restricted themselves to the remarkable sights found via a microscope, Levin had also peered at

Mars through a telescope. "My cousin had taken a course in astronomy at college," he says. "I got to take a look at Mars, and was fascinated."

Gil Levin's proposal blossomed into what was initially christened Gulliver. He originally envisaged a piece of sticky string that would shoot out from the spacecraft and, when retrieved, would be analyzed using the Labeled Release method (described in chapter 4). "I took a string and dragged it across the storefront and another down the street," he recalls. "When we tested them, sure enough, we found evidence for bacteria."

Levin was given a grant of $125,000 to build a working prototype but was called into NASA Headquarters for a quiet word. "I was an engineer with no PhD," Levin says. "I was told that I would have to get a doctorate if they were going to select me as a co-experimenter."

This was a foretaste of the bickering, informed by residual academic snobbery, that would accompany the *Viking* biology investigations. So Levin went back to school. He earned a PhD in environmental engineering while steering Biospherics Inc. onto new paths. "We developed new technologies in space biology, water and air pollution as well as oceanography," he says. This gave him a certain financial independence, which ensured that he was well placed to be chosen for the *Viking* mission to Mars.

Behind the scenes, though, the *Viking* biology team was not happy. Levin says he found working with his cohorts difficult. "It was an uphill battle for me as an engineer from a small company working with Nobel laureates," he says without obvious resentment. "I always thought that we would work on each other's experiments and help each other. Unfortunately, we never did." In fact, rancorous personal antipathies spilled over into the working relationships of the *Viking* biologists. "Basically, they all hated each other," says one participant who knew all the key players. "There were two Nobel laureates involved, and I got the impression some of the others wanted to join that illustrious band."

The mutual antipathies reached a crescendo even before the spacecraft left Earth. Test samples taken in Antarctica that had been declared sterile by some were subsequently found to contain microorganisms. Part of the reason Levin believes that life was detected on Mars comes from those very same Antarctic samples, which had been taken to test the efficacy of the *Viking* biology experiments.

Sometime after the *Viking* landings, Levin remained frustrated by the confusing, often contradictory data returned from Mars. Worse, he didn't agree with the consensus that had evolved. The party line was that the soils of Mars were so highly oxidizing that they had removed all signs of organic materials that would otherwise have been present, including those deposited from meteorites that had landed on the planet through-out the millennia. Rather than accounting for some of the unusual soil properties, Levin felt this explanation was seriously deficient.

"I had tried to duplicate the chemistry and physical properties of the Martian surface," Levin says. "I could get the acidity right, but not the thermal profile, in duplicating my Mars results."

It was only in looking through the thesis of a PhD student who had done most of the development work on the *Viking* gas chromatograph mass spectrometer (GCMS) that he realized something significant. As they were being prepared for the flight to Mars, both the chromatograph and his Labeled Release experiment had analyzed the same Antarctic samples in 1974. Roy Cameron of JPL and two assistants had gone down to the Dry Valleys, where they had sealed some soil samples for use by the *Viking* instruments as part of their test procedures.

After *Viking*, the standard line was that the mass spectrometer had found no organic matter on Mars, without which no life could exist. It was very hard to explain even the lowliest forms of life if there were no equivalent to their biochemical bodies. So Levin's Labeled Release results were taken as ambiguous, even though they had essentially tested positive on Mars. That is, the uptake of the nutrients in his sample technique had been assimilated by possible microbes. Levin was astonished to find that the reverse had been observed on Earth.

"The GCMS had tested the Antarctic samples and found no organics in the classic method," he says. "I went through our files and found we had detected living organisms in the same sample that the GCMS found no evidence for organic matter."

Checking on the sensitivity of each instrument, Levin was further amazed. Whereas the Labeled Release was capable of finding as few as fifty cells, the GCMS would require 10 million to 100 million cells to work. "Both instruments were right," Levin says. "The GCMS detected nothing, and our instrument did detect something on Mars. There's no conflict if you understand the sensitivities."

In other words, Levin believes he found life on Mars. His instrument was more than sensitive enough to have detected the presence of microbial life from the uptake of nutrients provided to the samples taken by *Viking*. The leader of the GCMS team, Klaus Biemann at the Massachusetts Institute of Technology, was unwilling to recheck his instrument against Levin's in the presence of known amounts of bacteria. "He said the working prototype was in pieces and could not be assembled," Levin says.

Nevertheless, Levin has persevered. Today, he points out that he has run his experiment in terrestrial soils and with microbial cultures, both in the laboratory and in extreme natural environments. "No false positive or false negative result was ever obtained," he wrote in October 2019. "This strongly supports the reliability of the [Labeled Release] data [on Mars], even though their interpretation is debated."

Still sprightly and indefatigable in his nineties, Gil Levin positively vaporizes at the merest mention of the usual explanation of why *Viking* found no Martian organics—that the surface of Mars is so violently superoxidizing it destroys any complex molecules on the surface. The standard theory is that hydrogen peroxide trickles down to the surface, where it reacts with the same bleaching propensity employed by teenagers in a thousand bathrooms.

Levin is having none of it. He was a coinvestigator on an infrared spectrometer flown aboard *Mariner 9*, the first probe to orbit Mars in the early 1970s. "I rechecked those data, as we had an excellent opportunity to detect hydrogen peroxide," he says. "We didn't find any evidence."

Today, Gil Levin would like a panel of experts to look afresh at all the data—his own results and all the more recent findings that have buttressed the notion that Mars could have supported life. "Such an objective jury might conclude, as I did, that the Viking Labeled Release did find life."

* * *

Ten years after the *Viking* landings, at a commemorative conference held in Washington, DC, Gil Levin and his colleague Patricia Ann Straat announced that there was another distinct possibility: that time itself was an unknown factor in any observations of possible biological activity. Martian metabolism rates might be ticking over so slowly there would have been no signs of organic detritus, they claimed. That was why none were ever found by the mass spectrometer.

They also beat the drum that any scientific models that tried to account for the chemistry of the surface could not be explained properly by the consensus view. In 1986, when Levin publicly concluded that it was more likely than not that his Labeled Release experiment had indeed discovered life on Mars, he had begun the exhausting task of examining the ten thousand *Viking* lander pictures to look for any kind of evidence for life. With the correct color calibration, he did find some rocks that showed changes over the years, similar to the way in which lichens behave on rocks on Earth. At that same conference, he and Straat showed pictures of one particular rock that seemed to have changed color. The greenish patches, they said, were similar to lichens that have a symbiotic relationship with algae and fungi (and have been found to thrive in the Dry Valleys of Antarctica).

Originally, Straat didn't completely agree with her older colleague. As more information about Mars has come in from newer missions over the years, she is now inclined to agree. In July 1976, she had exclaimed, "Oh my God, it's positive" when the initial run of the Labeled Release had been carried out. Fifteen sols (Martian days) later, that same instrument examined the first heat-sterilized sample. The results were negative. And that, as Straat has remarked, was when the controversy really started.

When they later heated another Labeled Release sample to 50°C, the earlier positive response was reduced. "That was fairly strong evidence that the active response was biological," she has said. In other words, whatever was present in the samples had been "killed off" at these higher temperatures. While this finding could have been due to some unknown chemistry, nothing has ever been proposed to account for it completely.

"The caveat is the lack of organic molecules," Straat says. That still remains a puzzle in the *Viking* findings. Ironically, though Curiosity has more recently found complex organics, they are not simple ones, like alanine or glycine, for example, which life can use to synthesize proteins. Unequivocal confirmation that the *Viking*s did find life on Mars would be very simple. A more sensitive version of the Labeled Release experiment, with greater capabilities for organic analysis, should be flown. A smaller, lightweight instrument has been developed and tested. "It could readily be turned into a flight instrument," Gil Levin has pointed out.

So far, their best efforts to persuade NASA have fallen on deaf ears. In the nineties, Levin was told that the agency did not have $25,000 to do an engineering study nor amend any of its new tranche of spacecraft. A Russian experiment that should have landed on Mars in 1997 to investigate the subtleties of the soil's oxidation, in which Levin was an investigator, failed when the *Mars 96* spacecraft crashed into the Pacific Ocean.

As a final word, though, Patricia Ann Straat is clear. The continuing controversy is exciting. "We might believe that we discovered life on Mars," she says, "but we won't know the true answer for a long time. I would like to see more life detection experiments sent to Mars soon to either prove it or disprove it."

* * *

In the summer of 2008, the White House, in the final months of the George W. Bush presidency, was briefed about the latest scientific results from Mars and the implications for life. Weeks earlier, the *Phoenix* lander had come to rest on the plains surrounding the north polar cap. Since then, rumors had abounded from the project. In early August, the venerable industry bible *Aviation Week and Space Technology* reported that the spacecraft had discovered something significant.

Because of this, NASA released its tentative findings early, which in some ways turned out to be a red herring. The discovery concerned "the potential for life" and, specifically, something called perchlorate. It was so obscure that one of the researchers involved later said, half in jest, "all of us had to go and look it up." It is a salt, technically, a chlorinated hydrocarbon, which has been found in the Atacama desert in Chile, for example. It is known that ancient organisms there have been able to adapt perchlorate for their energy needs. So, once again, it wasn't life on Mars, but something much more subtle.

Phoenix had grabbed two teaspoons of soil, which were analyzed by a mass spectrometer—determining chemical composition by weight after burning the samples in a furnace—to reveal the presence of compounds down to ten parts per billion. Despite the coruscating cold, *Phoenix* discovered perchlorate was present down to a half percent, "a fair amount" in one estimation. The implications for life were curious. If Martian microbes could live off the chemical energy released when its chlorine is highly oxidized to form chlorides, perchlorate would provide a new way of searching for life on the Red Planet.

"Its presence is good news for the possibility of life on Mars but very, very bad news for humans," says NASA's Christopher McKay of the discovery. Perchlorate's very toxicity is such that it will cause untold problems when astronauts land on the Red Planet. Breathe it in accidentally, and lung damage might result. But were two teaspoons of polar soil representative of the rest of the planet?

Up to this point, previous rovers had discovered elemental chlorine on the surface. The Mars Odyssey mission had mapped it all over the planet. It would be another four years before direct confirmation came when Curiosity discovered perchlorate directly in Gale Crater. Its presence means that another riddle has been staring us in the face for over forty years.

* * *

The consensus from *Viking*, which Gil Levin openly questioned, was that the complete absence of organics in the Martian soil meant no life on Mars. This finding in itself was not especially surprising. The Martian atmosphere lets in unfiltered ultraviolet light, which bleaches the surface. This kind of blanching would easily remove any organics and, thus, any extant life forms or evidence for past life that lived close to the surface as sampled by *Viking*.

Yet perchlorate represents a missing link. While it could be a "fuel" for microbial life, it has other uses today, including rocket fuel or accelerating the burn of fireworks. This latter finding may have led to an unintended consequence for *Viking*, researchers first reported in the summer of 2018. The *Viking* biology experiments used an oven that heated the samples to drive off their constituent gases (a forerunner of the ones aboard *Phoenix* and more recently Curiosity). Specifically, when the Martian soil was heated to a blistering 932°F (500°C), any perchlorate present would have *caught fire*. As a result, the more delicate organics in the soil would have been incinerated.

Yet they wouldn't necessarily have completely disappeared. The laws of chemistry are immutable no matter which planet you are on. When perchlorate is burned in the presence of oxygen, it produces a molecule called chlorobenzene, which contains carbon. In 2013, Curiosity discovered chlorobenzene in the soil within Gale Crater. So where did the carbon—within the benzene rings—come from on Mars? The answer to that question has opened a whole new vista with regard to the *Viking* results.

Chlorobenzene is a much hardier molecule, technically known as an aromatic organic compound. It is made up of a ring of carbon, hydrogen, and chlorine atoms (all of which have individually been mapped on Mars). It also seems pretty certain that these same elements were already present in the *Viking* data. *Viking 1* measured chloromethane at Chryse Planitia; in the colder, more northerly landing site of Utopia, *Viking 2* had discovered dichloromethane. In the 1970s, their presence was believed to be a by-product of the solvent used to scrub the landers before launch. To date, Curiosity has also found other "oxy-chlorinated species" on loose sand and in very ancient rocks. Even then, they have only been indirectly inferred, but they do seem to be present.

So begins a new chapter of this scientific detective story across space and time. The chlorobenzene in the old *Viking* data was a biochemical clue that nobody had thought existed. Now a group of French and American scientists are reexamining the information returned by the *Viking*s. "We see a higher abundance of chlorobenzene at the highest temperatures and in a sample protected from radiation," says study leader Melissa Guzman, a doctoral student at the LATMOS research center in Paris, "which is compatible with the release of organic compounds trapped in minerals."

It's not a slam-dunk conclusion for life, however. Another possibility is the carbon in the *Viking* samples came from accidental contamination before the spacecraft left Earth. That, as we saw earlier, was the primal fear of project officials in the 1970s. "Because we know there are instrument components with potential organic contaminants," Guzman adds, "we cannot fully rule out that the carbon component is coming from the instrument." But the signal returned from Mars was very different from that which would have been released by the *Viking* "mass spec" oven. In other words, if there was contamination from any of the instruments, it would have had a completely dissimilar appearance.

Guzman's research is one line of inquiry that is shining new light on old data. Christopher McKay has also been involved in looking through the records of the first successful landers on Mars. In research published in 2011, his team used soil from Chile's Atacama Desert and mixed it in with some perchlorate in the laboratory. After turning up the heat on the sample, an otherwise elusive chemical genie appeared. There it was: molecules of chloromethane and dichloromethane, chlorinated compounds that would be left over after perchlorate is heated—exactly as was seen on Mars forty years ago within the *Viking* GCMS instrument.

* * *

This latest chapter in the story of life on Mars is a fine example of what might be termed "planetary archaeology." In keeping with international agreement, all original data from space missions is stored at a specially dedicated facility at NASA's Goddard Space Flight Center, on the outskirts of Washington, DC. There were two separate elements to the *Viking* GCMS data. One was, in essence, a spread of colors from the gas chromatograph in graphical form. The other were the spectra from the mass spectrometer itself.

In those days, computers filled whole rooms and the analysis was as much an art as a science. So Melissa Guzman has assumed the role of a detective. The person who wrote the original software died, as did Klaus Biemann, the principal investigator. "All the similar computations we now perform use software," says Guzman, who works on the extended Curiosity science team. "He could do it in his head." On his death in 2016, several obituaries referred to Professor Biemann as the father of organic mass spectrometry, an extraordinary scientist who pioneered the subject and made it possible to make such measurements on Mars.

"They were more like artists," Guzman says in admiration. Many *Viking* veterans have referred to Klaus Biemann as a maestro, gracious and charming, who could occasionally be prickly, as he didn't suffer fools gladly. His pioneering legacy continues. "Even today with the databases we have, I will never find anything in my data unless I do a directed search for it," Guzman says.

Much of the raw information was unreadable and unusable. In the Goddard archives, there was *Viking 1* data stored in boxes marked *Viking 2*. There was little by way of supporting information. Guzman says archivists at both NASA and MIT have bent over backward to help. "We still don't know where the chromatographs are," she says. The correspondence files are not clear as to where copies of original material were sent, if at all.

Guzman and others have painstakingly restored as much information as they can. This has been made possible by the explosion in modern communications and what it is usually called citizen science. "We have connected with two individuals who successfully decoded the digital data," Guzman says. Amazingly, the original data have been formatted so they can be read on a desktop. "We've modified it and put it into the software which we now used to process SAM digital data," Guzman says, referring to the latest information coming down from Curiosity.

It is still early days for their reconstruction work. Their objective, Guzman says, is to "constrain the source of this carbon." They want to unequivocally discover if it is coming from the Red Planet or from the GCMS instrument. "What I want to know is what is the carbon source of this chlorobenzene that we see in the *Viking* data," she says. "Then we can know for sure what we have seen."

Perhaps, in a very real sense, this will be *Viking*'s ultimate legacy. Since the 1970s, organics have been confirmed in the Martian soil. Equally significantly, fluctuating levels of methane have been sniffed in and around the Curiosity landing site. Future missions are all designed to search for organic materials and ancient signatures of life on the surface. For the first time since *Viking*, they will use the latest technology to search directly for the precursors to life. Not only will they sense just how significant perchlorate is in this search, they will also narrow down much of the perplexing chemistry of the surface soils.

9

REACTIONS

I n the late winter of 1995, two middle-aged NASA scientists, David McKay and Everett Gibson, happened upon something peculiar in their Houston laboratory. Where once their work had been top priority —they had both cut their teeth analyzing Moon rocks returned by the Apollo astronauts—times had changed. The Meteorite Receiving Facility where they worked was now very much the poor relation. Building 31 at the Johnson Space Center was, as a writer from *Texas Monthly* would note, where "poor morale was manifested in peeling linoleum floors, dim lights, and in at least one building, payphones shrouded in cobwebs."

When JFK had given his famous "We choose to go to the Moon" valediction at nearby Rice University in September 1962, the young David McKay, then a postdoctoral student, was in the audience. His colleague Everett Gibson later joined NASA to look at the first Moon rocks. McKay, quiet and diffident, contrasted with the taller Gibson. It would be easy to characterize both by careers long past their best, but that was not the case. Both were highly experienced geochemists at the very top of their game.

They had spent the best part of a year examining a very odd meteorite that, they realized belatedly, had come from Mars. They were using a scanning electron microscope, which senses the way individual electrons are bounced off a minute target. Much more precise than using light, it allowed them to zoom in at over 100,000 times magnification. The individual electrons paint a highly detailed picture, akin to flying over an alien landscape at an atomic level.

McKay had spent innumerable hours poring over the images. One night, he and Gibson alighted on an area that looked a little different. Immediately, something caught Gibson's eye on screen. "We found a segmented structure almost 300 nanometers in length," he recalls. It was about the size of a single bacterium. "It appeared to be made up of a dozen segments with a head and a tail."

In the quiet of the lab, the two Texans looked at each other, knowing full well what the other was thinking. It sent a shiver down their spines. *This simply wasn't possible*, they told themselves. *Our eyes are deceiving us.* But they weren't. Close by this feature, which they soon christened "the worm," they happened upon a cascade of objects—later termed "the swimmers"—which looked as though they, too, had come from structures within a cell.

But what were they? Some form of contamination? Some strange fracturing within the rock? A spurious feature from the perspective of what they were examining? Or could it really be what it looked like: microfossils, the remnants of a once thriving microbiological colony on Mars?

Sometime later, there was a far more telling reaction from Gibson's wife, a bacteriologist. Unwinding one evening after working late, her husband had left some of the pictures of the swimmers on the kitchen table. On his return, he found her mildly curious. "What are you doing with these pictures of microbes?" she asked.

Eighteen months later, the image of that same segmented "worm" became front-page news around the world.

* * *

The announcement of this discovery in August 1996, hailed by President Bill Clinton as "surely [one] of the most stunning insights into our world that science has ever uncovered," was proof of something significant. The question of whether there is life on Mars is not just about otherwise obscure chemical reactions. It tells us a great deal about human nature. NASA's announcement that the meteorite may have evidenced ancient life from Mars took in accusations of politicking and under-hand behavior. Indeed, some of the intellectual undertow and arguments that followed were among the most vitriolic ever heard. In some ways, the announcement provides an object lesson in how *not* to reveal that life has been found on another planet. Indeed, the actual timing of the announcement had more to do with farce than anything else.

Originally, the announcement was slated to coincide with the Republican National Convention, 1996 being an election year, but there was a leak. An industry journal, *Space News*, ran a small item about the possibility of life having been found. According to some, that tip-off came from the vice president's office. Others have suggested a truly mind-boggling scenario: that it came from the bedroom whisperings of a Clinton aide, who had been ensconced inside a Washington hotel room with an expensive escort.

Later, the escort telephoned a supermarket tabloid. Life, she says, has been found on Neptune. And then, all hell broke loose.

* * *

Meteorites are the flotsam and jetsam in the seas of space, the largely forgotten detritus of our planetary system's messy birth. These rocks have ricocheted around for untold millennia at the behest of gravitational forces between the larger bodies of the solar system. Most were formed in the reaches between Mars and Jupiter and are calling out from the ancient past. Some, however, are even more curious.

Before the space age, a trio of meteorites had been known to be very different in their structure and chemistry. They were generally referred to as "Snicks," from the acronym of the places where they had been dis-covered: Shergotty in India, Nakhla in Egypt, and Chassigny in north-eastern France, hence SNC. When the Nakhla meteorite had landed in June 1911, it may have killed a dog, though this fact has been disputed in many accounts. More recently, a number of other meteorites were also suspected to have come from the Red Planet.

Dedicated teams have landed at the South Pole since 1976 to look for meteorites. They are easier to find out on the ice sheets than any-where else on Earth. More than twenty thousand have been discovered.

Of these, fifteen have been determined to be Martian. Another eighty-five from other collections have been identified as coming from Mars.

Until the early 1980s, there was only ever a suspicion that a handful of oddities were Martian. After *Viking*, one particular rock, usually referred to as the Elephant Moraine meteorite, more formally EETA79001, was found to have bubbles of Martian air trapped inside it—comparing its properties to measurements made on the surface of the Red Planet nailed its origin as Martian. The next crucial step was taken by someone who, like so many others, had started out looking for something else entirely.

* * *

Meteorites from Mars were the last thing on Monica Grady's mind when she returned from backpacking around Europe the summer of her graduation in 1980. Calling her parents in Leeds from a phone box at King's Cross Station, there was a pleasant surprise. She had been accepted for a doctorate. On that bright morning, Grady had little sense of just how important her work would become in unraveling the possibility for ancient life on Mars.

Well-regarded and fêted—awarded a Commandership of the Most Excellent Order of the British Empire—Monica Grady is today a professor of planetary science. She does not seem particularly bothered by media hoopla, as befits her no-nonsense Yorkshire heritage. Growing up the eldest of eight children, she was fascinated by the limestone scenery of the nearby moorlands. "I'd always liked rocks and was thrilled by their poetic side, you might say," she says.

As a geology undergraduate at Durham University, she examined samples returned by the Apollo astronauts. "They were absolutely wonderful," she recalls. "Six microscopes had been set up for us in the lab. It was a really big deal, and there was a hushed atmosphere in anticipation. It was almost like a religious event."

After pulling her proverbial socks up, she obtained academic distinction. That offer of a PhD place, she later learned, involved her supervisor, the late Professor Colin Pillinger establishing a new department in Milton Keynes. Given her earlier epiphany, she wanted to look at lunar samples. In the event, she was given meteorites to examine instead.

"We were going to be the reputation builders," Grady says, referring to the Open University. In due course, in the early 1980s, they did. Love also blossomed among the spectrometers, for she soon married Ian Wright, another researcher in the team. "I think it started when he was in his shirt tails, trying to get some equipment to work," she adds with a smile.

The group's interest in Mars had to do with timing and happenstance. The famous Nakhla meteorite was already in the British national collection. As noted meteorite researchers, Grady and her colleagues were also able to look at samples of the Shergotty and Chassigny meteorites. They were also in a good position to examine the Martian meteorites collected from Antarctica.

Collectively, meteorites from Mars have a considerable story to tell about surface conditions and, more importantly, the role of water throughout the Red Planet's history. The discovery of carbonates embedded deep within some of them shows that carbon dioxide and water did indeed react together throughout Martian history. Strangely, this sort of carbonaceous material in the form of organics was notable by its absence on Mars when the *Viking* measurements were taken.

In some sense, it was no surprise that Grady and her colleagues were the first to find them. The Open University team had at their disposal a powerful technique known as step combustion which they could use to examine this perplexing carbon. By carefully controlling the atmosphere in which a sample was burned and the temperatures at which this was done, they could learn, from both the gases given off and the carbonaceous remains left behind, just what physical processes had altered it.

In the case of carbon in meteorites, however, there is a catch. The technique can only be done if the sample is small yet representative of the greater whole. The reasoning is simple: if you take a whole meteorite—known as a bulk sample—and burn it, then it would be very difficult to disentangle just what sorts of carbon chemistry have been involved in the rock's subsequent evolution, after it left Mars.

Only the tougher rocks ever make it all the way from the Red Planet. They have been blasted off the Martian surface at hypersonic speeds. They then float through space for millennia, only to make a fiery entry through Earth's atmosphere. Small wonder most of the Martian meteorites are hard, crystalline rocks that were formed by volcanic processes in the ancient past. Out on the Antarctic ice, all sorts of contaminants from everyday life will tend to seep into any rock that has lain there for any length of time. This contamination will make it very difficult to disentangle just what sorts of carbon chemistry have been involved in the rock's subsequent evolution. So the trick is to do the combustion technique step by step.

Researchers take a small sample, add a controlled amount of oxygen, burn it at a set temperature, and then examine the chemical entrails. This allows them to unravel the role of carbon in stages, like reading the subsequent chapters in a whodunnit. Different forms of carbon burn at particular temperatures: graphite burns at far higher temperatures than organic forms, for example. So by looking at which of these different forms are present and then disappear step by step, researchers can infer many of the important chemical processes that have taken place.

In other words, this technique allows the disentangling of complex carbon chemistry into a coherent picture of how its host rock evolved. The Open University team were pioneers in this burgeoning field. "At the time when people were analyzing for carbon, most people would take a piece of rock and burn it to 1,000°C," Monica Grady says. "You'd get the whole lot, terrestrial contamination, everything, all rolled into one."

As time went on, they refined their techniques considerably (and have subsequently miniaturized such equipment to fly into space). At the start, they were taking steps in increments of 100°C, but this was later reduced to 25°C or even 10°C. Depending on what they were looking for, these refinements allowed them to deduce the subtlest details of carbon in the handful of meteorites suspected to have come from Mars.

"We thought that it would be interesting to look at this oddball bunch," Grady says. "We wanted to look and see if you could relate the amount of carbon to the amount of metamorphosis they had undergone. In particular, we wanted to see if you could find out about the temperature at which these things had physically changed."

The Open University team had another ace up their sleeve. A new mass spectrometer, designed by Grady's husband, Ian Wright, helped them examine whether any possible carbonates were in any meteorite samples. The team then decided to go for the "big daddy" of the Antarctic meteorites, EETA79001, in which that infamous bubble of Martian gas had been trapped.

Using the step combustion technique and teasing out individual molecules from mass spectrometry, the scientists in Milton Keynes were surprised to find a relative abundance of carbon isotopes, each of which has a distinct story to tell. The way they are changed depends on the processes they have undergone. With intense boiling, lighter isotopes will be preferentially lost to space; conversely, during condensation, they will remain. The exact ratios reveal whether the sample has undergone condensation, evaporation, or distillation. "If we didn't know anything about these meteorites," Monica Grady wryly recalls, "we'd say that they'd got carbonates in them."

It was as unequivocal as finding a fingerprint at a crime scene. Around this time in the late 1980s, other researchers steeped in the science of petrology—geologists who look at the origin, structure, and composition of rocks—had come to the same conclusion. When petrologists looked at the Elephant Moraine samples using their own different techniques, they too found carbonates present. But it all seemed too bizarre to be really true.

When the Open University team published their findings in *Nature* in July 1989 in a paper entitled "Indigenous Organic Matter in EETA79001," they hedged their bets. The organic matter, they reported, was "indistinguishable from terrestrial biogenic components." By looking at the decay of radioactive elements within them, it was clear that this particular rock was less than 200 million years old. They raised their heads above the parapet by concluding that "the implications for studies of Mars are obvious."

Today, Grady laughs at the memory of this typically British, understated equivocation. "There is a simple way of producing organic matter, and that is biology," Monica Grady says. "There are other ways, but we decided not to be so bold."

* * *

Early in the 1990s, it had become clear that there was also another very odd rock from a more recent haul of Antarctic meteorites. Discovered in

December 1984 in the Allan Hills, east of the Transantarctic Mountains, it had been cataloged as ALH84001.

Out "on the blue," as the Antarctic ice sheets are called, the meteorite looked green and was larger than most other samples taken. All were packed in containers and kept frozen, though this one was marked "Yowsa! Yowsa!" in the accompanying notes, for reasons that will soon become clear. First it was taken to Hawaii and then to Houston and the Meteorite Receiving Facility, where a nitrogen-filled vault (which had the advantage of keeping the samples dry) awaited it.

In 1993, examination of a routine sample from ALH84001 suggested, from its oxygen isotopes, that the meteorite was Martian in origin. Measurements from radioactive decay of elements like rubidium and samarium showed it was older than EETA79001. It had been blasted from the surface of Mars fifteen million years ago. It had landed on Earth 13,000 years previously.

Early in 1994, the story moved back across the Atlantic. Over the years, the NASA team had forged close links with Britain's Open University group. "We had been looking at carbonates, and it contained all these carbonates," Monica Grady says. "We could tell it was a very, very unusual meteorite."

Samples from ALH84001 were shared with the group. When Grady and her team examined them, they were astounded. The samples were "riddled with" unusual patches of carbonates, appearing beautifully yellow with dark edges under scrutiny of the microscope. "When I first looked at them, I was amazed," she says. "I had never seen anything like it."

These features were tiny: the carbonate globules were roughly the same width as a human hair. They were nothing if not aesthetically compelling. But how could they have formed? The rock was obviously volcanic, but it was clear that at some point it had been altered by water, probably with carbon dioxide dissolved in it, along with the presence of sulfur and other salts. Water would wash through the rock and break down the minerals contained within it. Secondary minerals would then be deposited along with the carbonate globules.

At this time, the main thrust of the British work was to untangle the carbon in the carbonates that had likely formed on Mars from any terrestrial contaminants as well as other indigenous carbon in order to shed light on the curious chemistry that had created the carbonates.

"Not many people had our sensitivity for looking at the other carbon isotopes," Grady says. "We looked at how the carbon had been reduced."

From the observed isotopes of carbon and also oxygen, even the most skeptical among the group were now convinced that their composition was not terrestrial. What they were seeing was unequivocally from Mars. The NASA researchers, in what Grady recalls as a "sheepish phone call," came up with similar findings. "We'd both got the same sort of results," Grady recalls. "They were quite unusual, as we had found unusually high amounts of carbon."

It made sense for the two teams to combine their efforts in writing up their research for *Nature*. Crucially, the Brits and the Americans were looking at the samples of ALH84001 in slightly different ways. Everett Gibson's postdoctoral researcher in Houston, Chris Romanek, had hand-picked the grains with a needle after observing them through a microscope. The British team looked at the whole rock and dissolved a sample to see how quickly it took for the acid to react. That gave the "flavor" of whether it contained calcium or magnesium carbonates.

By effectively checking on each other's findings, the Brits and Americans knew that their results were correct. The orange globules at the center of the carbonates seemed to contain a mixture of iron, magnesium, and calcium carbonate. "It's all very well to say you've 'got a carbonate and that it's formed by CO_2 under the atmosphere of Mars,'" Grady says. "You want to know how much carbon dioxide was dissolved in the water and what the temperature of the water was."

The NASA team had already determined that the rock formed at a low temperature. The British team, using a different approach by plugging numbers into a computer model, came up with their own best guess. The rock had formed above the freezing point of water and below its boiling point. There was also something distinctly odd. The amount of organic chemistry within the carbonates was unusually high, more so than could be explained by contamination of any sort. Even the most badly tainted meteorites on Earth had very much smaller amounts of carbon within them.

"What leapt out from the beginning was there was a lot of carbon in the meteorite, but we decided not to say it was indigenous," Grady says today. "We said that there are a lot of organics which can't be explained by terrestrial contamination."

Of the three British researchers, Grady was, and remains, the most skeptical toward any possibility of life. Yet she knew perfectly well what their results implied. At the time, none of them wanted to court controversy. Immediately afterward, the British group changed direction in their research. Strange as it now appears, there seemed little point in dwelling on the indigenous carbonates in meteorite ALH84001.

Even with the benefit of hindsight, Grady is wryly amused by what happened next. "Things go nice and quietly until something horrendous happens," she says, thinking about a rash of telephone calls from the media a couple of years later concerning a rumor about NASA finding evidence of some sort of fossils from Mars.

* * *

In Houston, the NASA scientists persevered. David McKay was the driving force. Working in tandem with Everett Gibson, he was determined to unravel the complete chronology of the bizarre carbonates riddled throughout the Allan Hills meteorite, ALH84001. "We did not begin trying to find life in this rock," Gibson later explained. "We began to look at the chemical and mineralogical evidence."

Within the carbonate globules, they discovered alternating bands of black and white structures at a very fine scale, quickly christened "Oreo rims." The NASA team examined their carbonates and the organic materials that coexisted along with them. For that, they used the scanning electron microscope (SEM), normally found in electronics laboratories to look for faults in printed electronic circuits. Colleagues at the Johnson Space Center had a brand-new, more advanced SEM that was primarily used to examine parts of the Space Shuttle. It was capable of magnifying 100,000 to 150,000 times.

"It was just amazing what we could see," McKay says. Features that had only been vaguely suspected were revealed for the first time in stark and staggering clarity. Yet it remained difficult work. Rather like flying over unexamined territory, it could be two to three hours before anything of note could be found, if ever. It wasn't just a case of pressing a few buttons and sitting back, hoping to hit gold—or even fool's gold, pyrite, which had been found in a number of the Martian meteorites. In this case, their attention was grabbed by signs of magnetite, a magnetic mineral made of iron and oxygen that they knew was normally produced only by bacteria.

Taking a long hard look at all the other minerals within the carbonate globules, known as "unique phases" by geochemists, the Texans were fascinated by the Oreo rims. It was clear that some sort of complex chemistry had given them their distinctive coloration. The white parts were primarily composed of carbonates, while the black contained an abundance of iron minerals and hints of sulfur. "When we find some of these phases together," Gibson says, "it strongly suggests biological activity."

From the beautiful orange globules they had also seen, it was obvious that there had been water percolating through the rock. At the time the globules formed, there were likely liquids flowing through the Martian crust that had become saturated with carbon dioxide. This caused the minerals to precipitate out. On Earth, precipitation like this occurs at the edge of geysers, where, significantly, there can be entrainment of organisms into mineral phases. It is a quick way of burying terrestrial microbes alive by locking them into the minerals.

The Oreo rims hinted that something similar could have happened on Mars billions of years ago. They contained not only magnetite but also contained iron sulfides like pyrite and even gregite, which is also typically produced by terrestrial bacteria.

Gregite set another temperature constraint on the chemistry that might have taken place. On Earth, gregite would decompose at temperatures above 250°C. For it to have survived on ancient Mars, conditions would presumably have had to be significantly cooler. Terrestrially, gregite and similar minerals are associated with biological activity, providing whatever microorganisms are in close proximity with a unique energy source that they could metabolize.

In the outermost part of the Oreo rims, there appeared clear evidence for the effects of running water through the pores of the rock. At the greatest magnifications, McKay and Gibson found within the carbonate globules some strange, oval-shaped structures that appeared curiously elongated and looked vaguely biological. Using the scanning electronic microscope, these "swimmers" were all of two hundred nanometers long and ten nanometers wide.

Both McKay and Gibson were well aware that these strange shapes could be artifacts of the examining procedure. Yet they were quite unlike anything the two had ever seen. Over the course of his career, David McKay has peered inside the structures of at least fifty thousand separate extraterrestrial rocks. So they kept looking, making sure what they were seeing was not where they might have put too much gold on the sample to let the individual electrons flow.

There was also the possibility of terrestrial contamination. Could there have been Earthly organisms that somehow got mixed in with the sample? McKay and Gibson learned that similar segmented "worm structures" had been found in terrestrial rocks. But McKay in particular had never come across anything remotely similar after looking at lunar samples for twenty years.

There were clusters of objects that startled them further: the wormlike features at a tiny scale. It looked as though they had crawled over the microscopic cliffs within the rock's fractures. Other features looked as though they had undergone division, like cells do, with telltale clover-leaf structures, reminiscent of the process of mitosis that replicates cells in biology. "It was just incredible and exciting," McKay recalls. "I mean, we really wanted to jump up and down when we saw those."

Yet their smallness surprised them. "We knew that they were smaller than what we'd see with normal terrestrial bacteria," Gibson says. "And that was one of the problems that bothered us."

Nevertheless, they soon learned there was a flourishing field of study in nanobacteria, biological cells that are ten times smaller than most cells on Earth. They were exactly the same size as the objects they had found in the Martian meteorite. McKay and Gibson were pleasantly surprised to learn that such tiny cells had been found miles underground in the Columbia River Basalts in Washington State. Not only did these cells seem to feed off hydrogen for their energy source, they were on the same scale as those seen in the meteorite.

Such information only encouraged them to be bolder in their thinking. Their next step was to consult a colleague down the hall, who, in the

time-honored way of all important discoveries, thought her two distinguished colleagues had gone completely off their rockers.

* * *

These guys are nuts. That was Kathie Thomas-Keprta's first reaction after
she had been invited by her ostensible boss, David McKay, into the secret
of what they had discovered. McKay took her to his office where Gibson
was already waiting, closed the door, sat her down, and revealed that they
thought they had found evidence for ancient life on Mars.

They showed her some slides and kept mentioning "fossils." McKay
said that they really needed her expertise. She mumbled something about
being too busy but knew she had little choice. Thomas-Keprta was a contractor to NASA—she worked for Lockheed Martin—and thus far had
been an expert in examining cosmic dust. It seemed impolitic to argue.
She left McKay's office stunned, thinking they were either playing an
elaborate practical joke or had gone bonkers. That night she went home
and said as much to her husband.

In early 1995, Thomas-Keprta was in her late thirties. She had accumulated twelve years' expertise with a different sort of microscope that
had a much higher resolution. Transmission electron microscopy is the
high-tech equivalent of holding something up to sunlight to see what is
inside. In this case, it involves illumination at the elemental level as electrons are passed through a sample. Diversions from their expected path
reveal subtle details about their microscopic structure and their mineral
composition.

It was only when she looked at the features within the globules for
herself that Thomas-Keprta was gradually won over. The amazing colors
and remarkable vividness of the carbonate globules is an experience that
few have had firsthand. For her it was a revelation.

Thomas-Keprta would often joke that the "Thomas" part of her
name was the important one. She was a doubting Thomas. "As time went
on, I became more and more convinced that the chemistry and the structure of the magnetic minerals found in the carbonate globules provide
good evidence for ancient life on Mars," she later recalled.

Her expertise was the next step in the discovery process.

McKay and Gibson knew that they had merely scratched the surface. Scanning electron microscopes can examine only the outermost layers and will only reveal what elements are present. The Houston team needed to probe deeper and use the transmission technique to look at the minerals. Without that, a significant piece of the puzzle was missing. The Oreo rims were not the only feast for Thomas-Keprta's eyes. Toward the center of the globules, there appeared to be curious pancake shapes in which she found tiny grains of magnetite, plus grains formed by pyrite, made from iron and sulfur.

These things are tiny. So tiny, in fact, that about a billion could fit onto a proverbial pinhead. Within the globules were chains of magnetite grains that are known to be associated with bacteria. The chemistry of these features suggested an environment that was good for them, too. They faintly resembled bacteria.

Magnetite is a common enough mineral, but the grains inside the globules were very unusual. Not only were they extremely small, they were nearly pure, chemically speaking. They seemed to have no structural defects. At the time, the only known way of producing such pristine crystals was by the action of bacteria.

Well, perhaps, Thomas-Keprta thought. Magnetite could be produced nonbiologically, too. In fact, magnetite had already been found in other meteorites. It was the shape of the crystals in this particular one that was so perplexing. Some appeared tear-shaped, others cubic, and many appeared side by side. That was hard to explain as the result of a random chemical process. Some terrestrial bacteria cause magnetite grains to line up along the magnetic field lines when they are produced internally. They effectively use these grains as a compass needle to navigate through Earth's magnetic field.

The flawlessness of the crystals, with virtually no visible imperfections, implied they had been formed within the bodies of bacteria. Bacteria can control the shape of the magnetite used to create their propulsive motion. By contrast, as Thomas-Keprta knew, when magnetite is produced inorganically, it is often mixed in with other elements that happen to be in close proximity.

There were other pointers toward biology. The fact that the magnetite grains seemed to coexist with pyrite was very unusual. Both are highly reactive and will, at the drop of a hat, change into other forms of mineral. Finding a reduced form of iron (an iron sulfide like pyrite) as well as its oxidized form (an iron oxide like magnetite) was also atypical. Chemically, this finding hinted at a "disequilibrium," where some unusual external agent had had its chemical hand on the tiller.

At this point, the Houston team enlisted the help of Hojatollah Vali, an Iranian-born mineralogist at McGill University in Montreal. He was given another sample and used a completely different technique to examine the grains. This involved freezing and thawing the sample to open up cracks within it so that new, clean surfaces were exposed and scanned afresh. He too found structures some 700 nanometers in length that appeared curved and fragmented. "I think we then began to have some real confidence that what we were seeing was real and in the sample," David McKay says.

Vali was puzzled too, but for another reason. The carbonates hinted at low temperatures, yet the magnetite and pyrite would form together only at higher temperatures. At face value, it was a bizarre chemical paradox, the equivalent of finding a hot ice cube. Unless, of course, something else had forced them to coexist at lower temperatures. The easiest explanation was bacteria.

More importantly for the geochemists, the carbonates were associated with the reduced form of carbon that is characteristic, on Earth at least, of biological material. "The relationship of all these things in terms of location, found within a few hundred thousandths of an inch of one another, is the most compelling evidence," McKay said later.

Their preliminary analysis had already taken the best part of 1994. Thereafter they spent another eighteen months trying to eliminate any other rational explanations, including examining the detailed molecular structure within the globules.

Three further samples—given the names Mickey, Minnie, and Goofy—were sent to Professor Richard Zare at Stanford, with whom they had previously collaborated on other meteorite samples. The Zarelab, as it is known, had a laser absorption mass spectrometer capable of detecting one part per trillion (or one part per billion of a part per billion), almost down to the level of individual molecules.

That meant they could look for Polycyclic Aromatic Hydrocarbons

(PAHs). Zare's laser can be tuned to excite specific organic molecules. In this case, the specifics they were interested in were little hexagons of carbon around which various hydrogen atoms were dotted. Within these PAHs, their constituent atoms are arranged in an aesthetically pleasing hexagonal ring. That is the polycyclic part. The aromatic part comes from just a peremptory whiff of such molecules as benzene and toluene.

PAHs are often found when incomplete organic combustion has taken place. On Earth, they are associated with oil, tar, asphalt, and coal deposits, which are usually the result of rotting biological matter. They are also found in mothballs—naphthalene—and within the gunk around burnt hamburger.

Here, the decomposition of biological matter was a possibility that commended itself. The PAHs found inside the meteorite contained three or four benzene rings, plus an occasional extra carbon or hydrogen atom. Zare's British-born postgraduate student, Simon Clemett, did the analysis. "I sent the samples to him under crazy names so he didn't know what he was analyzing," Kathie Thomas-Keprta wryly recalls. When completed, both Stanford researchers agreed they were not terrestrial contamination.

"The PAHs belong to the rock," Zare concluded. "The question is, where did they come from?" The presence of carbonates implied lower temperatures, and if that was the case, "then it's hard to imagine that they weren't produced biologically."

The concentration of the PAHs within the globules, the magnetite—produced when bacteria consume sulfur—and the presence of the micro-fossils suggested something stronger than unusual forms of carbon. "The coincidence of all these data suggest that there's a possibility that it could have been produced by primitive life on Mars," Zare says of his reaction at the time.

So the Houston researchers decided to test the water. In March 1995, Kathie Thomas-Keprta felt confident enough to present preliminary findings on the PAHs at an annual meeting in Houston. She was aston-ished to be asked point-blank by a reporter from the *Houston Chronicle* if what she meant was that she'd found life. "Absolutely not," she replied. Everett Gibson, when also questioned by the press, added that the com-plex hydrocarbons were not related to life. "It is too early, and not all the pieces are in place yet," he said at the time. It was a close call.

Over the next year, they refined their work. The extended team submitted their findings to the leading American research journal, *Science*, in the spring of 1996. After a few revisions, McKay and Gibson received confirmation that their paper, entitled "Possible Biogenic Relic Fossils in a Martian Meteorite," would appear in the August 15 issue.

* * *

In early August 1996, David McKay took his annual summer vacation. He and his family made their way to central Texas, where they would camp along the Frio River. Just in case, he took a beeper. After a few days, he was relieved that there were no last-minute panics. At the end of the first week, he decided to call into the Johnson Space Center. He found out the news had been leaked to the media, that publication of the *Science* paper was being brought forward, and that NASA had issued a statement and a press conference had been called for the next day. The agency's surefooted publicity machine went into overdrive, preparing press kits that included animation sequences and videos explaining their scientists' exhaustive analysis.

McKay left his family on vacation and rushed to San Antonio, where he jumped on the first available flight to Washington, DC. In the capital, he was joined by Gibson and their coauthors. The Clinton White House had already been alerted. Vice President Al Gore had apparently asked, "You mean a group of *government* scientists have done this?"

At NASA Headquarters, they were grilled by senior managers for three hours. Their number included Administrator Daniel Goldin, who was electrified by what they had found. During their presentation, he made twenty-seven pages of notes. At the end, though, he lightened up a little. "Can I give you a hug?" he asked an astonished McKay.

At a packed press conference held at NASA Headquarters on August 7, 1996, McKay and Gibson emerged blinking into the limelight. They described in exhaustive detail the intricate steps they had taken to uncover what they believed was unequivocal evidence of ancient life from Mars. It was always the images of the "worm" and the "swimmers" that garnered the greatest attention.

At 2:00 p.m. that Wednesday afternoon, Goldin kicked it all off under the klieg lights. "NASA has made a startling discovery that points to the possibility that a primitive form of microscopic life may have existed on Mars more than three billion years ago," he said. "I want everybody to know that we are not talking about little green men."

Then came what was later called their "Pearl Harbor moment"—or at least, a day of infamy. Each member of the team described their painstaking part in this extraordinary detective story. As lead author, David McKay, normally reserved and hesitant at the best of times, appeared nervous. He was surprised by the media attention and dismayed that their conclusions were largely ignored.

"None of these observations is in itself conclusive for the existence of past life," McKay emphasized. "When they are considered collectively, we conclude that they are evidence of primitive life on early Mars."

Perhaps the most remarkable aspect of this story came from the fact that, if they had indeed found traces of life, it would have been the oldest seen anywhere in the solar system at the time. What was strange, though, was how quickly and publicly the scientific community became polarized by their findings. As McKay was at pains to point out, they were bound to be contentious, yet throughout, his team were as open as they could be. "If we're wrong, we'll admit that we're wrong," McKay said a few days later. "I think our data will hold up."

More than 1.2 million people clicked onto *Science*'s website on that first day. By the time McKay returned home to Houston, he was welcomed by a barrage of vitriolic emails: "Shame, shame, shame! This is cocaine and Crayola science!" read one. "A bunch of catchphrases and suppositions purporting to be fact," ran another. Less easy to ignore was the statement from a well-known scientific authority: "This is half-baked work that should not have been published."

Battle lines were being drawn. Since 1996, the weird rock has gotten even weirder. The jury is still out concerning the evidence both for and against ALH84001 containing ancient life. The more it is examined, the less clear everything becomes.

To some extent, the findings remain inconclusive. Experts in many fields have refuted a number of the arguments that supported the conclusion that life had been discovered. To date, nobody has come up with a single fact devastating or conclusive enough to demolish or prove the central hypothesis. "ALH84001 is really a test for all the scientific community," Everett Gibson has stated in public on several occasions. "Can we decode the information in this sample, which is of an unknown locale on Mars; can we interpret the record it has brought to us?"

To some, the carbonate globules and their organics might not need microbes. The "worms" could have been the result of uneven patches where they had been prepared for electron microscopy. The presence of the PAHs is also seen by many as a red herring. Polycyclic Aromatic Hydrocarbons have been detected in just about every meteorite ever examined. It is hardly surprising that they have been found on the surface of Mars. The NASA team have more or less acknowledged this. "The presence of PAHs does not mean you have life in a sample," says Everett Gibson. "What it does, if proven to be indigenous, is show there's complex carbon chemistry that is retained in that sample."

The PAHs could quite easily be formed inorganically on the surface of Mars when carbon dioxide reacted with hydrogen in the presence of magnetite. Another school of thought is that the bulk of the carbon is terrestrial contamination thanks to the seepage of Antarctic meltwater over the 13,000 years the sample was stranded there. Many of the observed hydrocarbons are typical of long-term background contamination. There are sufficient PAHs floating around in Antarctica on the wind that, when dissolved in the ice, could stick onto the carbonate globules themselves.

The signal of the indigenous "carbon" is difficult to distinguish against the overwhelming number of possible contaminants. "There is no doubt in my mind at all that part of this organic carbon is definitely Martian," notes Monica Grady.

The arguments have continued. "We certainly have not convinced the community," David McKay lamented on the tenth anniversary of the announcement in August 2006, "and that's been a little disappointing."

* * *

Indeed, the jury is still out. Today, the remaining central argument—concerning the shape of the magnetite crystals—rests on whether they are biological or not. In other words, is there unequivocal proof that bacteria caused their unusual shapes? Everything hangs on the word of experts who have spent years looking at the shape of these magnetite crystals and enjoy arguing with each other.

In the mid-1990s, there was no known way to create magnetite crystals *without* biology. Now, though, other mechanisms to do so have been discovered. Given that the host rock was excavated by impact from ancient Mars, it is significant that shock waves can create magnetite crystals. Certainly, there would have been a powerful blast that sent the rock on its way. In 2005, it was suggested that it had been liberated from Eos Chasma in the giant Valles Marineris, the result of primordial bombardment. "It's a bit like being shown a photograph of an eye," notes one critic. "It's quite a stretch to say that it belonged to a Hollywood movie star."

Finding magnetic compounds could be misleading in itself. The presence of magnetic materials could have resulted from strong magnetic fields early in the history of Mars. "We had no idea of the magnetic field variability in the history of Mars," Everett Gibson has said, "especially at the time when these carbonates formed in this rock."

The fact that the magnetite grains appear to be nicely aligned—in one photograph, almost point-to-point—seems to be nothing more than an optical illusion. As several people have pointed out, bacteria aren't like polite tourists who line up to have their picture taken. To use the Transmission Electron Microscope, the object of scrutiny is tilted in such a way that unrelated objects could appear to be joined when they were not. "There is no reason for Martian bacteria to have these things inside them anyway," Monica Grady says. "It couldn't help them."

In any case, some of the structures of the crystalline shape suggest that they have been deposited at a very high temperature, which would have killed off any biological activity. Some of the magnetite crystals are elongated and have a structural defect known as a screw dislocation, which hints that they are formed at higher temperatures. Others are barrel-shaped and seem to have lined up on the carbonate minerals, a method of formation that suggests they are inorganic.

Nevertheless, Kathie Thomas-Keprta has reported that a quarter of the grains are the exact same size, shape, and structure as magnetite grains made by bacteria. In 2000, further work was published in a specialist journal in which she found six hundred magnetite crystals that looked identical to the ones produced by microbes. "If you did not know this rock was from Mars," she said, "you would conclude that it contains evidence for past life."

Though she implored critics to look carefully at the evidence, one came back with the retort that it was "Custer's Last Stand" so far as life was concerned. Ironically, one person who had assumed the role of Lieutenant General Custer in the last years of his own life was David McKay, who died in February 2013. Until his passing, the lead author of the original work was still pointing to the magnetite findings as significant. "The shape of the magnetite grains is still rather distinctive," he said in 2006. "If it were found on Earth, it would be a very strong biosignature."

In a development that sounds like something from *Macbeth*, a second team of NASA researchers created a perfect alignment of magnetite from nonbiological processes. That one of them was McKay's brother gave the subsequent debate a delicious twist. "He got a little testy about the results we were getting," Gordon McKay said of his sibling. "What we have shown is that it is possible to form these things organically." Yet David McKay insisted they were not the same shape. His brother's samples had been synthesized from ridiculously pure ingredients.

Today, the debate is more about the coincidence of the carbonates and magnetites and which came first. Some authorities believe the small magnetite crystals formed from the partial thermal decomposition of the host carbonate. "Their origins may be unrelated," Kathie Thomas-Keprta says. "From the perspective of the carbonate, the magnetite may have formed somewhere else."

Yet her most recent work with others suggests that the magnetite could have formed by neither thermal decomposition nor shock waves. The magnetite crystals could have been "brought in from somewhere else, suspended in the ground water, and then were added to the carbonates as they crystallized," Thomas-Keprta says. While it doesn't prove the biological argument is correct, it probably represents their last word on the subject.

Another analysis by Ed Scott and David Barber, both experienced crystallographers, centers on the magnetite crystals having a "topotactic" relationship with the surrounding carbonate. *Topotaxy* refers to the parent material determining the orientation of the product. Their reexamination of the coincident material has led them to the completely opposite conclusion. "This demonstrates that the embedded magnetite crystals formed within the carbonate crystals by diffusion of atmosphere and loss of carbon dioxide," they state.

Many believe this is the final nail in the coffin for the biological origins of the fossilized material. According to Professor Barber, who was a pioneer in the application of transmission electron microscopy to high-temperature rocks in the 1960s, the findings are akin to a kind of scientific Rorschach test. Someone with a biological sciences background will give undue weight to shapes and sizes. "This also showed early on in the incorrect identification of the carbonate 'worms' as microbes," Barber concludes.

* * *

As one scientific paper notes, there are still "animated discussions" about the findings inside ALH84001. "It is very difficult to take a view on a lot of these findings unless you know a lot about such specialized crystallography," says Professor Charles Cockell, an astrobiologist at Edinburgh University. "The fact that people who *do* know about these things still argue with each other is probably all you need to know. There is no definitive shape where people will say 'this is absolutely biological.'"

One tangible benefit from the meteorite debate is the genesis of the new field of astrobiology. The weaving together of several different strands of evidence, which Everett Gibson says have supported the original hypothesis, has seen laboratories around the world working together to find a way out of these kinds of scientific riddles in the future. Certainly, the missions to Mars launched in 2020 will use a similar approach, for the first time in forty years, to look for organic material and signatures for ancient life directly.

The debate about the meteorite will probably continue until Martian samples are brought back to Earth either refute or strengthen the arguments. Otherwise, some feel their colleagues will get ever more hung up on pointless minutiae concerning this particular and peculiar rock. "And I am not sure it will ever be resolved unless we get to examine more Martian meteorites in context," says Cockell.

To some, the arguments represent diminishing returns, a reductio ad absurdum with no way out of a particularly complicated labyrinth. Certainly, additional samples associated with the carbon inside meteorites will be required. Possible biological signatures may be found in further investigation of salt crystals, fresh basalt, and clay samples. "If there was life on Mars," Charles Cockell adds, "there would be other types of minerals and organic remains that would also reflect biology."

Twenty years after the original announcement, Kathie Thomas-Keprta provided a summary that could act as a leitmotif for the next generation of missions to Mars. What exactly constitutes life? "At the most fundamental level," she says by way of conclusion, "we still don't know whether the difference between animate and inanimate is simply a difference in kind or degree. In absence of such a definition, the search for evidence for life on Mars is plagued by ambiguities."

* * *

Generating publicity is the first thing on Veronica McGregor's mind every morning. A former CNN reporter, she is now the news and social media manager at the Jet Propulsion Laboratory, a poacher turned gamekeeper with a keen sense of news values. "When I started here," she says, "we were only targeting reporters." As the influence of traditional media has waned, with layoffs and print newspaper closures, the laboratory's big fear was not reaching the public. Even worse, taxpayers would no longer care about landing on Mars. "We were worried about it," McGregor reflects, "and it turned out everything had changed with social media."

With a host of new missions now heading to the Red Planet in the immediate future, all with instruments capable of detecting biologically significant activity, the question of life will be brought into sharper focus. If word comes of a "major discovery," it will hit social media first. Small wonder that space agencies like NASA and the European Space Agency are deliberately wooing a younger audience by posting regular updates from their spacecraft on Mars. Social media will fan the rumors ahead of time. Generation Z is more likely to take notice of #LifeOnMars and #GetYourAssToMars than traditional headlines.

Until the early 2000s, most information had been in the form of dispatches sent out to journalists. Now, online updates, released through videos and various social media accounts, keep the world informed. By their very nature—such releases tend to come in staccato bursts—they often create a false impression. Not so much "fake news" but, as McGregor acknowledges, providing no coherent narrative to connect the dots and show how coherently organized NASA's Mars exploration program is.

"There were certain times when we would worry about [this]," Veronica McGregor says. "There are a lot of missions in this Mars pipeline. And how do we explain to the public what this one's doing that the other one didn't do, and why we have to do this next one, and why it is so incremental."

JPL's push toward social media came after the Curiosity landing in August 2012, in large part due to worries about this kind of public fatigue. The one-shot approach of earlier missions such as *Viking* and *Pathfinder* often obscured the need to go back, especially when no life had been found. Many younger followers today are learning about the Red Planet for the very first time. McGregor says their work has needed to shift between explaining the discoveries and carrying out education. On Twitter, however, they have found that the public "are going to stay with you for the entire life of the mission."

A perennial hurdle is the reluctance of the people doing the actual work. In the 1960s, for example, mission engineers complained when the first TV cameras were placed in the Space Flight Operations Facility. Scrutiny made many uncomfortable. More recently, the same thing happened when McGregor wanted to place a live-streaming camera to watch over the Curiosity rover's assembly. There were protests about privacy (so the camera was held in one fixed spot, with ample room off-screen for team members to relax) and international security (so the camera was turned on only after the proprietary rover technology was put under wraps).

McGregor's team, working in another building, hosted a series of weekly live chats. During one, a viewer asked the rover team to wave at the camera. McGregor quickly gave a "corporate" response. They couldn't be disturbed. But to her amazement, in the clean room, white-clad spacecraft engineers turned to face the camera and began waving with their disinfected gloves. "So they had somebody on the team, who wasn't working on the floor, [who] was looking at the chat and called in to them," McGregor concluded.

Later, she learned that the assembly team simply loved the attention. A mission to Mars represents the ultimate selfie. It was, she says, "like watching the rover grow up." As this book is being written, the authors—and anyone else, for that matter—have been able to watch Perseverance, the follow-up to Curiosity, take shape.

JPL's first social media campaign was created for *Phoenix*, which landed on Memorial Day weekend in 2008. McGregor started it after seeing fears posted in online forums that anyone traveling would miss news of the landing. It was a big risk: at that time, Twitter was a tiny startup and not the vast engine of social and political machinery we know today. But, as McGregor recalls, *Phoenix*'s Twitter following "just blew up," since many more people than just the diehard fans joined in. "It was the fifth-most followed Twitter account in 2008," she says. "It preceded any celebrity accounts and a lot of other stuff. It actually came at a time when everyone was saying Twitter was going to die by the end of the year."

Since then, tweeting has gone into overdrive. Today, the laboratory reaches out to twenty-one million followers across all social media platforms. The Curiosity landing was live-streamed in the middle of the night when traditional news media were off the air. "It was just a great time for stories about Mars," McGregor says, "as well as our other missions."

Regular open houses at the lab and other NASA centers are routinely packed thanks to the greater exposure from social media, particularly after *The Martian* hit movie theaters in 2015. Just before then, a Pluto flyby captivated the imagination of the public, marking exactly fifty years since the first successful Mars mission, *Mariner 4*. Much of this interest has also been extended into the real world by holding regular meetups, first for Twitter followers, then social media followers in general. McGregor said the friendships that formed at these offline meetups forge connections that nobody anticipated. "They help each other out in their lives in pretty difficult situations. It's great. It's just amazing. So I think we're seeing the value of building that community and asking them to help us reach even more people."

10

SIGNATURES

Adam Steltzner is gung-ho, and it's only eight o'clock on a golden Californian morning. Smiling as he makes his way back to his desk, coffee dispenser in hand, he makes a minor mid-course correction. "I'm closing the door just so I don't bother everybody with my yakking." Open-shirted, relaxed, the chief engineer for NASA's Perseverance mission talks excitedly with his unique machine-gun delivery. Here in JPL Building 321, he is celebrated less as a fixer of problems than as a motivational speaker for the hundreds of engineers who work for him in the organization. Motivation, he says, is one of the fundamental pieces keeping them inspired to do great work.

"Everybody needs to somehow be personally invested and connected to our joint effort," Steltzner says. Given that he is now in his mid-fifties, most of his staff are younger. In the eight years since he became well-known, there has been another generational shift at the Jet Propulsion Laboratory. Many of his greater team who have come onboard since the Curiosity landing were in high school at the time of that achievement. No doubt a number of his younger charges remind him of his former self.

"There's a type of lazy, very bright person who really needs to be loaded up a lot with challenge," Steltzner says. Some clearly have the same independent streak as the boss. It is, he says, just a matter of working out how to harness such individuals to work together with one "singular, tough goal" foremost in all their minds. That is, "making it to Mars," a mantra they have all made their own.

In the years ahead, the JPL engineers will have to do it not just once but many times. Perseverance represents the firing of the starter gun to return samples to Earth, which will require at least two further missions in the coming decade. "What this means for each individual is very personal," Steltzner says. "I can't actually even unpuzzle it. But I can help search a little bit." That also means not corralling his team too tightly, or else they wouldn't get the job done in their own unique way. "I'm secretly psychoanalyzing," he adds, leaning over the desk. "Or maybe my colleagues would say not so secretly psychoanalyzing! I try and understand what makes them tick. What would really fire them up?"

The best engineers thrive in such a high-risk environment. The trick, Steltzner says, is to understand what will get his colleagues going. It could be responsibility, excessive workload, or simply playing their part in JPL's rich heritage of flight projects, making the impossible come true. "It's really individual per person, case-by-case basis," Steltzner says. "You're looking to make an environment where they feel safe in bringing all of themselves to the problem."

Thinking outside the box remains high on the list. That has certainly been the case as they have worked out how to build another machine to look for the signatures of biology that may still be present on the Martian surface. For the first time in forty years, a NASA mission will consider the question of life directly.

"It's been daunting," Steltzner admits. "It will [need] three missions to get the job done." Even the familiar aspects—the entry, descent, and landing, otherwise known as "the seven minutes of terror," from which he graduated cum laude—remain "a life and death challenge." There is a thin line between success and failure.

For all its outward similarities, Perseverance is not just a rerun of Curiosity. It will also land on six wheels, use nuclear power, and employ the highest resolution camera, with the widest field of view, ever sent to the surface of another world. At a cost of another $2.5 billion, it will employ the same chassis, power distribution, and communications system as Curiosity. The money saved in development has been lavished on a suite of seven instruments that are either new or improved designs of the ones that have been scrutinizing Gale Crater for the last eight years.

What Steltzner terms "one of the beauties of Perseverance" is the acknowledgment that there is a frustration inherent in their planning. The engineers have to estimate what instruments any future missions should carry before new discoveries are made. "We're guessing," he says. Perseverance will fly with the best instruments to get the job done. But before too long, another discovery will come along that demands yet another advance. "It is a guess at the future," he says. "It also represents the hard work that we did in planning." But they need to be flexible, or they miss an opportunity "to be competitive, to be innovative."

* * *

Somewhere near the top of Adam Steltzner's list of iconoclasts is an out-spoken flight systems engineer who has helped scope out Perseverance's suspension. Rich Rieber's conversation is laced with self-mockery and knowing irony, yet there is no doubt how bright and focused he is. His arrival at JPL came through "not paying attention in class." Frustrated and bored during a lecture, he was chatting on his laptop with a friend who had been recruited by the laboratory. He complained that he'd never got a chance to interview because he didn't know the JPL recruiters were coming to his California campus until it was too late.

While he was online, his friend quickly put in a word. Suddenly, Rieber's cell rang. The recruiter wanted to see him soonest. So Rieber made an excuse and left the class. He raced home to find a suit and smoothed the fabric and his hair so he didn't look "too disgusting." He talked his way into a JPL position based on his passion for machining.

Machinists are the unsung heroes involved in every space mission. They are the skilled workforce who rarely emerge into the limelight but actually fashion the machinery that eventually flies into space. Each piece of hardware they craft is no less a bespoke jewel than the items for sale on Rodeo Drive. The machinists have to develop their own lathes and microscopic cutters to make them.

Rieber loved it. "I spent more time in the machine shop than the machinists who ran it," he claims. At JPL, it is customary for entry-level engineers to work on different projects until they find their métier. When Rieber was assigned to a flight, he became responsible for communicating the scientific needs to some of the engineers. "It was very exciting," he says. "I was literally handed the keys to a half-a-billion-dollar space-craft at twenty-two, twenty-three years old."

Such a responsibility prepared him well for his subsequent career. "It was great. It was super fun." Since then, he has returned to the test bed where each prototype is tested by systems engineers like himself. "I'm the dumbest guy here," Rieber says. His first "encounter" with the Red Planet was working on the InSight mole that failed its preliminary design review because he misread a temperature threshold. Days of anx-ious phone calls followed.

"So, my other favorite saying is RTFM," Rich Rieber deadpans.

Every mission is different. There is no all-inclusive manual for con-structing machines to go to Mars. For all his knowing self-mockery, Rie-ber has been working on something his predecessors inspired when they added the Morse indents that spelled out *J*, *P*, *L* in the tire tracks on Curiosity. That act of subversion led to what Rieber calls the greatest Interplanetary Easter Egg Hunt. He smiles gamely despite the grief it actually caused. "Our overall philosophy for Perseverance was to not have problems with the tires."

This has meant completely redesigned wheels. Rieber is delighted at what his colleagues have achieved. "They knocked that out of the park," he says. "I don't know if you've seen our new tires? They're my sick rims, I say. Pimp my ride." His only regret is that they didn't put spinners on them. "I'm a little upset about that. You can't get everything."

*   *   *

Those selfsame sick rims will play a crucial role not just in collecting the samples taken by Perseverance but also in ferrying biologically interesting material back to Earth. Its engineers need to guarantee the rover's mobil-ity so it can take them to the "fetch rover," which may not arrive for a good few years. That all-important goal, toward which NASA and other space agencies are now working, cannot fail because of faulty tires.

Perseverance is 12 percent heavier than Curiosity. That means greater traction is required to keep it moving. The extra mass also causes addi-tional problems for the overall wheel design. As a test engineer, Rieber says his role is simple: "I'm paid to break stuff." That means if it fails under Earth's gravity, it will do the same on Mars. So he and his colleagues got rid of the "chevron" tread they had on Curiosity.

"If you think about it," Rieber says, "you've got this one stiff part of the tire. A hundred eighty degrees opposed to that is where you're going to get the most flexion." And at that point, the wheel wear will always start. "So now we have tires that are durable. We don't have to worry about pointy rocks." The replacement tires are more flexible. They can support the higher loads, conform to the terrain, and offer a better grip. The wheel skin is thicker. The inner diameter is fashioned from an epoxy paint that is not brittle, so it won't crack, and the whole pattern is stiffer than the rest of the tire.

At this point, Rieber gives vent to a wondrous stream of consciousness about "Fletcher interface flanges" and "circumferentially asymmetric" wheels, before he suddenly stops. He points to a photograph of the outer tread. "Nerd alert here," he says with a grin. "That was where it was most likely to fail."

They have also made Perseverance more intuitive to operate. Its onboard computer is more powerful and runs new, improved algorithms to guide its movement. They are using a simpler operator interface back on Earth—"like a video game"—for the duty drivers. "We are doing image acquisition and the processing while we're driving," he says. "So you can walk and chew gum at the same time."

With this new driver-friendly graphical environment, a drive that used to take two hours to plan can be done far more quickly. "Some dude did a drive plan in two minutes," Rieber says. The onboard camera system can track smaller features, and they will attempt to travel four times faster. "Curiosity drives one meter, it stops for four minutes, images, thinks, drives another meter. So that's about twenty meters per hour. We do that at about a hundred."

For him personally, Rieber is resigned to ridicule for his insistence that the Morse code from the tire treads be removed. Like his Caltech colleague who demoted Pluto from its planetary status, Rieber is amused that he might well have made himself unpopular. "My name was on the engineering change request to ask to eliminate that whole [Morse] pattern." And then Rieber flashed a final affable grin at the younger of the authors during his bravura performance in a JPL boardroom. "You can say you met me," he deadpanned, "and I'm a jerk."

* * *

Perseverance is being aimed at a bullseye some thirty miles across. Jezero Crater holds the key to understanding exactly what happened in some of the oldest epochs on the Red Planet. Yet the engineering challenges are formidable. Jezero lies on the western edge of Isidis Planitia, a basin north of the equator gouged out by an ancient impact. Already a grave-yard for two European-built landers, the region was once thought out of reach. Now, thanks in part to what is called Terrain Relative Navigation, which has reduced the landing "footprint" by half, it is entirely possible.

The aiming is crucial. Project officials don't want the follow-up vehi-cle—which will return the samples Perseverance will collect—to waste time in fetching them. In this regard, the rover also benefits directly from the "seven minutes of terror" last time around. It won't be a shot in the dark. Vital engineering data was collected during Curiosity's descent with the aim of helping any follow-ons arrive safely.

Real-time telemetry, transmitted as Curiosity descended, measured atmospheric density and high-altitude winds in detail. Perseverance will once again use a combination of parachutes, a landing platform, and a big old cable to come to rest on the surface. This time around, how-ever, it will attempt a precision landing. What the engineers have termed "Range Trigger" will effectively bring the vehicle in closer to its desired aim point. "It could shave off as much as a year from the rover's commute to its prime work site," the mission website observes.

The "trigger" refers to the exact moment when the parachute is popped open. In the past, this was always performed as early as possible; now its release will be dictated by how close the descending vehicle is to the desired landing site. If Perseverance is overshooting, the parachute will be deployed earlier; later, if falling short. To determine the right moment, its onboard "brain" compares the rapidly approaching surface to an onboard map assembled from high-resolution images from orbit.

As it descends, Perseverance's real-time position relative to the surface will be accounted for constantly. Once it touches down, there will be sound as well as vision: a microphone will pick up any audio from the surface. And the new, improved camera system will, to paraphrase David Bowie, have the clearest view. The panoramic camera atop the rover's mast will have a new zoom function far more powerful than Curiosity's.

Indeed, the *Z* of Mastcam-Z stands for "zoom," meaning it can see greater detail and detect more interesting rocks from afar. It will provide the geologists with context, the all-important geological perspective, to let the rover operators work out where they should look for the most interesting rocks. To collect a scientifically worthy set of samples, the geologic context has to be understood. That translates into locating a habitable environment where life might have taken root. Jezero Crater should provide a hole-in-one so far as life is concerned, in yet another geological wonderland that promises many treasures in the years ahead.

* * *

If the engineers are excited by Perseverance, so too are the scientists. Jezero Crater is unique "because it has a beautifully preserved delta," in the words of a mission geologist who helped choose the landing site. Dr. Briony Horgan, an assistant professor of planetary science at Neil Armstrong's alma mater, Purdue University, is excited by what the crater will tell her about the very ancient past and a number of the enduring mysteries of the Red Planet. "How did Mars go from this relatively inert planet with flowing rivers and lakes to the world we see today?" Horgan asks. "We also want to understand ancient rivers and how their water flowed across the surface."

In Jezero, the delta would have deposited sediments "and at some point overflowed to the other side of the crater." Though Perseverance will land on the crater floor, the hope is that it will eventually pass through the breached wall to examine exactly what happened when the water spilled over. As the rover makes its way through the delta, it will build up a picture over many years, layer by layer, of what shaped the surface over time. The story of what geological processes have been wrought over many billions of years will unfold.

Compared to Gale Crater where Curiosity landed, "Jezero is a lot older," notes Briony Horgan. "It's getting closer to 4 billion years, a far earlier period of Martian history."

Outside the crater rim, what geologists call "ejecta"—"huge blocks of the ancient crust" thrown out when the crater was excavated—will likely be the oldest material ever targeted by a Mars mission. "We haven't actually seen these super-old rocks from Mars up close," notes JPL geologist Abby Fraeman. "And so, they're going to tell us something totally new about what the environment of Mars was like around that time." As this was when conditions were most conducive for life to have emerged, she adds, "it'll be interesting in terms of understanding possible origins of life."

Sampling the rocks thrown out by the Jezero impact will shine a light on the subsequent decline of the Martian magnetic field and its atmosphere. The deteriorating conditions, "in some way or other, will be tracked by these really early rocks," says Briony Horgan. In this case, the older, the wetter. Where everything has been drying out in Gale Crater, conditions would have been much wetter for longer in Jezero.

Extensive minerals seen from orbit may be the most exciting of all. As the water collected from the delta, any available organic matter would have been deposited there. If life had taken hold, the remaining sedimentary layers should be rife with signatures of the biological processes that may have once taken place.

* * *

Jezero Crater also presents a curious paradox. Though carbonates are rarely observed from orbit in infrared light, they seem highly concentrated in this particular crater. "This region of Mars is unique because of this big, widespread area of carbonate-bearing rocks covering the surface," says Horgan. "We really don't see that anywhere else on Mars."

This, above all else, commended Jezero as a target. The ubiquity of carbonates suggests there was rapid deposition and precipitation of minerals, which were trapped in the water that once flowed through the delta. If there were microbes, they would have collected where the water evaporated or else at the bottom of the lake, where they had been deposited. Though the microbes would now be long gone, their signatures might have been chemically entombed. "We can actually see the delta from orbit," Horgan adds. "You can also see evidence for shorelines."

Dr. Horgan was lead author of a scientific paper published in November 2019 that showed something akin to a "bathtub ring" all around the crater. What she termed more formally as the "marginal carbonate-bearing region" will have preserved the carbonates, "and quite a lot is associated with these ancient shorelines." For biology, this could be a watershed moment.

Such material will preserve the changes in the local environment as the water was altered due to the changing atmospheric conditions. The shoreline of the lake will faithfully reflect what happened, as it is "a little higher than the breach where water flowed out," in Horgan's estimation.

What she and her colleagues aim to do in Jezero is unravel the exact details of how the minerals were deposited. The excitement comes from the fact that on Earth, carbonates act as "great fossilization agents." They are involved in laying down coral reefs, seashells, and—as would have been more likely on Mars—some form of stromatolites, ancient fossils embedded within regular layers of sediments at the bottom of these lakes. The Perseverance scientists hope to refine estimates of how much water was involved in the deposition process in Jezero Crater.

If life ever started on the Red Planet, then it could have survived and migrated deep underground. So, as an indicator for life, mineralization is a good bet. Ultimately, the greater question is, why is there so much carbonate deposited in Jezero Crater? Briony Horgan laughs as she replies, "That's the million-dollar question about Mars."

* * *

The way Perseverance will look for telltale signs of mineralization is very different from Curiosity. Foremost is the fact that the Sample Analysis at Mars (SAM) chemical oven is not being flown. That means a change from what might be termed the "bulk sample" approach to a greater emphasis in making observations down to the microscopic scale. "The SAM approach is super sensitive," says Professor Ken Farley of Caltech, project scientist for the Perseverance mission. "It is an exquisitely sensitive instrument, but it loses all spatial information." So, although Curiosity can detect minute traces of materials, its instruments cannot specify exactly where they came from within a microscopic sample. That is what Perseverance will fill in.

In total, the new rover will carry twenty-three cameras, six more than Curiosity, but most of them will be used for navigation and hazard avoidance. The rest will be used to create ever more detailed maps. Areas of interest will be observed from the far distance down to the microscopic level with successive cameras as the rover gets closer and closer. All the cameras will observe extra wavelengths to get a better spread of spectral information on the chemistry and extent of the minerals suspected in Jezero. Mastcam-Z will allow them to create "beautiful maps" of the very diversity of the rocks from visible light into the infrared. "That can help us tie the orbital data with the images we'll have on the ground," Dr. Horgan says.

Already, the geologists are planning "first order" traverses for the rover, depending on where Perseverance comes to rest. Even before landing, they have created a patchwork quilt of squares—roughly half a mile on a side—to identify the best areas to sample. Once they have observed the most interesting rocks, the various cameras, particularly a close-up camera, will be able to see "the small-scale context of what's going on," in Horgan's phrase.

An improved robotic arm will use ultraviolet and X-ray spectrometers to tease out new details. A weather station will return real-time measurements of temperature, wind, humidity, and dust. Perhaps a good old dust storm will kick up to reveal why local storms sometimes turn global, still an enduring mystery about today's Martian climate. Finally, a ground-penetrating radar will unpeel the layers of the surface just ahead of where the rover moves. The aim is to identify the best places to sample where the water once flowed, by examining the delta and "looking for biosignatures in this ancient water environment."

* * *

Biosignatures are the characteristic products of biological activity as opposed to those from obscure chemical processes. Hunting them down is Perseverance's express purpose. "They could be concentrated in forms that could be microbial in origin," Briony Horgan explains. "They would be apparent in surface textures that preserve them."

Some biosignatures might be more clear-cut than others. The discovery of a DNA molecule would be an obvious signature, whereas the existence of a free amino acid—which may make up some of the constituent strands of such genetic material—is not. Even amino acids can be synthesized chemically under the right conditions and have nothing to do with biology.

The problem from the outset is that biosignatures of the simple, cellular variety are not easy to find, certainly not with the naked eye. If they were, then they would likely have already made themselves known at the nine sites on Mars where landers have come to rest. And yet, it is only in the last few years that organics and perchlorates have been found. The Red Planet, as several people interviewed for this book have said, has lots of surprises still in store even today.

Hence the need to zoom in even closer. A much clearer biomarker would be a cell wall, because cells are the very basic biochemical engine upon which life is based. For this, certain chains of "fatty" hydrocarbons known as lipids would commend themselves, as they are likely to be long-lived (certainly, this is the case for lipids that have been detected within the terrestrial record of rocks). "Some biosignatures are more robust," notes astrobiologist Charles Cockell at Edinburgh University, "and would be more indicative of biology than other types of organic material."

* * *

If the Martian methane that has been headline news in recent years could be shown to have been produced biologically, then it, too, could be considered a biosignature. The only problem is that the trace amounts observed are so minuscule that it would be difficult to make such a judgment call on Mars. Narrowing down what one researcher calls "the crazy, insane source" in biology will be very difficult indeed.

The most likely culprit for biologically produced methane would be methanogens, microbes that release methane as waste. On Earth, they often thrive where oxygen is lacking. Even if the Martian methane is released overnight, as seems to happen in Gale Crater, it could still be billions of years old. The exhalations seen by Curiosity could have come from where water is scraping against ancient sediments somewhere deep below the surface.

"It's exciting regardless of the source," says Dr. Claire Cousins, a planetary scientist at the University of St. Andrews in Scotland. "It shows Mars is not geologically dead, a new finding in itself."

So far as biology is concerned, an unequivocal biosignature would come from accurately measuring what is known as fractionation within any Martian methane—that is, the relative abundances of the two most common isotopes of carbon, known as C12 and C13, which are attached to the four hydrogen molecules that make up methane (CH4). Life on Earth tends to use the lighter form and less of the heavier one. In this case, there being a greater fraction of C12 used by microbes compared to C13 would be a biosignature.

While Martian microbes might use the isotopes differently, if the numbers were significantly similar to the ratios seen on Earth, it might indicate the presence of some sort of biochemical activity. However, most atmospheric scientists believe this will be a very tough measurement to make. "We're not going to get the isotopes from methane on Mars because the abundances are too small," says John Bridges of Leicester University, who is on the science team for Curiosity. In time, improved instruments may be able to observe the methane with the detail needed, but the current rover is hampered in what it can do.

Certainly, measuring these sorts of carbon isotopes on Earth is useful and revealing. The extent to which the known ratios of the various carbon isotopes change enables a clear picture of the physical processes that have acted upon them. On our home planet, they can be measured accurately and in context, where the exact ratios can pin down the differences between biological and nonbiological sources, too. (Hydrothermal activity might also be another source for methane. Again, it can be produced both biologically and nonbiologically in such circumstances.)

"On Mars, we haven't got that same level of understanding," says Dr. Cousins. "So it's not just a case that you can measure one isotopic ratio"—such as the all-important C12 and C13—"and say, 'Ah, this must

be life,' because of a single measurement. It's never going to be as simple
as that until we get that same level of context on Mars that we have on
Earth."

Another pointer might come from examining the fractionation of
carbon in the largely CO_2 atmosphere, but that is still tricky to do even
though there is more of the gas available. Over time, the spectrometers
aboard the Trace Gas Orbiter may get a better look at the methane fractionation as they record ever more significant amounts of information.
Ultimately, the answer may only ever come from a sample of Martian air
returned to Earth and, even then, only if Perseverance is lucky enough to
catch any fleeting release of methane. As yet, none have been detected in
and around Jezero Crater.

Isotopic fractionation would also be a good marker for the activity of
cells in a rock, and not just of carbon, together with organics and micro-
fossils. To make these measurements will require examining the rocks and
surface soils along various different yet complementary lines. And that
is precisely what one instrument aboard Perseverance will do. In looking
for potential biosignatures, its principal investigator says, "it's going to
be very interesting because we're trying to find life as we don't know it."

* * *

While Adam Steltzner and the JPL engineers have been preparing for
the "seven minutes of terror," the principal investigator for its most significant life-detecting instrument says that by the time it starts returning
information, they will have already endured *seven years*. That is how long
they have been working on the instrument known as SHERLOC. For
Dr. Luther Beegle, this has given him ample time to develop a coping
mechanism. "Because of the stress, I get up at 4:00 a.m.," he says. "I
decided that it would be a good idea to go running."

The road to the surface of Mars has been an endurance test. Immediately after the announcement of the Allan Hills meteorite results in
1996, JPL started looking at life detection on Mars for the first time since
Viking. In 2013, the instrument that will now fly on Perseverance crystallized as a concept to which Sherlock Holmes has lent his name. The
fictional detective's words are guiding exactly how they will search for

past signs of life: "Once you eliminate the impossible, whatever remains, no matter how improbable, must be the truth."

The acronym SHERLOC stands for Scanning Habitable Environments with Raman & Luminescence for Organics and Chemicals. It is designed to ferret out evidence for organic and other molecules that may or may not be related to ancient life. These particular biosignatures include organics associated with minerals that have formed in water, as well as layers of organic minerals themselves. "We're going to go out and look into a rock," Beegle says. "We're basically going to abrade into a rock, and we're going to analyze what we find."

The SHERLOC instrument sits on the "turret" at the end of the rover's robotic arm. It is boresighted along with an X-ray spectrometer and another hand-lens, high-resolution camera. Each of these instruments scans the same tiny area in different ways. Working together, the texture and chemical imprints of material that may have been left behind by ancient life will be observed in exquisite detail. Each provides supporting information to the other instruments.

SHERLOC works by firing a microscopic dot of ultraviolet laser light at a minute section of a selected rock. The laser beam scans across the section of rock by use of a mirror. As it does so, the instrument makes a microscopic map of minerals down to fifty microns, half the thickness of a piece of paper. As the laser tracks along, the strength of the reflected signal depends on the individual grains and composition of the surface. At this microscopic level, even the tiniest variations in the texture of the rock will change how the light is bounced off.

SHERLOC has been designed to work best with complex rings of carbon atoms, which often form the biochemical backbone of interesting organic molecules. It's not just the rings: it will look for the long chains of hydrocarbons known as aliphatic, typically seen in membranes of cells on Earth.

How the instrument works represents two investigations for the price of one. After the laser light hits the surface, "we are looking for two different effects," says Luther Beegle. Some molecules glow or, technically speaking, fluoresce. "The fluorescence will come from organic materials that have ring structures associated with them," says Beegle. In particular, what are known as polyaromatics—including three amino acids—should make themselves known. All are important in life processes on Earth.

The laser has been chosen with this kind of fluorescence in mind. The laser is tuned to a precise value of 248.6 nanometers and what is known as an "excitement wavelength," that is, the particular wavelength of ultraviolet light (not visible to the naked eye) that is specifically characteristic of the organics they are looking for. "Anything with a ring is going to fluoresce really well," says Beegle. "So we'll see spectra from those fluorescences if there's organic material there."

Given how fine the target area is, slight misalignments or minuscule changes in temperature might alter the sensitive readings. To make sure they know what they are seeing is real, JPL has come up with a very bold plan: the rover is taking a well-characterized sample of Mars back to where it came from to use as a calibration target for the SHERLOC laser. In this case, the meteorite known as Sayh al Uhaymir, which crashed into the Oman desert in 1999, is the first sample of Mars ever returned to the Red Planet from Earth. The readings streaming back from Jezero will be compared to "controls" of what remains back on Earth. That way they will know their readings on Mars are essentially correct.

* * *

In the life detection business, everybody loves Raman.

The SHERLOC laser also uses a technique developed by one of the more remarkable pioneers in physics, the Indian Nobel Prize winner Chandrashekhara Venkata Raman. Professor Raman came up with a powerful technique to differentiate between chemical structures at an atomic level. Raman spectroscopy, as it is known, can detect individual atoms and measure their different combinations. SHERLOC will use the first Raman spectrometer to land on Mars, designed specifically to look for other forms of organic material.

Raman scattering is the more formal, technical name given to how ultraviolet light is subtly altered as it is bounced off the surface of a target sample. It will indicate the presence of minerals, particularly those that have evaporated from salty water and any left-behind organic compounds. "We'll look for things like any of the organics that we wouldn't necessarily see with the fluorescence," explains Luther Beegle, "other amino acids, some lipids, and things like that."

This means searching out a smorgasbord of interesting compounds and minerals, whose combinations of constituent elements can be identified. "We'll do things like sulfates, phosphates, and different types of mineralogy from the Raman spectra," Beegle says, "and we'll get them at each individual point we're looking at." Until the laser is actually shined at the sample, nobody knows where the strongest signal will be returned. So the Raman spectrometer needs to be as flexible as possible to search out the strongest reflected signal on the scale of microns. Autofocus— the savior of so many wedding photograph albums—is the name of the game.

To aid in its pointing and grabbing of these reflected signals, SHER-LOC has a scientific sidekick named Watson, whose somewhat contrived acronym stands for Wide Angle Topographic Sensor for Operations and eNgineering. It is essentially an improved version of the Malin hand camera flown on Curiosity. Once again, Watson is not just there for pretty pictures. It is essentially a zoom lens, capable of magnification down to thirty microns as it examines the surface at a different "spatial scale" (down to three microns per pixel). It will be able to identify targets within the laser sampling and where they came from. "When we come across an outcrop," Beegle says, "we'll use Watson to take images to look for things like layering, grain sizes, and other patterns."

The robotic arm also has a coring instrument at its end. This will be used to abrade any rock with a tool that breaks down the surface material. SHERLOC can see only the outermost layers of the surface—the upper two hundred to three hundred microns of material—and they may want to penetrate farther. "We want to get under the weathering layer a little bit, and we want to get underneath the dust," says Luther Beegle. "And then we will use what's called the contextual imager"—another small camera, which has a resolution of ten microns per pixel. "We'll be able to take a picture of what we've just abraded or even a natural surface if we're going to go down that path, depending on how things work."

SHERLOC then uses the laser to scan across the surface that has been observed by the contextual imager. Across a spot one hundred microns across, this microscopic camera will show exactly how the organics, minerals, and chemistry are distributed. "Those maps will determine where that rock came from," Beegle says. Using the X-ray instrument and the various cameras, some of which can see into the near infrared, they should get a full chemical analysis on the same material. "That's something that's really powerful, because these different techniques have different benefits to them," Beegle concludes. "It'll help us identify rocks that have potential biosignatures. That's really what we're looking for so that we can catch them."

* * *

As it moves through Jezero Crater, the Perseverance rover will take samples not just of rocks but of what one scientist involved calls "the sand and air." All will be carefully put aside in protected "caches," which, in Adam Steltzner's words, involves "carefully, individually sealing them in hermetically sealed 'super-clean' vessels." Ahead of the launch, the engineers have had to be more serious about cleanliness than anybody has ever been before. In fact, Steltzner terms these vessels the cleanest pieces of hardware humans have ever constructed.

Perseverance has to wait for what he calls "the return ticket," a "fetch rover" to gather up all these "cached samples," which will then be fired off the Martian surface. During that time, the scientists will be able to debate the merits of each hermetically sealed sample. Ultimately, what will result is an interplanetary lottery. "We are taking more sample tubes than we can bring home," Steltzner explains. "That was absolutely my intent, specifically for the following reason. The science debates should be rich, complex, deliberative, and take as much time as they need to."

It is only when those samples are about to picked off the surface of Mars that the debate about their relative merits needs to have come to a conclusion. "So we have space for twenty or thirty samples," Steltzner adds. "What we don't want to do is have to debate the value of a sample before taking it."

It is a case of grabbing first, as quickly as possible. Back on Earth, the scientists can then take as much time as needed to figure out what is actually in the tubes. "That's why we over sample," Steltzner explains. "Then that's why you allow that debate to go all the way up until it's time for them to come home, which will be several years."

There are limits to how and what can be sampled in Jezero Crater. A suite of samples that can help answer remaining riddles will be needed. "We have so many fundamental questions about Mars, the evolution of the solar system, and life," says Dr. Briony Horgan. "We can't just do that with one sample."

In this regard, Jezero will provide an embarrassment of riches. Ahead of the Perseverance landing, the scientists have a notional idea of what they would like to sample. As they debate where to take samples from, the question remains of how much time to devote to any particular spot.

NASA has already said that the highest priority samples are the ones that are astrobiological in nature, meaning something in which potential biosignatures may be found. Areas of interest include the mudstones that accumulated at the bottom of the delta. "That's where you might expect organics to be concentrated," says Dr. Horgan. "We would like multiple samples of the carbonates, especially if we see evidence that they were deposited in the lake."

Sediments generally, across the delta, may have different sorts of textures within them. According to Dr. Horgan, they will attempt to grab the most interesting. "That will tell us a lot about the water flow, how long it did, what the chemistry of the water was," she says. Of particular interest will be "the watershed": when the water started to flow and what resulted in the lake. "If there were any microbes living in the lake," she says, "the organics should be concentrated there."

One abiding mystery is the actual age of the material in the crater. "We don't have rigorous dates for everything on Mars," Horgan says. In particular, she wants to know the exact age of the lava that spilled across much of Jezero Crater and "capped" the floor. "We have some idea, but we don't really know how old these lava flows are," Horgan says. "We need to characterize them, and so we're going to grab some samples."

At this stage, the geologists hope to exit the crater to take samples from outside the rim. NASA hasn't decided yet whether they can, but that is desirable for the geologists to round off their exploration. They want to make a beeline for the breach, as one possible landing site for the sample return mission is outside the crater. "We'd love to get bedrock samples along the other side of the rim," Horgan says.

The rover will not last indefinitely. Nor will the return mission have the luxury of waiting on the surface. "We're trying to make it a decade from going in 2020 to return," Adam Steltzner says. Their aim is to "close the loop and bring the samples back to Earth as soon as we can."

For now, keeping the caches on the surface requires extremely pristine sampling techniques. Steltzner says, concluding as only he could conceivably put it: "We're just going to engineer this shit."

* * *

Organics are not organisms. There is a world of difference between the biochemical precursors for where life may once have existed and the actual presence of life itself. Biology is not a surefire guarantee when you stumble across organics, though they show that the ingredients are present. And the converse is true: it is very difficult to explain life without them.

Luther Beegle is at pains to point out that SHERLOC is not capable of detecting life directly. "We do not believe a single instrument could find life on Mars," he has said. "There is so much uncertainty as to what life on Mars might look like that it's hard to imagine a single instrument unambiguously identifying it."

Under present-day conditions on the Red Planet, life "as we know it" would be very scarce indeed—certainly below the limits of detection of any single instrument. But because of its greater spatial resolution, meaning it can see so much more at a very fine scale, SHERLOC should be able to discriminate between indigenous material and organics that have been delivered from space over geological time.

Meteorites have been landing on the Red Planet for billions of years. They have performed the interplanetary equivalent of gardening the topsoil. Some of the organics close to the Martian surface may result from meteoritic impact. SHERLOC will provide a clue as to where they might have come from.

The synthesis of biologically significant compounds is not just restricted to planetary surfaces or special conditions. Prebiotic chemistry might well provide the mechanism by which the precursors of life were delivered to both Mars and Earth, too. Indeed, if life is discovered on the Red Planet, the organics should tell us straightaway if it was related to life on Earth.

When they get up close and personal within a rock, Luther Beegle and his colleagues hope to identify exactly what kind of organics may have been deposited within them. "If the meteoritic infall went into a sandstone or a mudstone," Beegle says, by way of an example, "we would expect a certain pattern associated with that, which is kind of more homogeneous across the entire rock we're looking at." Since a number of people on the SHERLOC team are meteorite specialists, they already have extensive experience of examining Raman results from within meteorites they have examined here in their laboratories. "They understand what those spectra would tend to look like," Beegle says. "So from that experience, that's why we do the mapping to begin with."

When a meteorite impacted, its organics would tend to be spread across the surface they will be examining. "If we saw something that's homogeneous," Beegle says, "we would assume that's kind of meteoritic. If we see something that's really, really spotty, we would think that would be more ingrained into what has come from Mars."

Making absolutely sure will require the taking of many different measurements at the same time. "You just don't look at something in a space by itself," Beegle says. "We'll be able to look at the images, we'll be able to look at what [the X-ray instrument] is seeing, and we'll be able to differentiate everything." All of this information will come from the same sample, "and that's what makes this approach really, really powerful."

* * *

To the average person in the street, many of the results from SHERLOC will no doubt seem obscure. What results in the spectra generated are "lots of peaks," in the wry words of one astrobiologist. Their extent and shape depends on how the constituent molecules of samples are bound together.

Physically, over geological time, organics tend to degrade. They eventually turn into kerogens, a kind of graveyard molecule of undefined, high-molecular-weight organic material (which ultimately ends up as graphite). Luther Beegle terms it "disordered carbon"—as it has essentially been broken down over the ages—that produce in Raman spectrographs what might be termed "twin peaks."

Known as G and D bands, they represent, in Beegle's estimation, "two different ways the carbon or kerogen hangs out." They should be visible in the results from SHERLOC. But the peaks' position and what spectroscopic scientists call their "shoulders"—the actual shape of their outline seen in spectrograms—vary considerably. Although they can be produced by biological entities, they can easily be misidentified. Similar spectral shapes can also be produced nonbiologically. "So if you saw a G and D peak in a sample on Mars," says Professor Charles Cockell, "you wouldn't really know if it was biology."

SHERLOC will, however, produce very distinctive spectral peaks with DNA and lipids, fatty acids that are good indicators for the "machinery" of cellular activity (though they can also break down as a result of nonbiological processes). Luther Beegle and his colleagues can't wait to dig into the various spectra, much in the manner of scientific autograph collectors. "We would look for different concentrations of minerals of organics as well," he says. "If we see a lot of polyaromatic hydrocarbons, and we don't see any aliphatic hydrocarbons, we would find that interesting. But I don't know if we would target those as potential biosignatures. Once again, it depends on what the signature looks like."

In interviews, Beegle has also made it clear that if they do come across something unusual in the soils, they could easily identify it. "However," he has cautioned, "we expect life on Mars, if it is present today, to be extremely rare." Some forms might not be so identifiable, and ultimately, he errs on the side of caution. "It is a real challenge to tell the SHERLOC story without generating sensational headlines about 'life on Mars,'" Beegle has written. As Carl Sagan famously remarked, extraordinary claims require extraordinary evidence.

Ultimately, Perseverance represents exploration at its finest. "A lot of our work depends on what we're going to see," says Beegle. "We're not exactly sure. If we knew what we were going to find, we wouldn't spend all this money and anguish doing it."

If that is not the best encapsulation for what the next American mission will do on Mars, it surely comes close. If exploring the Red Planet has been about obtaining the clearest view—from telescopes to orbiters to landers—this latest work is ushering in the microscopic era with these exquisitely fine-scale "mineralogical maps." "It's all about looking at something [minuscule]," says Luther Beegle, "especially these maps and determining whether or not that what they show could be created abiotically or biotically or what the probabilities of it are. And then we'll go from there."

Certainly, to find definite proof of life on Mars will require many lines of inquiry with different instruments. SHERLOC is the first step on the long journey ahead. "I do think Mars is one of our best bets in trying to understand the early evolution of life and maybe the origin of life," says Dr. Briony Horgan. "We're hoping Perseverance will be the golden key that will get us that information, but it might be a longer process."

For an engineer like Adam Steltzner, who has devoted the last twenty years of his life to thinking about Mars, that will be the ultimate payoff. "I think that there was likely life on Mars at one point," he says. "Whether it still exists or not, I have a hard time knowing. My guess would be yes. Why do I feel that way? Well, we know that when life was just getting into a hold here on Earth, the conditions to support life were back on Mars. It's a very interesting question about how once the conditions are acceptable, what's the probability that life starts? That's a fun, interesting question that's almost unanswerable."

A pause and a smile.

"But we'll do our best to find out."

* * *

Within days of Perseverance lifting off from Cape Canaveral, a smaller, but comparatively powerful, launch vehicle should have followed suit from the steppes of Kazakhstan. But with just weeks to go before launch, the painful decision was made to delay Europe and Russia's own plans to land on Mars. What is known as the ExoMars mission will now have to wait another twenty-six months before it can be launched in the fall of 2022.

It is the latest in a series of delays and misfortunes to befall the project.

Originally, there would have been two missions with a rover each (one from the United States and one from Europe). ExoMars has been scaled back so that there is now just one European rover, which will carry some US instruments. (The same NASA team that built the SAM oven on Curiosity is involved with this European mission.) In the wry estimation of the ExoMars project scientist Jorge Vago, "We've been rolling with the punches."

If the politics were labyrinthine, the vagaries of sending missions to Mars are even more capricious. At the beginning, all the omens were good. ExoMars's first element, the Trace Gas Orbiter, was successfully launched on March 14, 2016, carrying with it a tiny "technology demonstrator" to land on the surface. Named Schiaparelli, it originated with an American notion "so Europe could develop [the] competency" to land on Mars. The orbiter successfully reached the Red Planet seven months later.

Just before it did, on October 19, 2016, Schiaparelli was released. Minutes later, controllers in Germany lost its signal as it descended. A few days later, its crushed remains were spotted on the surface by NASA's Mars Reconnaissance Orbiter. Six months later, the European Space Agency released a report about what had happened. When the parachutes deployed, the descent stage swung like a pendulum. This swamped its sensors so that the onboard computer became disoriented. Consequently, the internal reference reset itself. By then, Schiaparelli thought it had already come to rest on the surface, so it did not fire its retro engines. What resulted was—in one participant's estimation—"a failure to detect a technical failure."

The European Space Agency (ESA) and Roscosmos pressed on. Over the next two years, the next mission came together. A Proton launcher was earmarked to dispatch a Russian landing platform from the Baikonur Cosmodrome to carry a European-built rover. Its aiming point is slightly more northerly and closer to the central meridian than Perseverance. Yet despite their best efforts, a host of technical problems have now delayed liftoff until September or October of 2022.

"This is a very tough decision," said the director general of ESA on March 12, 2020, "but I am sure the right one."

The most serious issue concerns passage through the Martian atmosphere. To avoid the extra complication of the NASA sky crane system, the Russian-built lander is much smaller and weighs about 660 pounds (300 kilograms). It has been designed to come in on an innovative parachute system. It needs two parachutes, each of which pops out with its own drogue (a feeder parachute that starts the slowing down). As the lander, known as *Kazachok*, encounters the outer layers of the Martian atmosphere, it will be traveling at just under four miles (six kilometers) per second. A thermal heat shield will slow its speed to roughly twice the speed of sound.

Then it is double or nothing. The first parachute is designed to slow the lander down to below the speed of sound. After that, the largest parachute ever designed for Mars, with a diameter of 115 feet (35 meters), will take over and slow it down for a safe landing. To do that, it actually needs two parachutes, each with its own drogue, so there are, in fact, four firings of mortars to release them. All passed muster in 2018 when they were tested from a helicopter at roughly half a mile above Earth.

Subsequently, there have been serious problems with high-altitude tests. Dummy ballast, of the same mass as the vehicle that will land on Mars, was dropped from the stratosphere (from roughly seventeen miles or thirty kilometers' height) after release from a large helium balloon on August 5, 2019. The main parachute failed to open, and the whole system crashed into the Arctic tundra near to the test site in Kiruna in northern Sweden.

Tears in the fabric were later spotted in the television images that accompanied the descent. Tearing had also taken place during an earlier test. On that occasion, at the end of May 2019, both parachutes failed. Three months later, the larger parachute failed and the whole package came down on the smaller parachute designed for the supersonic portion of the descent on Mars.

In the fall of 2019, it was all hands on deck. Working with NASA and JPL, further parachute tests were carried out by the Europeans. The deployment of parachutes is often more of an art than a science. Initially, it seemed the problems were with the ejection mechanism, but final high-altitude tests to qualify the parachutes were pushed back to March 2020, which would have been too close for comfort for launch that summer.

A series of other difficulties have compounded matters. A solar panel needed to be replaced due to a fault. Important software had yet to be completely tested, with bugs found in the software overseeing the critical descent. The decision to delay was purely for reasons of safety.

Europe and Russia have decided to err on the side of caution. The rover will be kept in Italy, maintained at strict levels of cleanliness, as are all landers being sent to Mars. This will cost more money, but the agency is not asking for any increase in budget. ESA says it wants to ensure a successful mission, yet it is not clear how well individual components and delicate instruments will survive until the fall of 2022 and whether they will have to be replaced.

* * *

Technical failure remains the unspoken elephant in the clean room. The odds have not exactly been stacked in favor of either Russia or Europe when it comes to Mars. Of the ten missions officially announced by the Soviet Union since the 1960s, only three were partially successful. The rest failed completely, and this tally does not count the many more that were never officially admitted.

Despite these vexations—which, to be fair, are part and parcel of any space program—the ExoMars engineers remain upbeat. One factor that has commended its landing site is its lower elevation, roughly a kilometer (a half-mile) below the mean surface level of the planet. That extra slice of atmosphere, "that kilometer lower will give us more time," says Paul Meacham, the lead systems engineer for the rover at Airbus in the UK, which could mean all the difference in the world.

The stationary surface platform is expected to operate for about a year. The *Kazachok* instruments will take pictures of the landing site, watch the local weather, probe the internal structure of Mars, and take measurements of the atmosphere. They will also look at the distribution of water ice under the surface and measure levels of radiation in and around the landing site.

Kazachok's in situ measurements will be carried out in conjunction with the Trace Gas Orbiter, which will serve as a communications relay for the lander and rover on the Martian surface, as it was supposed to do with the failed Schiaparelli demonstrator. Though ESA's Jorge Vago says,

"It will be difficult to resolve the question of life on Mars with a single mission," ExoMars will dig deeper than any previous lander. The project scientists hope they will find greater concentrations of organic material that have accumulated over time and not been damaged by radiation, the highly oxidizing soils, and wind. The planned mission—if the warranty holds—is seven months (218 sols), and "all that time should be nirvana for all our sample collection."

* * *

The European rover has been named after Rosalind Franklin, a chemist and crystallographer who was overlooked in the credit for having the critical insight into identifying the double helix, the elusive yet crucial component of DNA that passes on genetic instructions. Finally, Franklin will take her rightful place in the pantheon of explorers who have made the search for life on Mars possible.

Though smaller than Perseverance, the Rosalind Franklin rover has any number of innovations of which its builders are inordinately proud. Foremost is the fact that its body has been fashioned from carbon fiber. This means it is lighter and stronger than any previous rovers sent to Mars, helping it to cope withthe freezing cold of the Red Planet. The central body is roughly the same size as Spirit and Opportunity, and it weighs about the same. The rover sits on a lower, wider chassis that carries six sturdy wheels to rove across the surface.

No matter who designs it, the basic principles of moving on Mars remain the same. In that sense, the Airbus engineers have had to take into account "the pointy things," as one onlooker terms the rocks. Rosalind Franklin uses a much more flexible spring steel wheel system to provide greater suspension.

After launch, its six wheels are bolted down during the flight to Mars. En route, the health of the rover will be checked: software will be updated, communications checks will be carried out, and the all-important state of the batteries will be monitored. When it arrives on the Red Planet, the rover will stand to attention in the sense that its mast, solar panels, and wheels will all be unfurled.

To guide it off the landing platform, there are ramps and rails. This means the Franklin rover won't have to perform the first three-point turn on another world, which would have required extra redundancy in the steering motors. "We don't have to do anything too complicated," Paul Meacham says. "Essentially, you are able to drive backward or forward in a straight line."

Controllers will see which direction provides the best exit point. After being guided down by the rails, Rosalind Franklin will roll off the ramp. Its cameras will inspect the general layout of the terrain and carry out "a bit of engineering analysis to see how things work."

Traction is similar to Spirit and Opportunity. Rosalind Franklin has six wheels arranged in pairs but with each one capable of independent motion. Each wheel pairing is mounted on a pivoted bogie, which allows for better ability in steering. Because it is smaller, the rover can also do a smaller point turn when it is on the surface. If an obstacle presented itself, or a crevasse opened up, the driving team doesn't have to worry about a turning circle (as you would in reversing a beach buggy).

One important innovation is the equivalent of a shoulder joint. It allows the rover to effectively stand up, using this clever joint, to allow all six wheels in turn to function as legs. The first pair move forward, then the back two stand upright. Instead of rotating, the wheels stay put on the surface. The shoulder joint allows the rover to "walk" across the surface, albeit very, very slowly.

In simple terms, it gives the drivers a leg to stand on. If, as happened with Spirit, it gets stuck in sand, the Franklin rover can use these "legs" to get out. The shoulder articulation allows for what is called "wheel walking." It alters the rover's height, changes the traction, and keeps it out of trouble, even though to some, it looks comical and slightly sinister, like a slow-motion version of *Dawn of the Dead*, as it drags itself across the surface.

* * *

In the spring of 2023, the little Cossack that can—the literal meaning of *Kazachok*—will come down in a low-lying plain close to the Martian equator. Oxia Planum represents another of the older surfaces on the

Red Planet. It stretches from the edge of the southern highlands of Mars where they open out onto the northern lowlands.

According to Jorge Vago, the surface is characterized by complicated geology and "a relatively featureless plain" that is dominated by certain clays. "We see time and time again that the oldest deposits on Mars appear to be high in magnesium silicates, formed under somewhat warm conditions," Vago explains. "Both are consistent with the planet being young, and there was a lot of hydrothermal activity with circulation of hot fluids."

Mapping from orbit shows a number of interesting features in these plains. They contain certain minerals known as philo-silicates, whose chemistry suggests they were formed by the action of water that is chemically neutral (neither alkaline nor acidic). It seems likely that this whole area was covered in standing water, fed by flows that were rich in clays and minerals and that fanned out across the surface in the ancient past. As with Jezero Crater, suspected riches may be found in the form of biosignatures.

While Curiosity has come across clays only recently, in the winter of 2018–2019, Oxia Planum is expected to be replete with them. The plain is very large—many thousands of square kilometers (and miles)—covered by a finely layered sequence of clays. Unraveling the order of how they were laid down will reveal when there was sedimentary deposition and wetting. The key point is that the clays formed in situ and were not brought in from elsewhere. They hint at an intriguing past for the possibilities of Martian life.

The clays in Oxia Planum suggest two possible scenarios. They could have formed due to the long-suspected ocean that covered whole swaths of the northern hemisphere, or what Jorge Vago calls "a humongous lake," into which volcanic ash fell after the greatest period of geological activity in Martian history. Another possibility is the water bubbled up from below to alter the basaltic minerals to form lake beds.

The relative smoothness of these largely featureless plains commended it as a landing site. A number of geologists believe the whole area was under water for a significant amount of time. "If we can show that Oxia was at the bottom of a sea," Vago explains, "then that will be proof there was once an ocean on Mars."

While the jury is still out—"Some days I'm convinced by the ocean hypothesis," says one study participant, "other days, I think it is really silly"—there are at least five suspected shorelines that run through Oxia Planum. Those ocean edges drawn up from the original *Viking* data do not seem to be there in the more recent, higher resolution images. Others believe past shorelines are still visible.

All throughout Oxia Planum, there are more recent outflows of lava; from crater age estimates, they are just 2.5 billion years old. Certainly, the mission planners hope to avoid these areas, where there may be mesas and cliff faces "which would be very difficult to get off." Though they are seen at the outermost edges of where *Kazachok* is being targeted, ESA officials want to avoid them.

Old means gold, so far as the search for life is concerned. "We are really going for these old surfaces which contain clays," Jorge Vago says, "as the goal of the mission is to look for distinct traces of life. We are chemistry driven and we want to look for biosignatures."

* * *

Oxia Planum should certainly *look* very different. Detailed mapping from orbit, down to the scale of ten inches (twenty-five centimeters), has revealed that it is fairly unique in terms of inherent coloration. "It shouldn't be as orange as some of the other places we have landed on," Jorge Vago says, "because clays are sort of gray with a slight tinge of blue."

Since they were deposited by ancient flooding, dust will have blown in from elsewhere and accumulated, as other missions have revealed. So any differences will be subtle, with slight variations in color. Similarly, the deposits of clays will not appear dramatic. Close up, they will appear more finely laminated and finely grained, much like the landscape in Yellowknife Bay observed by Curiosity.

"These clays are really good at preserving ancient organic molecules," says Dr. Claire Cousins of St. Andrew's University, who is involved in the ExoMars mission. She will be using the rover's cameras to search out interesting variations in the surface minerals, so that rocks more likely to have harbored biosignatures can be targeted.

The clays in Oxia Planum include extensive sediments, where

prolonged water flows have created a remarkable hunting ground for biosignatures, "a sweet spot" in Jorge Vago's estimation. Conditions will have been preserved as far as possible from 3.8 billion years ago. This was when the Red Planet lost its magnetic field, its atmosphere thinned, and any water on the surface became less stable, either lost to space or frozen in place. There may have been icy coverings and glaciations, but how thick that ice was is still a matter for debate. "It could have been cold at the top, but kept warm—a liquid layer—by hydrothermal activity," Vago says.

The mission will show how clement Mars was in the ancient past. "If it's cold today," Vago explains, "it should have been even colder were it not for the fact the atmosphere was denser and there were more greenhouse gases."

Thereafter, the Red Planet changed significantly. Volcanism was at its peak in the nearby Tharsis region, spewing gases out into the atmosphere, which tended to be acidic. The outgassing changed the nature of both the water on the surface and the chemistry of the minerals that were deposited. Over time, this would have created hydrated sulfates (the sulfur would have come from the volcanic eruptions). The acidity would also have changed the temperature at which chemical reactions would have taken place. "There is less chance for life when you get into this 'sulfate' period, because the water was acidic," notes Dr. Peter Grindrod at the Natural History Museum in London.

For geologists, understanding the context of how this transition occurred is important. A planetary scientist like Claire Cousins would be happy to come across "a delicious mudstone," one that has exactly the right kind of chemistry to host and, more to the point, preserve organic biosignatures. "If the sediments are right, deposited in the bottom of a lake and not the kind of chemistry that destroys organic signatures," she says, then the Franklin rover will have found what they are looking for. But from all earlier experience, what is actually going to be found on Mars is usually very different. "I would say from how things have gone on previous missions," says Jorge Vago, "we're going to be surprised."

* * *

Although the landing sites for the next missions are different, they speak of an ancient past on Mars when life could have evolved. Given that they have different geological histories, though, will it be possible to extrapolate from just these two sites on Mars, Jezero Crater and Oxia Planum, for the whole of the planet?

"The way we've done geology on Mars is almost the other way around to what we've done on Earth," notes Dr. Grindrod, also an ExoMars investigator. "That is, you go into the field, figure out a very small area, then another, and then join them up to get a bigger and bigger picture."

To understand the whole of the Red Planet, geological context is important. "For Mars, we started out with low-resolution global pictures," Grindrod adds. "We have come up with a way that Mars formed and evolved: it was volcanic, then water flowed out, and nothing else has happened since then."

Thanks to at least nine landers and high-resolution images for the whole of the planet, those local studies have been fitted into a global picture of Mars. In that sense, the biological attractiveness of two landing sites adds variety thanks to their slightly different geology. Indeed, many scientists interviewed for this book who are involved in both new missions are confident that, if there is evidence for life on Mars, it will eventually be found.

It is quite a sea change from the doldrums immediately after *Viking*. That earlier pessimism has been superseded by a growing optimism. "We have a better handle on the chemistry and the environments on Mars today," says Claire Cousins. "We also now have a better feeling for what life could have taken hold there and the environment it could have lived in as well."

Both the Perseverance and the European ExoMars missions will come to land in what were water-rich environments in the dim and distant past. Biogenic elements could have combined in just the right proportions to have allowed life to evolve. With water, metabolism within cells would have been possible.

Out of the melting pot of planetary formation, the Red Planet shared the same sorts of elements as Earth did. Crucially, water flowed across the surface of Mars early in its history, a factor above all else that speaks volumes about the possibilities for life. Narrowing down the aqueous chemistry that might have created biology, or at least the precursors of life, is now the focus in the next decade as ever more complex missions will search them out on Mars.

"It's an interesting human study why these things come in waves," NASA's Christopher McKay reflects. "It's very easy to get carried away on the question of life on Mars. We need to sift through the facts of what we know and what we don't know."

11

RETURN TO SENDER

Professor Kenneth Farley is the first human being to ever date a Martian rock. Not in the sense of sharing a meal in an intimate setting, but rather determining remotely the age of a specimen on another world with extraordinary accuracy. It's all part of the day job for this Caltech professor, recently elected to the National Academy of Sciences for his preeminence as a laboratory-based isotope geochemist. Performing such a measurement led to him becoming project scientist for the Perseverance mission.

Ken Farley's fervent hope, as with so many others interviewed for this book, is to make similar measurements with Mars rocks here on Earth. That will banish any lingering errors in the delicate measurements that he and others are capable of carrying out. "On Curiosity, I led the investigation to make the first radiometric age determination on another planet," Farley says. "And I think we were reasonably successful." It showed the rock was 3.85 billion years old, but the error in that measurement was plus or minus 250 million years either side of that. "Reasonably successful," he is at pains to point out, "isn't the same as actually knowing the age with the kind of precision that we need."

Hence the desire to bring samples home.

Using Curiosity's chemical oven, the Sample Analysis at Mars (SAM) instrument, Farley and his colleagues measured abundances of a form of potassium that had decayed into argon. The tracing of this decay of radioactive elements, technically known as "geochronology," was performed in a sample that was taken from a mudstone in Gale Crater in 2013. "It is perhaps among the most demanding of measurements that you can make," Farley says, "and the furthest from being really successful on a space flight."

Typically, this involves observing a very low abundance of the material being examined. "Those measurements are tricky," Farley says. "It is just analytically difficult." Worse, they are often too demanding for the current "capabilities" for robotic missions. "There are a very large number of questions about Mars that we struggle with because we don't have absolute ages."

From his own and others' existing work, geochemists can see the tremendous value of working on actual samples of Mars in their terrestrial laboratories with the kinds of instrumentation that will nail down those numbers. "When samples come back from Mars and we are able to actually get radiometric dates," Farley says, "it will be a huge advance."

For many scientists interviewed for this book, that is the moment that everything goes into overdrive. As far as looking for life on Mars is concerned, some even say there is no point trying to do in situ examinations until we bring samples back. Remote sensing, even with powerful instruments, has its limitations, as even the success of Curiosity shows.

"We are at the stage where we clearly have detected Martian organic matter, but it is unclear whether it is biogenic or not," Farley says. "It's kind of the fundamental question that makes organic molecules super interesting." In other words, how much of a tracer of life are the organics measured on Mars? Even with the latest "wet chemistry" innovations that are improving the SAM instrument's ability to search for organics, the science teams have not been able to make further progress in Gale Crater. "The organic matter has multiple sources that are hard to separate from each other," says Farley. "When we have those samples back on Earth, we will approach the same question using a diversity of instruments."

* * *

That day is coming. No sooner will the Perseverance mission land than preparations will begin in earnest for its follow-up. In 2026, another JPL-built lander will land in Jezero Crater, carrying with it a European "fetch rover." This rover will need to have enough available space to collect all the cached samples and load them into the new lander that has brought it to Mars. Also onboard will be an ascent vehicle, which will return those precious samples back to Earth.

To achieve all this requires new technology and advances in computer control autonomy. Small wonder that the cost for returning samples from Mars has grown steadily over the years. It is probably beyond one country's largesse, even that of the United States. So the next steps will require international collaboration. Despite its own failures in landing on Mars to date, Europe is optimistic. The European Space Agency (ESA) believes it is up to the task.

It will be sending an orbiter to work in tandem with the sample return lander. The problems Europe has had with the Red Planet have not been replicated elsewhere in the solar system. ESA has successfully sent missions to Mercury and Venus as well as a couple of comets. "Had we not delivered those things, we could not say to NASA we had the confidence to pull this off," says Sanjay Vijendran, ESA's lead engineer for Mars Sample Return. "Our orbiter also has to be in place and ready to provide data relay for the US lander mission in 2026."

The "architecture" of how the various elements will fit together is still being defined. Europe and the United States will jointly walk the tightrope that balances technical considerations and budgets. "A big part of what I am doing is identifying just what the individual steps are," acknowledges Brian Muirhead, a veteran JPL engineer in charge of the US contribution to Mars sample return. "We are very early in the development process."

Nevertheless, they have broken down what needs to be done into smaller, manageable elements. So far as possible, they will use tried and tested technology. NASA is having to juggle priorities in the run-up to an election in November 2020, despite its successes in recent years. After Perseverance, it has no further funded missions to the Red Planet. The intention is to proceed with the plans even if the money hasn't been formally allocated. Brian Muirhead says any uncertainties with the Mars sample return effort are "associated with the costs and schedules." Yet, for all the breezy hype, some of the exact technical details still baffle the brightest engineers.

Way before 2026, NASA will need a new Mars orbiter to relay all the communications between its existing and forthcoming missions. Its current orbiters are aging and nearing the end of their lives. Paradoxically, the continuing success of other missions, particularly Curiosity, is eating into the margins needed for communication. Some of the responsibility will be taken up by Europe, which has the political wherewithal—ministerial approval and money guaranteed for the next three years—to move to the detailed definition phase of what it will contribute to returning samples from Mars.

Foremost among them is the Martian minicab, which will fetch the cached samples taken by Perseverance. "The challenge for our 'fetch rover' is to get where the samples have been dropped in the time we have allocated," says Sanjay Vijendran. Then they have to get those samples into the ascent vehicle whose design, Brian Muirhead says, is also a tough technical problem.

Both space agencies are moving "quite aggressively at getting those missions underway in 2026." Originally, it had been thought this fetch mission would arrive later than Perseverance's warranty. But as Curiosity's longevity shows, it is not beyond the realm of possibility that its successor rover will still be working. The planners are considering that Perseverance might now interact with the 2026 mission. The new rover is likely to live up to its new name.

That means the project scientists are looking at a fully loaded sample return mission. "And so, you are now working under the gun that somebody's coming to meet you and you have to be there for the samples," says Ken Farley. Over the next five years, give or take a few launch windows and delays, the engineers will work around the increase in management complexity. "There are a lot of difference in cultures that make a technically very hard job even harder," notes Brian Muirhead somewhat diplomatically.

* * *

In the near future, Mars will be coming under the scrutiny of three other rovers looking for organics and, in time, biosignatures. Curiosity is still making its way up Mount Sharp in the middle of Gale Crater. After its earlier contamination problems, the SAM oven in its "wet chemistry" mode will start looking for organics in new ways. When it is launched in 2022, ExoMars will be using the "son of SAM" aboard the Rosalind Franklin rover to do the same in Oxia Planum. And by then, they will have been joined by a Chinese-built rover, about the same size as Spirit and Opportunity, which promises to be the country's first mission to Mars.

As ever with Chinese space activities, precise details are largely missing. That said, in November 2019, diplomats from nineteen countries were invited to watch a test of the airbag systems that will be used on the Huoxing-1 mission.

China is expanding its presence throughout the solar system, having carried out the first landing on the far side of the Moon in 2019. Now, it has its sights on Mars. "Exploring the Red Planet and deep space will cement China's scientific and technological expertise," said Jia Yang, the deputy chief designer of Huoxing-1. Known more formally as the Mars Global Remote Sensing Orbiter and Small Rover, it will be launched on a Long March booster in July or August of 2020. The lander will separate from the orbiter and come to rest on the surface by use of airbags. The orbiter will spend about a year in the region of Mars. The rover is expected to last for at least a Martian year. It is an impressive development given that the program started only in 2016. The instruments it carries include a series of spectrometers capable of looking at methane and some of its isotopic fractionation. The Chinese mission will also search for biosignatures in a similar way to Perseverance, and in time, ExoMars. The aim, Chinese news agencies have reported, is to return samples in 2030.

* * *

Even with all this unprecedented forensic examination of the Martian surface, the world should not expect miracles. No one instrument will reveal whether life might have existed on the Red Planet. "Biosignatures will always need to be corroborated by different methods," says Charles Cockell, professor of astrobiology at Edinburgh University. "There is no single smoking gun for finding life."

As the rovers make their way across the Martian surface, their instruments will target interesting features. A variety of cameras, working across many different wavelengths, will identify areas most likely to harbor biosignatures. As the rovers physically reach the most interesting rocks, observations at microscopic levels will zero in on the exact composition. Even with the highest-resolution images, there will still be a great deal of effort to determine if something in the rocks is indeed a biosignature.

"In particular, if you saw some sort of organic residue," says Dr. Claire Cousins of the University of St. Andrews, "it would have to be a whole suite of different things." Much of this evidence will have degraded and broken down over time, "and what we will be looking for are repeating patterns of organic structures," she says. The ultimate aim will be what Dr. Cousins calls the motherlode. "We would like to see not just organics alone above background level that make it hard to interpret, but a collection of organic compounds that we can analyze properly."

The different approaches by the American and European teams on SHERLOC and ExoMars work around some of the limitations that even such instruments as advanced Raman spectrometers have. "They are working in the best band for mineralogy," ESA's Jorge Vago says of their American equivalent. "They have tried for a band that is trying to do organic molecules." According to him, the problem with Raman spectroscopy when it comes to organics is, "if it says, 'I have organics here,' it is then impossible to know exactly where they are."

The Franklin rover will use the "son of SAM," a more advanced chemical chamber, derived from the one on Curiosity, that will employ a very different kind of laser for its mass spectroscopy. Though the European Raman will see different organics than SHERLOC, their different approaches essentially complement each other.

For all the fluorescence that both instruments hope to see, all that shimmers is not necessarily scientific gold. "One of Raman's limitations is that it is sometimes quite difficult to get detail about very complex organic molecules," says Charles Cockell. Unless there is a large, unequivocally biological molecule in close proximity to the laser pulse, it is difficult to know whether it is definitely a signature of some sort of life process or not.

Paradoxically, another potential problem for both Raman spectrometers is that on occasion the fluorescence could be overwhelming. When the light is bounced back at the detectors, it might well be akin to full headlight beams dazzling you in the rearview mirror at night. In that case, though, something interesting is there to cause such an effect. "We can turn down the current," Luther Beegle says. "We can do fewer laser pulses, and then we'll get a better signal from that. But if we max out on the signal because there's too many organics there, that's a good problem to have."

Once they have learned how to operate the instrument on Mars, the SHERLOC scientists have any number of tricks up their sleeve. "We have a lot of knobs," Beegle says. "We can turn down the current on the high-voltage power supply. We've built in a lot of ways to mitigate this problem." If the return signal was too bright, they could simply fire fewer laser pulses. At present, they are planning somewhere between one and ten pulses per spot they are examining at a particular current level.

* * *

When the October 2026 launch window opens, NASA's planned lander and Europe's orbiter will make their separate ways toward the Red Planet. The NASA lander is taking the scenic route. If it headed directly to Mars, it would arrive in Jezero Crater in the middle of the dust storm season in early 2027. As with previous US missions, it has to head straight down. The dust would add yet another level of complexity. Even if it survived, the lander would have to hibernate on the surface until the dust cleared.

The 2026 lander will need the support of the European orbiter, which, because of the built-in delay to the landing, also has the luxury of taking a leisurely lob toward its destination. The European mission employs a mixture of two established technologies to be sent to Mars. The first is solar electric propulsion, which uses solar energy to strip electrons from argon, with the resulting ions providing thrust. The orbiter will also employ a traditional chemical capability. This hybrid system will extend its lifetime around Mars. "Because the orbiter is using electric propulsion," Sanjay Vijendran says, "we can't get to Mars faster than two years."

Once the European orbiter arrives in 2028, its trajectory will be changed by the electric propulsion to ensure it will be there to support the landing. The orbiter will be used for communications with the fetch rover, which will be ferried down to the surface as close to Perseverance as is possible. Its job is to pick up the samples that will be waiting in and around Jezero Crater. In the intervening five years, the Perseverance rover will have been reaching out to candidate rocks, drilling into them, extracting samples, and placing them in tubes. Each sample will weigh about fifteen grams (half an ounce), and the tubes will be hermetically sealed within an hour to limit exposure to the Martian atmosphere or any potential contamination. The cached samples will sit and wait until they are picked up. So long as they are completely pristine, a few more years won't do them any harm.

Perseverance is considered the "caching rover." As Sanjay Vijendran notes, "it's not been designed to last as long as it would need to" if it were used to take some samples to the next lander. According to Ken Farley, "the warranty on the Perseverance rover" is a one and a half Martian years qualified lifetime. From the outset, some wonder whether they will have enough time to fill all the sample tubes. "We would need an extended mission to fill them all," says Farley. "But since you're not guaranteed an extended mission, our goal would be to work efficiently, yet not sacrificing the quality of the samples. And so, we are organizing the science team to work under that kind of pressure."

This introduces a variety of headaches. The rationale for leaving the cached samples on the surface is simple. If they are placed in "the belly of the rover," as Farley describes it, and Perseverance should actually die, the precious cargo would be locked inside. Similarly, if they drop all the samples on the surface and the fetch rover fails, they would likely lose everything. Although not designed to travel so speedily, Perseverance has to travel only half the distance of the fetch rover to meet the ride home. "Conversely, if all the samples haven't been dropped," Sanjay Vijendran says, "they have to be delivered by Perseverance or they cannot be accessed at all."

So there will be a little bit of both. Some samples will be left on the surface for the fetch rover to retrieve, while others will be kept within Perseverance. "If we have two working rovers when the lander arrives," says Vijendran, "it makes sense to use them both to deliver the samples."

This stratagem reduces the risks should one of the rovers fail. Beyond that, the longer the caches are left to their own devices, the more they are at risk of the Martian seasons, the extreme weather, and, of course, the dust. There is a real risk they might be hidden or difficult to find among increasing layers of dust deposited by the wind. And though the European fetch rover has been designed to work for five months, "the season has driven the overall mission design." The fetch rover must land and get its work done before winter sets in.

Winter kills. For the whole retrieval procedure in Jezero Crater, the freezing cold of the Martian winter, even in Isidis Planitia which is close to the equator, sets very fundamental limits to the rover's longevity. Temperatures will routinely drop below −100°C in the daytime in the dead of winter. "For the first time, we are designing a rover mission without any nuclear power onboard," Sanjay Vijendran says. Unlike the most recent NASA rovers, the ESA fetch rover will not carry a radioisotope thermoelectric generator to provide for its heating and electricity needs.

The decision not to use plutonium makes the rover less complicated and less expensive to build. Yet after landing in August 2028, the clock is running. "Once winter sets in," Vijendran says of the fetch rover, "it won't be able to function."

The baseline design is for 150 sols, a figure that includes what Vijendran calls "prepping the return." This means the rover will have to pick up all the samples, place them into a container on the lander, prepare the sample container for launch into space, and then fire the samples into orbit. That time period, he says, allows for considerable margin, but "if the rover doesn't do everything in that time, it's not the end of the world."

* * *

Whenever it lands on Mars, the Rosalind Franklin rover will also be on its own. What makes the European ExoMars rover stand out is what one participant calls "the dirty great drill," an ironic comment, given that it is probably the cleanest drill ever sent to the Martian surface. The Franklin rover will be capable of reaching more than six feet (two meters) below the surface, "so we can drill into solid bedrock," says Jorge Vago, and sift for "concrete evidence," for any signs of life on early Mars.

To date, all drilling on Mars has encountered unexpected problems, as the mole on InSight did. Yet Rosalind Franklin won't be piercing in the dark. It has forward- and rear-facing ground-penetrating radars, so "we can get depth perception that way." Its operators can anticipate any obstacles. The drill itself has been designed for the hardest rock imaginable, though the operators won't bite off more than they can chew.

Where Perseverance is abrasive, ExoMars is percussive. Its drill, unlike the American ones, "rotates, it doesn't hammer." To avoid mechanical or chemical problems, it will extract a core sample very slowly. The deeper the drilling, the more likely "we are sure that what we are looking at material that has been protected," explains Jorge Vago. That's because deep under the surface, the sample material will have avoided radiation and the scavenging chemistry of the surface. Outcrops of rock are the target. In particular, if the bedrock is exposed, the rover instruments are more likely to find the specific minerals the researchers know will be important. "And the sort of minerals that we think could hold biosignatures," Vago says, "are hydrothermal deposits or clays circulated by hydrothermal fluids."

The Rosalind Franklin rover has a slightly different way of looking for organics than SHERLOC. The pristine material is first examined with a camera system called the Close-Up Imager, which takes images in red, green, and blue. The close-up camera "also takes infrared spectral images" that observe everything at grain-size resolution. It also looks where they have been taken from. When the samples are then crushed and chemically analyzed, the results can be cross-referenced to where exactly they were removed from the surface. As noted earlier, when looking for life, an overriding factor is context: was that particular sample found in a geological environment where it could have plausibly supported life?

To answer that question, a microscopic camera system known as the Micro Observatoire pour la Mineralogie, l'Eau, les Glaces, et l'Activité, or MicrOmega (with a literary allusion in the direction of Voltaire), will scan the crushed material to see what sort of geological processes have acted upon it. "MicrOmega will be imaging the crushed mineral samples," says Jorge Vago. "Then within the rover we have intelligence to analyze infrared spectra." The infrared absorption bands reveal telltale signs of what chemistry is present. "And we can examine in detail the mineral grains from their composition." That will bring out subtle details of any suspected hydrothermal environment, such as clays and carbonates, "all of which point to past climate conditions and the possibility for ancient life."

To do so, Micromega's Raman instrument fires a laser beam of green light that illuminates different kinds of organic molecules than SHER-LOC. When they fluoresce, they indicate the presence of organics, though Jorge Vago warns that a lot of energy is required before the Raman lights up. This may make it difficult to distinguish the organics exactly.

Any "pigments," as Vago calls them, have precise patterns. In particular, MicrOmega gets a good look at what are known as the "heme" group. These are molecules in which atoms of iron bond with those of oxygen (as in hemoglobin, which gives blood cells their red coloration) and which are present in many biologically important molecules. Here they are more likely to be simpler and use the iron from the ferrous oxides in the Martian soils.

At the microscopic level, the crushed mineral grains appear more like mountain ranges because they are so greatly magnified. Their "range" is a few hundred microns across. The laser instrument fires at the crushed mineral grains at a certain height, then "moves along, then it will fire again at a place where it is much lower." Having an autofocus will mean all the difference in the world. "You don't know where the signal will be strongest," Jorge Vago says. "You need to have an autofocus to find that part."

What it should be able to do is find "pretty sophisticated molecules that are mainly phototropic," ones that respond to light at the parts-per-million concentration. If that is the case, it would hint that precursors to life were available on Mars. "We are talking about very primitive organisms that rely on chemistry for their metabolism," Vago says.

* * *

Europe hopes that the experience it will gain from the Rosalind Franklin rover in 2023 can help in the subsequent scoping for its successor three years later. Their foremost technical challenge is the need for speed. The fetch rover will have to beat the land speed record on the Red Planet by a factor of at least two and possibly three. "It will need to drive very quickly," says Paul Meacham, the Airbus systems engineer, "faster than any rover has traveled before." At the moment, the Airbus engineers, who are likely to build the fetch rover, believe this next rover will be quite a lot smaller than Rosalind Franklin. It might have four wheels instead of six.

That NASA "retrieval lander" can, JPL believes, be targeted with an accuracy of 4.7 miles (7.5 kilometers) to where the Perseverance rover will be waiting inside Jezero Crater. The engineers are allowing for a straight-line drive of just over six miles (ten kilometers) there and back to retrieve the samples. That means the rover will be capable of traveling a total of 12.4 miles (twenty kilometers) to fetch and retrieve the samples over five months. It will grab all of the available caches as quickly as possible.

The average daily drive will need to be of the order of "hundreds of meters." Even then, that refers to "blind driving," meaning there will have to be no stopping nor waiting for hazard avoidance. This also requires the fetch rover to be working perfectly and encountering no hazards in the way, which means there will be a need for a new, improved software package to control it. The fetch rover will have to navigate with far greater autonomy.

Paul Meacham says it would be tough to do what they want by scaling up the autonomy they are using on Rosalind Franklin. The speed, for example, means that the images used for navigation might become blurred. "There are all sorts of things [that come up] with the increased movement," he says. So a more powerful processor will be required. There is also a very different design ethos behind the fetch rover. "Unlike previous missions, when you don't get to a nice scientific target, you can get there eventually," adds Sanjay Vijendran. "With a sample return, you either get the samples or you don't."

These prized samples will be transferred from the rover to the lander. They will be collected in a container that will carry about thirty tubes. It started off as a ball shape—a sphere is dynamically easier to capture in orbit—but since then, it has "been squashed a bit," in Vijendran's estimation. The more cylindrical design allows for the length of the tubes to be accommodated better. Ideally, the tubes will be loaded one by one into a holder within the sample container, packed in tightly for the journey back to Earth.

While on Mars, the container will resolutely remain on the American lander. The fetch rover will hold each cached sample in what looks like a test tube rack, and the lander will do most of the hard work of transferring it to the container. The lander's sample transfer arm is about six feet (two meters) long, while the rover's isn't long enough for this operation. Once again, the loading will have to be performed autonomously. If it were directed from the ground, it would take at least a month. There simply won't be time for that on Mars.

* * *

"At some basic level, life on Earth is formed in little bags of carbon," Luther Beegle says. "And if you look at what happens in these extreme environments, the carbon tends to clump and the carbon tends to be together."

With SHERLOC, what they are hoping to find is a lot of organic material in the same microscopic area "as a mineral that is known to be fluvial"—that is, it has been deposited in water. "So, for example," Dr. Beegle continues, "if we saw a sulfate vein that had leached in from the ocean or the lake that we're going to go land in and there was a correlation between where the organics are and where the sulfates were, that would be really interesting." That would be targeted as a potential biosignature.

SHERLOC will sense a full spectral spread of silica-rich rocks, sulfates, nitrates, and phosphates, all of which are important in the creation of life. "We're basically targeting all of them," says Luther Beegle. "It's different concentrations for different types. We can see calcium sulfate, no problems, along with gypsum. We see phosphates all the time. We can see silica-based rocks, all the carbonates, all the calcites and dolomite, and evidence for nitrates."

They will also find any aromatic organics across each hundred-micron spot SHERLOC examines. "We will be able to detect about one in a million, and that's a really low concentration if you think about how much material is actually in a hundred-micron spot size," he adds. The camera associated with SHERLOC will get a better sense of where these longer chains, which are also crucial in organic chemistry, may be found.

"Most organics have a Raman signature that you'll be able to detect," Luther Beegle says. "At least, all the amino acids do, and a lot of the lipids do. So it's a two-step approach and with different concentration levels for the different measurements we're looking at."

In particular, they are hoping to find organics that have been formed within the all-important fluvial minerals. "We would definitely target those," Beegle says, to be cached and returned back to Earth. "If the organic materials were spread out uniformly over a rock, that would not be as interesting as if we find clumps of organics all over the particular rock, especially if we thought it didn't undergo any thermal transformations."

Though the SHERLOC team will be able to determine the geological context for each sample, many onlookers remain reticent as to its divining capabilities. "You are going to be very lucky to stumble across any of these features on Mars," says Charles Cockell, "unless you find extant life there, which is unlikely near the surface."

Indeed, most of the scientists interviewed for this book say that it is also going to be very difficult to find evidence for ancient life with a Raman spectrometer unless it's right in front of the instrument's nose. The X-ray spectrometer boresighted with SHERLOC should be able to look for differences between iron oxides and sulfur minerals. But it won't be able to distinguish whether they are biological.

* * *

The centerpiece for Rosalind Franklin is its largest instrument, a toaster-oven-sized laboratory that takes in the crushed samples from the surface. MOMA, the acronym for Mars Organic Molecule Analyzer, is the result of a marriage between a gas chromatograph mass spectrometer and an improved version of the SAM oven already working aboard Curiosity. As with SAM, it uses another laser to look at molecular fragments. This will give the instrument a broader range of possible organics to target compared to SHERLOC. Its gas chromatograph heats samples into a gaseous form and measures the exact spread of colors in their resulting spectra.

To do that, the crushed samples will be baked by use of thirty-two separate ovens that move on a small carousel. Each is about the size of a thumb and is designed for "a single-shot use" (and possibly might be used twice). The sample is sealed, heated to a blazing temperature of 900°C (1,650°F) if required, and then its vapors rise upward into the mass spectrometer. (That part is similar to the ones flown on *Viking* and more recently on Curiosity, the SAM oven.)

The heating in the ovens is, Jorge Vago says, a tricky thing. Technically known as pyrolysis, it can be achieved in steps within an inert atmosphere. Depending on which molecules are being examined, the individual steps are different. Each stage is in some way chemically aggressive.

"You modify the molecules that you are trying to study," Vago says, "but you tend to destroy them." The superheated molecules travel down a long tube coated with a special substance. This filters out various interlopers of little interest, because the coating preferentially favors some that flow more easily than others. Once they're stuck inside the trap's fluctuating electric field, they're in just the right spot for the detector to figure out their chemical composition.

The second mode of MOMA uses an ultraviolet laser for a more delicate form of molecular analysis known as desorption. It has never been used on Mars before. As well as the green Raman laser, this other laser fires a beam in ultraviolet light as part of MOMA. The samples are zapped for all of a billionth of a second so they aren't fragmented irreparably. The extreme oven heating mentioned earlier would easily destroy lipids, the building blocks of cell membranes. To search for such delicate structures, samples will be vaporized by this ultraviolet laser much as SHERLOC does, directed to a focus smaller than the diameter of a human hair.

On the first day it is used, the operators will find the best place to sample and then aim and fire the laser at it. There will be twenty Raman spectrometer spots and a couple of MOMA laser desorption measurements. "Normally when we get a sample, it first goes into a refillable container," explains ESA's Jorge Vago. "Then we look at it with MicrOmega, Raman, and then the laser desorption part of MOMA." All this involves the lasers firing into the sample, "then the 'proceedings' are analyzed in the mass spectrometer."

If something of interest is suspected, the ovens can be used to investigate further. "If we see something with the Raman laser, the next part is to use laser desorption if you want a careful examination of the organic fragments," says Jorge Vago. "If that still looks pretty good, then you can think of using an oven." Once the samples are dropped into the ovens, a dozen have the remarkable property of allowing the science teams to look for a twist in the origin of life—literally, whether the lipids and organics are left-handed or right-handed, since certain combinations are favorable for life. How it does so involves a remarkable and, to the uninitiated, convoluted way of narrowing down their structure.

* * *

Of all the steps needed to return samples of Mars back to Earth, the most complicated involves exactly how the ascent stage will work. "The launching off the surface of Mars worries a lot of people," says Sanjay Vijendran. "It is a totally new and difficult problem to crack."

For the first time ever, a rocket for the return will have to be taken along for the ride. Not only does it have to wait out the journey to Mars and the landing, it also has to survive the environment on the surface for the best part of a year. And then, when ignited, it has to work the first time and loft the sample into the right orbit.

Nobody has ever fired anything off the Martian surface. The ascent vehicle will head toward a waiting vehicle in orbit around Mars. How its samples are then transferred so they can be returned to Earth is another difficult step. The whole procedure will require capture in orbit and transfer into a containment capsule that has in no way come into contact with the Martian environment. Speed remains key. After racing to get the samples ready, the capsule will need to make the date dictated by the next launch window to return the sample back to Earth after being "caught" in Martian orbit.

"It's clear to me that one of the big challenges is to make a rocket that will function properly in the Martian atmosphere," says Ken Farley. "And the kind of challenges that exist are that the rocket has to sit on whatever platform it arrived on. It has to then go through the huge diurnal temperature fluctuations on Mars. So the propellant has to survive all that."

At present, JPL is looking to find a fuel that can survive with the degree of thermal control that they can reasonably apply. While liquid propellants are more stable at freezing temperatures, there is a greater difficulty in maintaining them. Solid fuel rockets are also tried and tested, but they will not work below the freezing point of water for any length of time. So various options, involving heaters powered by batteries or nuclear power, are being examined. Some sort of hybrid system, perhaps using new, untested lightweight fuels, had been considered. At the end of April 2020, NASA indicated that a two-stage solid rocket motor would likely fit the bill, based on proven technology and known thermal characteristics.

What goes back up from the surface will have exactly the same problems its carrier had when it went down in the first place. The Martian atmosphere is very difficult to navigate because of its unpredictability. Worse, the atmospheric resistance on the way back up to orbit will be hard to anticipate. Despite its comparative thinness, dynamic changes will make launchings as difficult as landings.

And then it is a case of "catch me if you can" by the European orbiter, which has to have been in a position to relay data from the fetch rover but also "needs to be in position to observe the launch of the ascent vehicle," Sanjay Vijendran says. These two demands dictate a compromise between what is ideal and what is pragmatic. The orbiter will need to be in an orbit low enough for the ascent vehicle to reach without it needing a ridiculously large rocket. On the other hand, it cannot stray too close to the planet or the Martian atmosphere will drag it down. The ascent vehicle will have to reach "an orbit that will be stable for at least ten years in case something happens to the orbiter before it catches the sample."

Such an orbit would give them another chance if this first attempt failed. The samples would still be waiting for another mission to come and retrieve them. Even if the engineers get the launch and rendezvous right the first time, the orbital capture represents another difficult milestone. When the precious samples are in range of the European orbiter, there need not be a docking in the traditional sense—"not as such," says one engineer. "That's going to be tough," says Paul Meacham of the capture. "No matter how you do it, it's difficult to achieve that sort of accuracy." And the whole sequence will have to be carried out automatically. "We can't be in the loop," says Dr. Vijendran of Earth-based control, "because of the time delay in signals."

The orbiter will be the active chaser. Long-range cameras will be used to look for the sample; then, as it comes in closer, the orbiter will maneuver itself to catch the sample canister. At present, the ESA engineers are considering a large, self-contained "box" into which the sample container will be captured. This collecting system cannot be contaminated by anything that has come into contact with the Red Planet. "You want to minimize any contact between the ascent vehicle or container with the orbiter," says Paul Meacham. "You want to keep it as simple as possible in capturing the sample."

Whereas the ascent will keep many people awake at night, other engineers don't seem to think this capture is such a huge deal, to actually orbit-match with the sample container. But what compounds the worry is making sure that nothing returns to Earth that has been in contact with the environment of Mars. "This is a big step that sounds very challenging," Ken Farley says. "It's what we call breaking the chain."

* * *

Life, as we know it, is a complex phenomenon. One of the fundamental, persistent difficulties is how to define exactly what it means. There is no universal, all-inclusive definition whose criteria can be entirely satisfied. Most biologists, however, would probably agree that life can be defined by its ability to reproduce and evolve via natural selection.

Even then, a more fundamental question remains. How did life get started? How did the chemical reactions required for life somehow join together to form a system capable of reproduction? What were the ingredients? And if life ever did get started on Mars, how would we know where to look for it? There are many unknowns in our current understanding of how life evolved on Earth, let alone Mars. Indeed, most of the competing theories—as the late Fred Hoyle once pointed out—are about as likely as a hurricane blowing through a scrap yard and producing a fully formed, gleaming jumbo jet as a result.

To be fair, the creation of life on either planet did not come from out of nowhere. There were at least three billion years of evolution. There is also a very clear and important role for randomness. The various combinations that triggered the biochemistry at the start, and then any subsequent mutations within biological molecules, were all random. The end result, however, was not random, as the process of natural selection came into its own. This effectively shepherded the combination of molecules that were able to reproduce and adapt to whatever environment they evolved in.

There is nothing random in the process of evolution. It is the production of the *variations* that is random. "What that means is, you don't go from a collection of molecules into a 747 completely randomly in one step," explains Professor Charles Cockell. "Evolution has never had to do that." Such long-term progress also explains how life is able to propagate. The crucial step is having a molecule capable of replicating itself. That could, for example, involve the joining together of the amino acids that form the first strands of an RNA molecule. RNA, ribonucleic acid, is a precursor to the "double helix" of its deoxyribonucleic form, DNA.

The creation of RNA would have been by trial and error over extended periods of time. Ribonucleic acid has a number of nitrogenous "bases," which contain amino acids in differing combinations that are slightly different from those used by DNA. The double helix represents the next step up, as it were, in biological evolution. Nevertheless, if a small piece of RNA was synthesized from the various amino acids just a few bases long, that would be the start of something much more significant. "As soon as you have one molecule that just replicates, the whole process is underway," says Professor Cockell.

Variations can then occur as those bases sometimes alter. Indeed, some will fall apart, and others will actually start to replicate slightly faster. This has been shown in the laboratory. RNA can be synthesized, which then makes molecules that can then make copies of themselves. At some point, metal ions can be introduced that cause errors in the replication. Over a period of days, molecules develop that can replicate faster than the original. Others fall by the wayside, dispersed into the medium of whatever is being used to cultivate them.

The local environments start to select new variations. Once there is a basic cellular structure in place, its complexity can be increased. Eventually, some sort of membrane—a cell wall, for example—allows these molecules to be encapsulated. "So you end up with increasing complexity and you start adding things into the membranes," says Professor Cockell.

"The molecules start producing with increasing complexity when you start to do that. Other molecules get stuck in the membranes and start producing more than one molecule."

A few threads of the constituent RNA may eventually result. These strands then start interacting with each other. Many of these reactions take place in minuscule fractions of seconds—some at the barely perceptible scale of a millionth of one billionth of a second, known in scientific notation as femtoseconds. "And when you think you've got a whole planet with femtosecond timescales over billions of years," Cockell continues, "the whole process becomes quite unremarkable."

* * *

As JPL and the European Space Agency work together, elements of the Mars sample return mission are changing almost on a monthly basis. If the lander seems to be getting bigger during the development process, the sample canister is getting smaller. "It used to be the size of a European football," says one amused observer. "Now it's been squashed down to the size of an American one—not surprising, really."

Breaking the chain adds another critical element. The part of the craft that has been in contact with the Martian surface "will never come anywhere near the terrestrial biosphere," in another engineer's estimation. Anything that has been exposed to Mars—its surface or atmosphere—will have to be sealed. That means, in Sanjay Vijendran's words, "you are always going to have an interface which is still exposed to the Martian atmosphere."

Once the container is captured, this exposed surface will need to be sterilized. What Dr. Vijendran terms "sterilization of the seal itself" will make sure nothing ever gets exposed on Earth. This is another technical headache that needs to be solved. It will have to be done in a confined space with limited power and carried out autonomously. It would only involve a small area. The designers are examining the best sealing and sterilizing techniques. This process includes looking at heat, radiation, chemistry, and plasma for quick and efficient sterilization.

So far, heat seems the best option. The external face of the container would be heated "for a few seconds to a few hundred degrees or so" once it has been captured in orbit. Heating has the advantage that the energy used will dissipate very quickly, ensuring that none infuses the sample container. Even a small rise in the temperature of the sample would make a mockery of the very reason for returning it to Earth in the first place.

Once the sample container has been sterilized and sealed, it will be ready for delivery back to Earth. For that, the container will be placed inside the return capsule, "which is attached to the side of the capture system." In outline, yet another lid will close the whole system off and maintains its integrity. Once all this has been achieved around Mars, the orbiter spacecraft will wait for the first opportunity to come home.

The sample needs to get back to Earth in one piece. The overriding aim is to make sure the pristine samples inside are not compromised during the journey to Earth and passage through our atmosphere. Of paramount importance is maintaining the returning vehicle's physical integrity, especially during the landing phase, when "the possibility for parachute deployment failure should be accounted for in designing a rugged sample canister," as one report had it.

For that, another mechanism will push the container into another hermetically sealed space within the returning vehicle. The main technical challenge is that the container has to be "orientated for landing," which, in the wry estimation of Sanjay Vijendran, means "it is going to come in like a Martian meteorite, hard and fast."

* * *

One of the most important steps along the way to creating life is the synthesis of proteins, large molecules that can contain up to a million individual atoms. Proteins are made of separate, smaller molecules known as amino acids. A protein's ability to perform different functions depends on its shape, and this in turn is dictated by the amino acids that form it.

As chemistry becomes biochemistry, these compounds become more complicated. They are capable of both copying information and replicating themselves efficiently. In this sense, the ultimate "informational" molecule is DNA, deoxyribonucleic acid, which is made of proteins and amino acids that carry the genetic blueprint for and, in so many ways, define each and every species.

If there is DNA in any form of life on Mars, one distinct difference may come from a discovery made by Louis Pasteur, the pioneer of vaccination and the heat treatment for liquids that bears his name. In the 1860s, Pasteur became curious when he saw dregs of wine left behind within their bottles. When he looked at their crystals under a microscope, Pasteur saw that they twisted the light in one direction. When the same crystals were synthesized in his laboratory, they twisted the other way. In time, it has been observed that this bias, which Pasteur christened chirality, permeates the whole of science. Nature is, you might say, even-handed.

Molecules can be either left- or right-handed. Indeed, some molecules can be mirror images of each other. In nature, proteins use amino acids that twist light to the left. In the case of DNA, twenty of the most common amino acids are involved (why those particular ones remains a mystery). Of the twenty, nineteen can be either left- or right-handed.

When amino acids are synthesized in the laboratory, there are equal numbers of left- and right-handed ones. Those that form DNA strands are left-handed. In other words, only biology distinguishes between them. On Earth, DNA prefers to use the left-handed version of the amino acids. On Mars, though, the preferred ones might well be right-handed.

It is a matter of chance which form of amino acids—whether left- or right-handed—is used to synthesize proteins. As a result, substantive differences between any amino acids found on Earth and on Mars will identify a truly alien precursor to life on the Red Planet. The only unequivocal biomarker, biologists believe, will come from a different suite of amino acids than those found on Earth. Probability dictates that the exact same group of amino acids won't be selected in exactly the same sequence elsewhere in space.

On the Red Planet, there may be left-handed amino acids with a slightly different structure. Right-handed amino acids will likely show that Mars had supported indigenous life. But this orientation of amino acids is ultimately dependent on chance. A slight surplus of right-handed nucleotides—the basic structural unit of nucleic acids such as DNA formed from proteins—could have been formed preferentially in an ancient pool of water. These uniquely Martian markers may have seeded any subsequent evolution of biological material on the surface.

As a mystery, it is worthy of Agatha Christie. The mirror—the symmetry between these right- and left-handed worlds—has indeed crack'd. Chirality also points to a way of distinguishing life on Mars that might have originated on Earth, or vice versa. (There is also the chance that life could have arisen from hitchhikers on meteorites or other interstellar material.) If there is such a thing as unique Martian DNA within an ancient microbe, chirality may well be the clincher.

* * *

Some of the sheen for successful sample return was taken off when, in the summer of 2005, a NASA spacecraft that had returned from interstellar space bearing samples of primordial dust crashed into the desert in Utah. Its parachute failed to open, and the spacecraft, called *Genesis*, slammed at full speed into the salt flats. In the amusing words of veteran space analyst James Oberg, it was proof of the space version of Murphy's law: "Every component that can be installed backward eventually will be."

In this case, the culprit was an accelerometer. It is a sensor that detects how fast the returning vehicle is moving. It was supposed to prompt the spacecraft to slow down by firing its engines when it recognized it was accelerating. But the accelerometer was installed the wrong way around. Instead of slowing, *Genesis* slammed into the desert faster.

So the first sample returned from the Red Planet will involve "ballistic reentry," which negates the need for a parachute that could fail. "It's actually easier to effectively crash it into the Earth," marvels Paul Meacham. "It sounds crazy, but if you have some sort of crushable material to protect the sample, it is easier."

A handful of other spacecraft have successfully returned samples from other remote outposts in the solar system. The main technical challenge is the need for extremely high reliability in the containment. For that, a passage through Earth's atmosphere that is as simple as possible will be required. "We need to be sure we have a high confidence we're not going to accidentally introduce anything from Mars into the Earth's biosphere," adds Brian Muirhead.

When and where the samples will land is not known at present. The likely candidate site will be the Utah Test and Training Range, which in 2023 will see the return of samples from the asteroid Bennu and the

Osiris-Rex mission. At the moment, the international agreement between NASA and ESA concerns only the flight elements; at a later stage, they will work out what exactly the return and recovery entails. But as one participant says, "It will be horribly political," and the feeling is that the returned container will first be taken to a US facility.

Despite the science fiction movie staple of Earth being contaminated, that is unlikely to ever happen. "It's an incredibly low probability that there are hazards on Mars," says Professor Ken Farley. "I don't wish to convey the idea that I believe that there are pathogens on Mars that could affect anything on earth. That seems incredibly unlikely."

* * *

Chirality implies a curious paradox at the heart of this quest. The conditions needed to incubate life are very different from those required to sustain it. Both Earth and Mars may well have exchanged material—"shared spit" in one astrobiologist's vivid phrase—on a regular basis in the past. Indeed, life that originally got started on Mars could have traveled across the celestial gulf and survived here. It is an idea with a noble heritage.

As long ago as 1871, Lord Kelvin suggested that the idea that "life originated on this Earth from the ruins of another world may seem wild and visionary. All I maintain is that it is not unscientific." Panspermia, this notion of prebiotic molecules and indeed biotic ones being wafted throughout the universe in interstellar grains and cometary bodies, is sometimes seen as the most elegant explanation for the origins of life. Ancient life on Mars and Earth could be related. "The reason that the solar system has life is because it had an incubator," NASA's Christopher McKay says. "That incubator could have been Mars. By necessity, these worlds are not sustainers."

The Earth would have been an ideal sustainer, because its oceans would have been cleansed of their original organics from the early heavy asteroid bombardment and any subsequent geological activity. After the impacts died down, Earth would have been in a better position than Mars to continue life's evolution. Certainly, during the heavy bombardment, the repeated rupturing of Earth's crust and the resulting explosions would probably have sterilized a lot of biologically significant chemistry.

As has been recognized in recent years, there has certainly been exchange of material from the two planets. A remarkable half ton of material lands on Earth from Mars annually in the form of meteorites. There would have been so much more in the ancient past.

It could be that there was an exchange of genetic material early on between Earth and Mars. "In some sense, it would be more staggering that Mars remained completely lifeless," says Professor Charles Cockell, "when we know it had liquid water at a time when we know Earth had life." Was genetic material ever transferred from Earth to Mars in sufficient abundance to have seeded either planet? In other words, just how different was any life that may have evolved on each? The two planets are about as biologically isolated as the continents are today on Earth.

"The evidence to suggest these differences will fall on the genetic sequencing," Christopher McKay says. "I would love to show substantive differences." Unequivocal evidence that life on Mars formed separately hinges on the twenty amino acids—nineteen of which are left-handed—that form the biochemical basis for DNA. Although amino acids are easy to make, nucleic acids are much harder to create. Wide-scale differences from planet to planet would be reflected in subtle differences between the DNA strands found on Earth or on Mars.

"My guess is the DNA will be the same as us," says McKay. "It won't be that hard to tell—there is a distinct signature to life on Earth." This means that any unequivocal Martian DNA might consist of right-handed sugars and left-handed amino acids. Should any future missions discover such features that are broadly similar to ones on Earth, but with substantive differences in their biochemical details, then we would know we had found life that was indigenous to Mars.

"That would be fascinating," McKay says. "It would be close enough to us so that we can make sense of it, but far enough away that it is a distinct, separate point in biology. That's what I'd hope for, but my objective reading of the evidence is that it is going to be the same."

* * *

Given that the returned samples from Mars will have been chosen for their biological attractiveness—biosignatures, involving "something super cool" in Jorge Vago's expression—extraordinary steps will have to be taken when they are examined on Earth. No matter how remote the chance they may contain organisms, whether alive, dead, or fossilized, contamination is no longer an academic matter. Protecting our environment is going to be a major initiative involving space laboratories all over the world.

"Planetary protection is essential," NASA's Planetary Protection Office notes, "[to] preserve our ability to study other worlds as they exist in their natural states, to avoid contamination that would obscure our ability to find life elsewhere—if it exists; and to ensure that we take prudent precautions to protect Earth's biosphere in case it does."

So far as Mars is concerned, it is a two-way street. One way runs in the direction of the Red Planet and the "forward contamination" of it and other solar system bodies: that is why the sample tubes, as well as the hardware and instruments, being carried by Perseverance will have to be ridiculously clean. "We don't want to put Earth stuff in the sample container," says Sanjay Vijendran, "or else you'll lose the scientific value of the samples."

So the whole business has already started in the clean rooms used to assemble the Perseverance hardware. "We have to keep everything clean until we get to Mars," notes Ken Farley. "Our big challenge is to keep the samples as free of terrestrial contaminants as possible." And that means not just keeping Mars out, but keeping any Earthly materials out, too, even when the precious samples are being examined after their travels.

"If you think for a minute, where is the worst place in the entire universe that we know about for looking for organic matter from another planet?" Ken Farley asks. "Well, the worst place in the entire universe is Earth because we are absolutely dripping with biogenic organic matter. So that kind of paints the challenge that we had in keeping everything clean."

If they expose the inside of the sample tubes they are using to the terrestrial atmosphere, that would cause way too much contamination. "So we have had to go through these really extraordinary steps to keep everything clean," he says. The sample tubes have to be completely clean of organic matter, cells, and microbes. "It has to be free of a lot of inorganic species that we might use, for example, in geochronology."

* * *

Suspecting that there might be chiral molecules on Mars and actually finding them are two very different propositions, partly because they are likely to be fairly elusive but also because of the curious chemistry of the Martian surface. In Jorge Vago's wry phrase, organic "gardening" done by meteorites is "going to be a pain in the neck for us. They are not the organics you are looking for."

Equally troublesome are the perchlorates discovered by the *Phoenix* lander in 2008. Indeed, their presence has had a significant impact on the design of all subsequent Mars missions. As noted, their signature was present in the *Viking* data and more recently discovered by Curiosity when indirectly observed as "oxy-chlorinated species" on loose sand and in very ancient rocks.

Perchlorates will have to be accounted for. Where interesting material has washed in across the surface, the perchlorate concentration will tend to be very low. Another interpretation is that as any water has subsequently evaporated, the concentrations, like salt left behind when boiling potatoes, may well have been preserved. If there had been flowing water over time, it would have tended to concentrate the perchlorates left behind.

Certainly, the concentrations of perchlorate are not expected to be the same everywhere on Mars. The question is, if they are found, are they ancient or modern? The feeling is that they are more likely to be more recent, geologically speaking. Chlorinated compounds, brought in by meteorite impact, are needed to form them. Their deposition would have been greater farther back in time, when there would have been larger doses of ultraviolet radiation available from the Sun. That would have provided sufficient energy to fire the very esoteric chemical reactions needed to create life.

If all goes according to plan, ultraviolet radiation itself will be used in Europe's forthcoming search for life on Mars. Within MOMA, the laser desorption instrument uses a very powerful ultraviolet beam to illuminate "the organics from the sample." In Jorge Vago's phrase, these laser bursts have "the glorious quality of not exciting the perchlorates." The zapping is so precise and quick—it takes all of a billionth of a second—that the target samples cannot be destroyed or oxidized by the perchlorates, "and that's very novel," Vago adds. "It's the first time that it's been done, and it will hopefully tell us a great deal."

At about 200°C (400°F), the perchlorates burn off. As they do, they take any interesting material with them. "So you run the risk of oxidizing the majority of the organics," says Vago. To get around that, if any interesting material is discovered, the ovens can be used as a next step in a clever way. The interesting samples can be "pyrolized," heated in a controlled environment, at a lower temperature. But, as he says with a smile, "It's more devious than that."

A dozen dedicated ovens—three groups of four, each of which contains a different "derivatization agent"—will effectively illuminate exactly what is contained in a particular sample. A derivatization agent "is something which attaches to the molecules you actually want to detect that are very fragile," Vago says.

Because their properties are known, these "der agents" enable the research teams to see what was there originally. One attaches to a molecule within the sample chosen for its suspected chirality. In the case of an amino acid, for example, it will actually cut the acid in a certain place. "Because we know the derivatization agent so well," Vago explains, "we know how it behaves, so we can reconstruct what it has actually become attached to."

There are three derivatization agents available for extracting chiral information in Oxia Planum. One works with the smallest molecules but has an unfortunate tendency to destroy the chiral information. Another is specifically designed for preserving the chiral information. The third is better for larger molecules. "Either you can absorb a chiral molecule, or you desorb it, specifically preserving the chiral node with the special derivatization agent."

Once again, maintaining the exact temperature is important. "If we get into trouble with the perchlorates," Vago says, "one of these agents gets activated at a much lower temperature than the perchlorates do." In other words, the organic molecules can be tested for chirality without perturbing the perchlorates and destroying their presence. The gas chromatograph mode of MOMA with the ovens is the best technique for looking for smaller molecules (with molecular weight of a few hundred) like amino acids and nucleic bases. The laser desorption technique can be used for proteins and more complex molecules, and "we can always use the laser desorption and the mass spectrometer, as you don't have to worry about the perchlorates."

Once the ovens are used up, that is not the end as far as extracting chiral information is concerned. Within the gas chromatograph there are "chiral columns" (four in total) where a chiral analysis can be done. "At any time, you can run your sample through a chiral column," says Jorge Vago; "if you are looking for chiral molecules within amino acids or sugars, they can be very fragile."

If chirality is there, MOMA will find it. If the target molecules are substantively different from ones found on Earth, they will have hit the jackpot: evidence for a different genetic evolution of possible Martian microbes in the ancient past or certainly microbial precursors.

* * *

The opening of freshly returned samples from Mars will be a milestone in human history. They will have to have been moved to a very high-security "bio safety" facility. While the scientists involved know that the likelihood of contamination is infinitesimally small, the public is another matter. "They've seen *War of the Worlds*," notes Ken Farley, "and they've seen all these terrible things that could potentially happen, like in the *Alien* movies."

As with cargo from ships returning from exotic destinations in the past, the samples from Mars will have to undergo quarantine. This might involve a "dispersed" center where the team would not come together until after the samples were returned, or amending existing NASA facilities, or even building a new, dedicated center. No current facility fits the bill exactly. The debate today is whether that first look can be accommodated in a facility like the Lunar Receiving Laboratory in Houston, which lacks expertise in microbiology, or perhaps somewhere like Fort Detrick or the Centers for Disease Control in Atlanta. While the latter have experience of containment, they have little curatorial experience.

"And so, these are the big challenges to figure out how you build such a facility," Farley says. "And it has certainly been the kind of thing that nobody has done before." For those expecting something like *The Hot Zone*, Farley points out that there is a major difference. "We don't really care if organic matter or terrestrial microbes are exposed to the Ebola," he says. "We do care if the Ebola gets out. We don't care if one of us gets in."

Even the most sophisticated of Earth laboratories will have to deal with the problem of quantities. The samples of Mars will be rarer than gold dust. The entire Mars sample inventory will likely consist of only 500 grams of material. It will, Ken Farley says, be partitioned out, gram by precious gram, to a variety of researchers all over the world. So there will be a huge push to become superefficient in the use of the atoms available to them. Such basic techniques are well known. "We're not going to have to invent something entirely new," Farley says. "We're just going to have to get a lot better at it."

This extends to the way in which researchers will attempt to look for biosignatures in the returned samples. "How do you do the same sort of superefficient kinds of analyses looking at organic matter," Farley says, "and trying to determine whether that organic matter is biogenic or not?"

* * *

As strange as it may seem, most people interviewed for this book think it highly unlikely life from Mars will be returned in any of those samples. "They will essentially be surface samples," says Sanjay Vijendran of the tubes cached by Perseverance. "That's the reason why ExoMars is digging down to two meters."

Which means that, if the drill on the Franklin rover works, it has a greater chance of finding something interesting. Others believe it may take even deeper drilling. Material that has preserved life's presence might have survived buried deep below the surface. "The search for life on Mars," notes one recent paper, "leads underground."

Luther Beegle was one of the forty-one coauthors of this paper published in *Nature Astronomy* in May 2019, which was unequivocal. "Go vertical and look for life. We may be able to find the Garden of Eden on Mars." Given that subsurface water is at least two miles below the surface, that is well beyond current technology on any planet at the moment. As the experience with InSight has shown, even drilling a few centimeters is difficult.

"On Mars, you've got very cold permafrost below like concrete,"

Charles Cockell notes, "so it's not going to be easy to drill into the sub-surface at all." Looking for fossilized remains will require a digging device capable of drilling for many miles under the surface. Even if you want to do that, with billions of dollars, rigs, and whole teams of drillers, it is very difficult to do the same on the Red Planet.

Perhaps, then, it may be better to frame the question in a slightly different way. Answering the riddle of life on Mars means narrowing down the more likely places where life might be found. As Curiosity has shown, ancient lake beds can preserve organics. "We've got places to look to find life in these kind of environments," says Professor Charles Cockell. "These findings don't show there wasn't life on Mars, because it could be located in some isolated oasis. On the other hand, it would be strange if there were vast lakes that extend across the planet and were totally devoid of life."

The corollary is that, if these most habitable environments that were lake beds and ancient salt deposits are examined and no signs of life are found there by either Perseverance or ExoMars, it would be a fairly strong indication there was no life on Mars. "You could never completely rule it out," says Professor Cockell, "but you could say, 'We've looked in these highly probable environments for life where we knew liquid water persisted for a long time, we find no evidence of life' and that suggests very strongly this was a lifeless planet."

* * *

The discovery of life on Mars will be a defining moment in this century. It takes a miraculous singular occurrence, life, to the irreducible universal statistic of two—initially, yet implying that the cosmos is teeming with organisms of one kind or another. Such a discovery would transform how we view ourselves and perceive our place in the universe, since we would know that life is possible in other locations.

If life ever got started on the Red Planet, perhaps it died out due to the overwhelmingly hostile climate conditions that developed. But, more intriguingly, it might have adapted to the declining climate. It could have started out in hot springs and then descended into deep aquifers, where it remains, waiting for us to discover it in the years to come.

In the next few years, we may find the next set of tantalizing clues that, in time, will bring us to the answer. There will doubtless be red herrings, false clues, and strange occurrences like those this book has recalled, but finally we will come to answer the greatest scientific question of the age.

ACKNOWLEDGMENTS

First and foremost, we would like to thank our respective partners, who have had to share the burden of our time away from more important familial duties—J and Sarah are, in every sense, coauthors. We also would like to thank our agent for this book, Humfrey Hunter, and all at Skyhorse, particularly our editor, Cal Barksdale.

As is clear from the text, we have canvassed widely and owe a special debt of thanks to the following for their help and assistance: Professor Jack Holt, Dr. Christopher McKay, Dr. Donna Shirley, Dr. Anna Horleston, Dr. Bruce Banerdt, Dr. Luther Beegle, Professor Charles Cockell, Dr. Abigail Fraeman, Dr. Ashwin Vasavada, Dr. Frances Butcher, Dr. Zach Dickeson, Dr. Jorge Vago, Dr. Ros Barber, and Dr. Melissa Guzman.

A special thanks to all the public information staff at JPL who helped with this book at a particularly busy time in the run-up to the launches in 2020. In particular, Elizabeth would like to thank Veronica McGregor, Andrew Good, and Mark Petrovich. They kindly took time out to escort her around the sprawling campus in golf carts, even squeezing in a visit to Perseverance (then under construction) in between hours of interviews.

* * *

Nick also wishes to thank:

Angelo Zgorelec, who gave him his first break; Robin McKie, Peter Beer, and Frank Miles, who taught him how to become a journalist in print, radio, and television; Duncan Mil, who always knew there would be life on Mars; and "that old gang of his," fellow newspaper colleagues from the Noachian era, most especially Nigel Hawkes, Robert Matthews, and Nick Nuttall, who were then on the *Times* in London; as well as contemporaries and companions on press trips in the ancient past: Henry Gee, Sue Nelson, Steven Young, Marcus Chown, and Dr. Stuart Clark as well as the late Steve Connor, whose passing has seriously diminished the world of journalism, let alone science writing. To Dr. David Whitehouse, he owes a particular debt of gratitude. A very fine author in his own right, David has cast his eye over drafts and been a constant source of good advice.

Nick also wishes to thank JPL and its many excellent people over the years who encouraged him—first as a kind of space-age Doogie Howser, when they gave him a summer job when he was still at school. In public information, he would like to acknowledge several people who have subsequently passed away: Don Bane, Jurrie van der Woude, Frank Colella, and Frank Bristow. More recently, Mary-Beth Murrill kindly escorted Nick and Sarah Booth around the lab. Over the years, he also benefited from endless discussions with people who had experienced the first Mars missions firsthand, not least the late Jonathan Eberhardt and Mark Washburn, author of a very fine book about *Viking* that got Nick started on his own road to the Red Planet.

Elizabeth wants to thank:

All her clients who have believed in her over the years as well as Algonquin College (where she teaches communications). Space.com, especially editor-in-chief Tariq Malik, who brought her to the International Astronautical Congress in Washington, DC. Space.com also supported her efforts to live at the Mars base in 2014. Another great supporter over the years has been the University of North Dakota's Department of Space Studies, which patiently saw her (an online student) through an MSc and a PhD that together took nine years. She has made a lot of friends and connections through that program and is now trying to give back to newer students. Elizabeth would like to thank her own mentors: Tim Lougheed, Kathryn O'Hara, and the (late) Peter Calamai, who unfortunately died shortly before work was started on this book. Her assistant,

Christina Goodvin, was very helpful in going through edits during the latter stages of the book's development.

Finally, she would like to thank her parents and siblings, who somehow understood her space-crazy ways all the way back to when she was a teenager and wanted to become an astronaut. Dare to dream.

And from both of us:

It is our dearest hope that somebody who reads this book will actually make the journey to Mars. We'll be with you every step of the way. And don't forget to write.

NOTES

Chapter 1: Frozen in Time

1 **contentious factions:** One of the more controversial discoveries was the announcement, two decades ago, by another NASA team led by the late David McKay (no relation), of supposed microfossils inside a meteorite from Mars that was retrieved from Antarctica. That particular story is covered here in chapter 9.

3 **mostly hidden from view:** Satellite data shows that of the 5.4 billion square miles (14 billion square kilometers) covered by the continent, 99.6 percent of that area is composed of ice.

20 **collar:** What this means is that both spacecraft can see up to 87° in both hemispheres.

22 **kind of icy sludge:** In the original *Science* paper, the word "lake" didn't appear. The term was later added in the press release that accompanied the story published by the American Association for the Advancement of Science. Some eyebrows were raised in the scientific community at this sort of glossing over of the ambiguities in the analytical and interpretational approach.

Chapter 2: Inside Out

25 **hearing of humans:** A human wouldn't hear a marsquake. The seismic waves that have been observed are too low in frequency for us to hear. Typically, a healthy young human can hear down to about 20Hz. This first event, and all the subsequent ones the seismometer team have observed, are well below 20Hz.

25   **Sol 128 event:** A day on Mars is thirty-nine minutes longer than ours, so it is given the name *sol* to distinguish it from the terrestrial day. The Sol 128 event marks the number of cumulative Martian days since InSight landed in November 2018. Compared to the detailed spread of the signals received since that date, this event is clearly the exception to the rule. "It was very strong in the high frequencies," explains Bruce Banerdt. "Usually the high frequencies get attenuated by their transmission through the rocks and you end up with mostly low frequency." In other words, the Sol 128 event contained a lot of high-frequency energy. "It was not very coherent in the sense that there wasn't a good polarization signal in it," Banerdt adds, referring to the way in which waves oscillate in more than one direction. "So the motion was kind of all over the place and random, so we weren't able to get a depth."

25   **magnitude 3 on the scale:** Geologists involved with InSight have created a number of different magnitude scales that have been specifically calculated with the Red Planet in mind. Broadly speaking, magnitude measured on Mars is very similar to magnitude on Earth. It is a measure of the total amount of energy released in a quake, and thus the severity of the shaking. The seismologists have to calculate this from the seismograms a little differently because of the somewhat different conditions on Mars.

27   **package of seismometers:** Though often called a French-built instrument, the "SEIS package" is more involved than that. The instrument development was led by a French team, but the completed instrument contains both French and British components. It is effectively two seismometers, each with three components. The primary instrument, the VBB, was designed and built in France and is tuned to listen to low-frequency waves; the second instrument, the SP, was designed and manufactured in the UK and is tuned to look at high-frequency signals. The two sets of sensors are all packaged together in SEIS. From an American point of view, it meant NASA's budget did not have to be expanded to develop a state-of-the-art instrument package from scratch.

28   **portent of war:** All ancient civilizations remarked on the maleficent aspect of this red-colored star that wandered so menacingly around the sky. Surviving texts from ancient Babylon refer to it as Nergal, the God of War, while similar records in Sanskrit refer to the same deity as Mangala. To the ancient Chinese, it suggested

fire and was called Ying-Huo. To the Aztecs, it was Huītzilōpōchtli. More familiar in Western culture is its Greek name, Ares, and the Old English, Tiw, which informs our name for the second day of the week. That we know Mars by its Roman name is but an accident of history, all the more curious for the fact that farmers in Ancient Rome prayed to it for the success of their crops. Today, spring officially begins in the month of March whose French name, *Mars*, is all the more appropriate.

29 **nearest equivalent mission:** The InSight spacecraft is based on the successful *Phoenix* mission that had landed at the north pole ten years earlier. By using the same basic design, risks have been reduced, as have development costs.

30 **officials were upbeat:** The InSight lander's parachute was, they were at pains to point out, stronger and could open at a higher speed. The spacecraft even had a thicker heat shield "partly to handle the possibility of being sandblasted by a dust storm," NASA had said.

34 **underlying noise:** Vibrations from the oceans and weather are, seismically speaking, deafening here on Earth. This continental cacophony is caused by faults in tectonic plates bumping and grinding. On Mars, researchers are discovering signals that could never be observed on Earth, as our planet is so full of noise.

34 **weather station ever built:** As a weather station, InSight is setting the gold standard. "We're taking some of the best atmospheric data at the surface that's ever been done on Mars," Banerdt says. A suite of simultaneous pressure, temperature, and wind data is recorded around the clock at high rates and with fine precision. "I think that's going to be really revolutionary in understanding Martian atmospheric dynamics and planetary atmospheres in general," Banerdt adds.

34 **Downslope winds:** Technically known as an adiabatic wind, they result when an air mass overflows a mountain and, as it descends on the other side, dries out and warms up. In the Alps on Earth, these notably warmer winds are known as foehn winds, and in the Pacific Northwest as Chinook winds, from the local indigenous people who first noticed them. In the thinner Martian atmosphere, they are much less notably warm but still apparent in the observations.

38 **other side of the planet:** InSight's longitude is 135°E, equivalent to a meridian that passes through the Pacific Ocean through Australia if it were on Earth. A network of seismometers would allow the origin of any marsquakes to be narrowed down more effectively.

38 2.4 hertz: Frequencies are measured by the unit named after Heinrich Rudolf Hertz, the first person to conclusively prove the existence of electromagnetic waves. They can be thought of as the number of times they are seen every second. 1Hz is one cycle per second, and as noted in the text this corresponds to the hum of electronics within the InSight spacecraft.

38 periodicity: According to Suzanne Smrekar, there are a number of possible explanations for the patterns that might underlie the seismic signals. Tidal stresses due to Phobos, the larger Martian moon, might in theory induce periodic stresses. Tidally induced quakes are common on our own Moon, but there the stresses are much larger. Phobos is tiny compared to the size of the Red Planet. Some marsquakes could result from thermally driven processes such as the freezing of an aquifer, which could create quakes. While it is an interesting idea, Dr. Smrekar cautions there is no evidence for it currently.

39 tides induced: The Martian moons, Phobos and Deimos, are both very small compared to Earth's Moon, but should be detectable from the almost imperceptible tides they generate on Mars. As of this writing, it will take many months and a great deal of data processing to bring out that signal.

39 mole has been ratty: *The Wind in the Willows* has had a disproportionate influence on explorers of Mars. In 1976, after *Viking 1* landed, a number of rocks in the immediate foreground were chosen as candidates to be moved out of the way. They were given the informal nicknames Mr. Toad, Mr. Badger, Mr. Rat, and Mr. Mole, the names of characters in the book by Kenneth Grahame. In the event, Mr. Badger was chosen.

42 liquid core: Because they do not travel through liquids, shear waves leave behind what is known as a "shadow zone" when they are observed in terrestrial seismometry. This is used to determine the radius of the inner and outer cores. To understand the exact conditions, account has to be made for the refraction of the pressure waves at the boundary of the outer core as well as its inability to transmit shear waves.

44 winter poles of the planet: In September 2018, another group of

JPL scientists reported changes to Earth's own spin thanks to ice loss in Greenland, the movement of glaciers, and the convection of heat within the terrestrial mantle.

45   **subtle wobbles:** This type of wobble occurs on Earth and is called a Chandler wobble, after Seth Chandler, the American astronomer who discovered it in the late 1800s. The exact cause remains puzzling, with little scientific agreement on its source. One explanation is that the liquid interior of Earth could play a part.

45   *Phobos 2*: As its name suggests, this Soviet mission had been planned to land instruments on the larger Martian moon. However, contact was lost before it could perform such maneuvers after entering orbit around the Red Planet. *Phobos 2* did not make enough systematic observations over a long enough period to answer the question of whether Mars has a magnetic field. It did amass hints that there was possibly some sort of magnetic activity, but the observations were either so fleeting or so contradictory that many declared them illusory.

46   **much lower orbit:** Though now a well-established technique in spacecraft engineering, Mars Global Surveyor was the first mission to use what is known as aerobraking at the Red Planet. As it passed through the outermost fringes of the Martian atmosphere, it slowed down by a slight amount because of atmospheric resistance at roughly fifty miles (one hundred kilometers) altitude. This changed its orbit without recourse to eating into large fuel reserves. Over a four-month period, the maneuver lowered the high point of the craft's orbit from 33,554 miles (54,000 kilometers) to near 280 miles (450 kilometers). While attempting to do the same with its nearest point—periapsis—the maneuver had to be halted in October 1997 because the pressure imparted by the atmosphere was causing one of its solar panels to bend out of shape (probably as the result of damage that had occurred at launch because it was not latched properly). A gentler aerobraking than originally envisaged was then carried out. Mars Global Surveyor didn't enter its operational orbit until the spring of 1999, when its high point was shrunk down to 280 miles. At this altitude, Surveyor circled Mars once every two hours and was in the best orientation for its mapping operations. This meant Mars Global Surveyor always crossed the dayside equator at 2:00 p.m. local Mars time when it moved from south to north, so the mission would see the same features

at roughly the same local time every sol and any changes would be more apparent.

47   **aurorae:** In late December 2014, the MAVEN spacecraft detected widespread auroras over the northern hemisphere of the Red Planet. The "Christmas lights," as they came to be known, circled the planet and descended so close to the Martian equator that, had they been observed on Earth, they would have extended down to Florida. More recent observations have shown that not only is solar material getting in, Martian material can get out. The areas where there is no underlying magnetized crust show greater escape of material. "If we look at the escape rate as Mars rotates," says Bruce Jakosky, "we see it change significantly as the planet rotates depending on whether these 'bumpy things' obstacles are pointed at the Sun or away from it."

48   **electrically conducting:** The Martian ionosphere extends outward from seventy-five miles above the surface of Mars out to several hundred miles. It is extremely tenuous and is made up of ions and electrons that are created by the action of the solar wind. Unlike Earth's ionosphere, the Martian ionosphere is not shielded by a strong magnetic field. The way in which the Martian magnetic field interacts with the ionosphere is not well understood, thus complicating the way in which magnetometers can piece together the underlying field.

51   **gravity field:** The gravity exerted by the Red Planet is weaker than on Earth as Mars is much smaller. As spacecraft orbit the planet, they can measure the acceleration toward the Martian surface thanks to the underlying gravitational field. They have found that it is not uniform due to localized concentrations of mass, such as the Tharsis volcanoes, as well as distinct topographic changes in the surface, such as larger impact craters. As the underlying topography has been measured by laser altimeters, the gravity field can be inferred more accurately. The greater accuracy will show exactly how the surface is deformed, for example, by tidal forces exerted by the Sun or Phobos. This deformation in the crust essentially reveal show "stiff" the underlying interior is and just how liquid the core is. Measuring such forces from the surface will enable InSight to contribute to a better understanding of the underlying gravity field.

Chapter 3: Curiosity

60 perilously thin: Conditions had to be ideal for the flight control-lers to be as certain as they could be that the rover would make it all the way down to where it was supposed to land. Consulting colleagues who work on the other NASA Mars orbiters, the Curi-osity landing team were provided with extensive thermal sound-ings—heat profiles of the atmosphere—along with images of any dust or clouds. That way they could better anticipate any potential problems along the path the vehicle would take.

63 Chris Isaak: A renowned musician who first came to public atten-tion for his crooning in the 1990s that was compared to greats of the genre like Roy Orbison. In the 1998 HBO series, *From the Earth to the Moon*, Isaak played astronaut Ed White.

65 The rover determined: This test was carried out by the CheMin instrument. The great benefit of the instruments aboard Curi-osity is that they can effectively check on each other's findings. When the SAM chemical oven later came online, it independently confirmed the CheMin result. SAM infers mineralogy based on the temperatures at which various components of the rock break down. It found that the clay released water at its characteristic tem-perature of around 700°C (nearly 1,300°F).

65 hundreds of scientists participating: By the end of 2019, Ashwin Vasavada says there are roughly one hundred engineers and five hundred scientists, a number of whom are based at JPL. Adam Steltzner has been quick to acknowledge all these unseen colleagues as "a huge team." In his estimation, there were over three thousand people at JPL who worked on Curiosity throughout its develop-ment and execution. "Thousands of others in over thirty-seven states and seven other countries all had committed their blood, sweat, and tears to the effort of building that rover," he has said.

68 Rubin's legacy: In neat symbolism, one of the researchers on the extended Curiosity science team is David M. Rubin of the Univer-sity of California, Santa Cruz, Vera's son.

71 tally of seventeen cameras: There are four front Hazcams, four rear Hazcams, four Navcams, two Mastcam cameras, MAHLI, MARDI (considered in chapter 5), and the ChemCam RMI.

71 total dose of radiation: In 2013, Curiosity's Radiation Assessment Detector (RAD) showed that a mission to Mars that stayed 500 days on the surface and just under a year—360 days—traveling to and fro would only rack up about 1.01 sieverts of radiation. That is comparable to the 1-sievert lifetime radiation limit the European Space Agency sets for its own crews.

74 **brought from elsewhere:** Even before this Curiosity discovery, organics have been found inside meteorites that have been blasted from Mars (a story told in chapter 9). Even farther out in space, in the dark dust lanes between the stars, radio astronomers have discovered biologically important molecules. They can literally "tune into" the frequencies at which these molecules resonate back on Earth. Each resonance is unique. The characteristic radio signals for over sixty compounds including ammonia, formaldehyde, and water itself have been discovered. All are used in the basic units of genetic systems. Their precursors—amino acids, purines, and sugars, among others—have also been seen in laboratory experiments which have attempted to create conditions where life could form.

75 **three billion years old:** This was partly based on past Curiosity measurements of the natural radioactive decay of elements found within the sediment. Over time, radioactive potassium decays into a neutral form of argon. So the more argon there is, the older the sample would be. The difficulty is that the sediment, or at least the rocks it was eroded from, first crystallized from volcanic magma very long ago, which will have changed the radioactivity.

78 **very sensitive detectors:** Two separate telescopes were used at Cerro Pachón in Chile and Mauna Kea in Hawaii. Infrared measurements have to be made on mountains because the rest of Earth's atmosphere "drowns out" any signals from Mars.

78 **patch tracking mode:** Mars Express is tiny. The more recent Trace Gas Orbiter is more than three times as large. As a result, the earlier spacecraft can be pointed to a spot on the surface much more easily, with less wastage of precious fuel.

80 **boundary layer:** In aerodynamics, boundary layers have a variety of meanings. Here it is taken to mean where the flow of surface winds can be shown to follow the topography of the surface. The actual boundary is formed where the airflow is no longer affected by the surface. It can move up or down depending on the relevant atmospheric conditions.

81 **Belgian-built spectrometer:** Known as NOMAD (Nadir and Occultation for MArs Discovery), it can measure the spectrum of sunlight across a wide range of wavelengths (infrared, ultraviolet,

and visible). This broad coverage enables the detection of low concentrations of the rarest gases seen in the Martian atmosphere.

Chapter 4: The Road to Utopia

86 visitors from Mars: So claimed one who knew von Kármán and his fellow Hungarians, the German physicist Otto Frisch. They were a "galaxy of brilliant Hungarian expatriates," as he termed them, whose number included Edward Teller, Leo Szilard, Eugene Wigner, and John von Neumann. All made fundamental contributions to modern science, not least in the years before and during World War II.

90 Deep Space Network: Originally, one of the "southern hemisphere" dishes was located in South Africa, but growing concerns over the apartheid regime in the 1960s meant that it was abandoned in favor of the Spanish station outside of Madrid.

91 exobiologists: The term *exobiology* was coined by Nobel laureate Joshua Lederberg, who was involved in the *Viking* missions. Subsequently, it has largely been superseded by the more general term *astrobiology*.

94 By 1967: Voyager had the misfortune to come to fruition during a year that was NASA's annus horribilis. The fire that killed the *Apollo 1* astronauts in January was the subject of a congressional investigation that some within the space agency thought was little more than a witch hunt. The more or less unrestricted funding that had accompanied the Apollo program was already being scaled back, and Voyager's appropriations hearings had the misfortune to coincide with that. The total budget for this Mars mission was roughly the same as the expenditure each month on the Vietnam war.

96 *Mars 2* and *3*: These Soviet spacecraft were altogether more ambitious missions. The landers they carried were designed to take measurements from the surface of Mars. With no way of altering their trajectory in the face of the dust storm, the landers were simply released on cue. *Mars 3* survived all the way down to the surface. After twenty seconds its transmissions ceased, presumably as it was blown around or its sensitive electronics became clogged with dust. The fact that a pendant of Lenin was delivered to the Red Planet was dutifully noted by the Soviet press.

98 George Low: Often described as one of the "wise men" of Apollo-era NASA, Low later headed human spaceflight within the agency. He had been a former deputy head of space science within the agency, hence the discussion with his old Rensselaer Polytechnic pal, Ed Cortright.

98   **success of the Lunar Orbiter:** Many of the engineers who were involved with *Viking* cut their teeth on the Lunar Orbiter spacecraft. All five orbiters worked and, for their time, produced the highest resolution images of the surface of the Moon. In the fallout from the Voyager cancellation, NASA Headquarters felt that dealing with Langley would be easier than dealing with JPL, based on recent experiences with the invariably independent contractor in Pasadena. As Gerry Soffen noted, Ed Cortright held a lot of sway with the NASA higher-ups in Washington, DC, and could see which way the wind was blowing.

98   **NASA Langley came out top:** NASA Langley established the Viking Project Office, which managed the whole project. JPL would continue to look after the Deep Space Network and would lead the development of the orbiters. Martin Marietta was the primary contractor; the company actually cut metal and built the lander. In time, the engineering teams would mesh, and nobody ultimately cared who paid their wages. As well as the designers of each experiment, there were Langley managers assigned in supervisory roles to oversee the design and manufacturing process, feeding back into the Project Office and the gimlet eyes of Jim Martin. Though Carl Sagan tartly commented that this was proof of "*Ad Astra Per Bureaucracia*," even he would not contradict Jim Martin when it came to choosing landing sites for the mission.

114   **all the *Viking* biologists moved on:** Unlike today's missions, where people appear and then reappear on later missions, *Viking* was a means to an end. The *Viking* scientist Ben Clark has noted that there were at most fifty involved with all the main experiments; today, as noted with Curiosity, there are roughly five hundred. There were few early career scientists involved in *Viking*. "Furthermore, many of the scientists had strongly established reputations to defend," Clark wrote in November 2019. "Humor was rare. For many this would be their one and only space mission."

Chapter 5: The Measure of Mars

115   **Malin now has a ringside seat:** Malin cameras ride on the Juno orbiter currently in orbit around Jupiter, the Lunar Reconnaissance

Orbiter, and the OSIRIS-Rex, where it has helped in landing the spacecraft on an asteroid called Bennu for a mission that will return samples to Earth in 2023.

116 **Mount St. Helens:** This large volcano in Washington state had lain dormant for more than a century, when some lava flows and unusual activity were observed in March 1980. Two months later, a spectacular eruption tore off the top of the mountain, turning the surrounding forest into dead ash and killing around sixty people.

117 **wicked sense of humor:** Dr. Paine, who passed away in 1992, is often cited as the last great NASA Administrator. In 1971, he resigned when it was clear that he could no longer maintain a political consensus as Apollo drew to a close. More piquantly, he was known as a "swashbuckler" in Washington. After the first Moon landing in 1969, he had enthusiastically endorsed plans to send astronauts to Mars.

118 **standard geological textbook:** *Earthlike Planets: Surfaces of Mercury, Venus, Earth, Moon, Mars* was a unique summary of knowledge from the first fifteen years of planetary exploration when it was published in 1981 (New York: W. H. Freeman & Co.). Malin's supervisor, Bruce Murray, had been involved with the early *Mariners* to Mars and Mercury and was famous for his sometimes antagonistic style. From 1976 to 1982, Murray was director of the Jet Propulsion Laboratory. When the galleys were being proofread, the author description beginning "Michael C. Malin is" was amended to read "a smart-ass graduate student" by a passing wit.

122 **susceptible to jamming:** A jammed tape recorder nearly scuppered the ability of the $1.3 billion *Galileo* spacecraft to record data as it orbited the giant planet Jupiter in the mid-1990s. Once the spacecraft was launched, the mechanism for the tape recorder, used to record its observations for later playback, literally jammed. JPL learned to avoid putting too much stress on the tape spool and transmitted data in more efficient ways. No data was lost in its phenomenally successful eight-year odyssey around the largest planet

Chapter 6: The Pathfinder

152 **budget was on a downward curve:** Although *Viking* cost ten times as much as *Pathfinder*, it was altogether more ambitious, involving two orbiters and two landers planned to last many years. One

former *Viking* scientist pointed out: "Four limousines cost more than a motor scooter, even a very fine one."

157 **sending of dangerous nuclear material:** Rarely mentioned in discussions about the launch of nuclear power sources into space is the remarkable story of what happened in the 1960s, when an early Nimbus weather satellite crashed into the Pacific after launch from Vandenberg Air Force Base in northern California. Its power unit was retrieved by Navy divers. So unharmed was this (admittedly heavily insulated) lump of plutonium that the whole item was placed on board the next in the Nimbus series, which was launched successfully into space.

158 **Rocky:** The prototype that morphed into Sojourner was known as Rocky 4. Rockies 1 to 3 were earlier definition studies. This rocker bogie motion has been proven to work well, and the Curiosity rover, currently trundling across Gale Crater, uses an advanced version of it.

160 **150 million:** Most of the hits came on the following Monday (three days after the landing) because, at that time, most people had access to the internet only from computers in their offices. In October 2019, project scientist Matt Golombek proudly pointed to a chart of website statistics that was still affixed to his office wall more than twenty years later. There was a peak on the day of landing, a drop, and then on the first day of the next week, people "had to go to work to actually start seeing the images." For Golombek personally, it meant that he was also invited back to his old high school for what was known as a "Dr. Matt Golombek Day."

164 **weathering:** This kind of silica enhancement is now known to be a global process on Mars as there have been interactions between the atmosphere and the surface over geological time. As a consequence of the changes to the Martian orbit discussed in chapter 1, the various climatic changes will have allowed thin films of water to be stable for geologically significant periods of time. Even this amount of water would be enough to alter what was happening underneath the rocks' surface and to create some weathering of the most susceptible minerals. As silica is the least soluble, it would tend to accumulate, and that is what has bumped up the silica content. Elements such as chlorine and sulfur would tend to be dissolved out and moved around the planet by atmospheric processes. This increase in silica content has also been replicated in

laboratory experiments, and the high silica content due to weathering has been verified by three subsequent missions.

165   **weren't going to take him on:** By 2001, Donna Shirley wryly noted, Administrator Goldin "was on his third president." The generally harmonious dealings between engineers and scientists who had worked on Sojourner was missing on some of the follow-on missions. Today, she notes that a sample return mission is going to be very, very tough to carry out, even with the advances in technology that have taken place in the years since.

166   **now in her late seventies:** Donna Shirley's website may be found at www.managingcreativity.com

168   **Wally Funk:** The story of this remarkable pilot is told in the recent book by Sue Nelson, the former BBC science journalist, while the greater story of the *Mercury 13* women is also being told in the new book by Rebecca Siegel, *To Fly among the Stars: The Hidden Story of the Fight for Women Astronauts.*

169   **all-female spacewalk:** NASA planned to run the first all-female spacewalk in March 2019 with Anne McClain and Christina Koch, two rookie astronauts. But the idea was put on hold after McClain, who had just completed her first spacewalk, found that the spacesuit torso size was not comfortable. She needed to downgrade to a smaller size in the off-the-rack spacesuits available. The design of these NASA Extravehicular Mobility Units, as they are more formally known, dates back to the 1970s and favored larger (i.e. male) astronauts at a time when crews did not include females. The EMU suit can swap parts for different body sizes, but modifications can take up to twelve hours of crew time to carry out. NASA determined it was best to prioritize other things and postponed putting the two women outside at the same time. The spacewalk went ahead with a man and a woman instead. NASA then sent up more spacesuit parts on a cargo mission and successfully ran the first all-female spacewalk in October 2019, with Koch and Jessica Meir.

Chapter 7: Waterworld

178   **similar to shorelines:** Their first paper on the subject, which was published in 1989, hedged its bets in jargon: "Transitional Morphology in the Western Deuteronilus Mensae Region of Mars: Implications for Modification of the Lowland/Upland Boundary."

183 Martian atmosphere is being lost: Various atmospheric gases provide useful information. In this case, Bruce Jakosky and his team examined the abundance of two different isotopes of argon, which is chemically unreactive. Isotopes are atoms of the same element with different masses. Since the lighter of the two isotopes escapes to space more readily, it will tend to leave gas behind which has been enriched in the heavier isotope. The Jakosky team used the relative abundance of the two argon isotopes measured in the upper atmosphere and at the surface to estimate the fraction of the atmospheric gas that has been lost to space. Isotope fractionation, as it is more formally known, has many uses in trying to understand conditions on Mars.

188 chemical ratios of hydrogen and deuterium: Roughly speaking, the D/H ratio can be used to determine where objects in the solar system formed, based in part on our knowledge of how much deuterium and hydrogen has been observed in other planets, comets, and asteroids. The amount of water and heavy water (which is formed with deuterium rather than hydrogen) contained in these bodies helps nail their origins. Compared to Venus, the D/H ratio on Mars is larger by a factor of five. First measured by telescopes from Earth, the D/H ratio was also observed in meteorites that have come from Mars. The meteorite measurements were very different, possibly because of biased sampling and because of unknown geological processes that altered them. Most recently, the SAM instrument on Curiosity has measured the D/H ratios in water released upon the heating of sand samples. In one series of measurements taken from a wind drift site, the vapor released contained more deuterium than that seen in Earth's water, i.e. the Martian water is heavier. This is to be expected. The lighter hydrogen atoms in the Martian atmosphere are escaping faster than the heavier ones. As two hydrogen atoms are chemically bound to one of oxygen to form a molecule of water, its loss to space forms a "two for the price of one" bargain so far as water loss is concerned. If the water had remained as vapor in the atmosphere, then it too would have been lost to space. Theoretical calculations show that the deuterium/hydrogen ratio would have been even higher.

Chapter 8: Claims

200 brightest for a century: Because the orbit of Mars is elliptical, oppositions—when Mars is on the same side of the Sun and at its closest to the Earth—vary in distance. At certain oppositions, Mars can be much brighter because the two planets are closest. The most recent was in 2018. These close oppositions occur roughly every fifteen years, so the next time that takes place will be in 2035.

200 two tiny moons: Throughout the eighteenth and nineteenth centuries, it was widely thought the Red Planet would have two moons. In Jonathan Swift's *Gulliver's Travels*, published in 1726, for example, the Lilliputians had already discovered them.

201 William H. Pickering: This is not the same Bill Pickering who was director of the Jet Propulsion Laboratory from 1954 to 1976. Lowell's associate had already "discovered" Martian lakes in 1892 and made many discoveries in astronomy. He believed he had found vegetation on the Moon and that there were insects in one particular large crater.

201 golden gate: A handful of other pioneers momentarily alighted from a train that had stopped at this same railhead in December 1913, among them Cecil B. DeMille. They had headed west to escape the influence of the Edison Company, which more or less controlled the technology of moviemaking. Rather than pay their tithe to Edison, these pioneers, bankrolled by an entrepreneur named Samuel Goldfish (later changed to Goldwyn), were looking to shoot a new Western. They stayed in Flagstaff only long enough to conclude that the scenery was too dull for the backdrop to their proposed movie. Both equipment and entourage were shunted to points west. The more appetizing scenery of orange groves in a little-known suburb of Los Angeles called Hollywood beckoned. But for DeMille not liking the Arizona scenery, Flagstaff might have become the moviemaking capital of the Western world.

210 We were devastated: Glenn Cunningham's testimony was typical of those who were involved. On a visit to JPL three months later, project officials were still visibly shaken. One engineer had tears in his eyes when Nicholas Booth talked to him about the loss over a drink.

212 Video resolution images were made available: This, and all the other images of the "Face on Mars," may be found at Mike Malin's website at http://www.msss.com.

217 detect hydrogen peroxide: More recently, observations from the Mt. Hilo Observatory on Hawaii have been made with nearly a thousand times greater sensitivity. "They can see down to the levels at which hydrogen peroxide is predicted to produce the Labeled Release results," Gil Levin said in the late 1990s. "And they find absolutely no evidence [for it]."

222 all original data from space missions: As a result of the International Geophysical Year that begat the opening of the space age, all space agencies were bound by international agreement to store and make available any of their scientific findings. The NASA Space Science Data Coordinated Archive (NSSDCA) was created to act as the permanent archive for all of NASA's space science missions. Its "active archives" provide access to data to researchers and, in some cases, the general public. NSSDCA also serves as NASA's permanent archive for earlier space mission data, some of which has not been fully digitized. Until March 2015, it was called the National Space Science Data Center (NSSDC).

Chapter 9: Reactions

225 nanometers: A nanometer is roughly 1/50,000th of the diameter of a fine hair strand. A micron is a millionth of a meter, while a nanometer is a billionth of a meter. A strand of hair is roughly fifty microns across, so what McKay and Gibson observed was a hundred times narrower than a human hair.

225 "What are you doing with": McKay and Gibson told this story to the NASA PR flacks who were preparing the two scientists for the press conference in August 1996. An equally telling reaction came from McKay's precocious thirteen-year-old daughter, Jill, when he had left some pictures on his desk for her to find and he teased her with a simple question: "What does that look like to you?" She frowned and said, "Bacteria."

226 Republican National Convention: In some quarters, even the timing of the announcement seemed suspicious. The meteorite paper was originally due for publication on August 16 in *Science* magazine. Reporters would have flocked to Washington, DC, for the press conference and filed their stories for their August 16 editions. But that would have coincided with the Republican National Convention in San Diego and, some have argued, would have diverted attention from it. Others have suggested that the

Democratic White House would hardly want the GOP convention to take attention away from an announcement of this magnitude.

227 **the late Professor Colin Pillinger:** Despite his preeminence in planetary science, which included election to the Fellowship of the Royal Society, he had not been offered tenure at Cambridge in the early 1980s. When the Open University courted him for a permanent post, he took the bait. His research group, including Monica Grady and Ian Wright, gladly decamped the forty-five miles to Milton Keynes.

228 **only the tougher rocks:** The rocks from Mars that are of more interest to geologists are what Steven Squyres, the former Cornell professor of astronomy, has termed the "soft, crumbly ones," the sedimentary layers that show telltale signs that the surface was clearly once altered by running water. The problem is that those kinds of rocks would never survive the trip if they were blasted off the surface of Mars. "We don't have any rocks like that in our collection of meteorites," Squyres told the journalist Miles O'Brien, "because they just don't make it, they don't survive."

232 **the observed isotopes of carbon and oxygen:** In particular, they found that the samples were enriched in Carbon-13 and deficient in Oxygen-18. The Carbon-13 findings suggested the fluids that had flowed through them had come into contact with the Martian atmosphere. The chemistry of the oxygen depended on the temperature at which the carbonates were actually deposited by the water. By dissolving the samples in acid, the researchers deduced the temperatures were fairly low, as was the oxygen composition. Oxygen, too, will have been lost to space over time, with its heavier isotopes staying put preferentially compared to the lighter ones. This has indeed been confirmed by measurements. Given that all these findings were, in some sense, out of this world, even the most skeptical among the group were now convinced that the observed oxygen compositions were not terrestrial.

232 **rock formed at a low temperature:** It was Chris Romanek, who was working with Everett Gibson, who initially identified the carbonates as being low-temperature deposits. Without that insight, the carbonates would have been thought to have formed at a higher temperature and therefore not of interest with regard to a life-on-Mars argument.

234 entrainment: Biological and other material can get entombed within minerals that precipitate out when water evaporates in a process known as entrainment.

235 nanobacteria: The size of the observed features in the Allan Hills meteorite would have also limited their biological significance. Two years later, a panel convened by the National Research Council concluded that the suspected nanobacteria observed in the Allan Hills meteorite were a hundred to a thousand times too small to be biological. That is, there was not enough room for them to develop the machinery of a cell in their proteins and DNA to allow for even simple metabolic processes.

240 always the images of "the worm" and "the swimmers": Ironically, when the Houston scientists had presented their findings to the journal *Science*, they got short shrift. "We delivered eighteen images with the manuscript," Gibson later said of their submission in April 1996. "The editors and our reviewers felt that the other data were more enticing than the images." Given the extraordinary nature of the findings, there had been an extraordinary peer review process. Normally, a couple of reviewers would be called in to assess any new findings. This time there were nine. According to some sources, the first draft didn't pull any punches. It was much stronger concerning the possibility of their having found life.

241 Battle lines were being drawn: Much of the contemporary news coverage as well as that in specialist journals noted some of the residual antipathies that seemed to pile up. Six months later, they erupted as a stream of vitriol that was faithfully recorded in a *Newsweek* article (February 17, 1997) headlined "War of the Worlds!" It chronicled a number of vituperative comments about the NASA team's work. One scientist referred to the Houston team as an inferior group of researchers wanting to set the agenda. One of the NASA group described a critic as someone who hadn't held a permanent job in thirty-two years. "What bothers me is the fact that claims of finding life are becoming more and more entrenched," said Professor Jeffrey Bada of Scripps, one of the more vociferous critics of the NASA work. "They are much more inflexible, and I don't feel that does science a good service. When Congress and the public see us squabbling, it's not good."

242 part of this organic carbon is Martian: More recent analyses have shown that the form of carbon observed also implies it was cooked during the meteorite's fiery entry through Earth's atmosphere. Tim Jull of the University of Arizona has shown that much of the carbon is the short-lived form of Carbon-14. That could only have come from when the meteorite would have been heated up as it passed through Earth's atmosphere. Comparing the atomic signatures of carbon in the rocks to those found on Earth, Jull has found much of the organic material in the meteorite to be similar to those found on Earth. Though there were some carbon molecules that were clearly extraterrestrial, that didn't amount to much in the cosmic scheme of things. "I don't think they require the intervention of bacteria to create them," Jull said at a conference in August 1999. "It is not a biotic form of carbon. It is a molecule that can be created by chemical reactions."

248 **online forums:** Specifically, Veronica McGregor namechecks the users of UnmannedSpaceflight.com, then numbering in their thousands, which represents people who are devoted and highly literate followers of the space program.

Chapter 10: Signatures

253 **Caltech colleague:** In his interview with Elizabeth Howell, Rich Rieber namechecked Professor Mike Brown, the California Institute of Technology astronomer who has found several dwarf planets about the size of Pluto in the outer solar system in the early 2000s. Brown was one of the leading advocates to remove Pluto from planetary status, a move that remains reviled by many, including Alan Stern, former NASA executive and the lead of a Pluto flyby mission called New Horizons. Brown cheerfully calls himself @plutokiller on Twitter.

257 **chemical oven:** Perseverance will avoid the problems with the perchlorates that have been discovered on the Martian surface. Using a "conventional GCMS (mass spectrometer) destroys the very molecules you are looking for," Luther Beegle says. It is suspected that this is what happened on the *Viking* missions.

258 **local storms sometimes turn global:** Because the Martian atmosphere isn't held as firmly as it is on Earth, the wind can whip up dust from the surface much more efficiently, which can lead to storms that eventually cover the whole the planet. This was the case with the extensive dust storm that started in the summer (on Earth) of 2018 and quickly obliterated the Martian surface from view. A

similar event occurred in 1971 when *Mariner 9* arrived. Though a wealth of data has been obtained in between those events, atmospheric scientists are still unsure what makes a regional dust storm grow into a much larger one. Not all localized storms do. Global storms tend to happen when Mars is closest to the Sun, when its southern hemisphere receives more sunlight.

260   common isotopes of carbon: Both forms of carbon have six protons orbiting their nucleus. Carbon-12 has six neutrons in that central core, while Carbon-13 has seven. The extra neutron affects its physical properties in a significant way so far as biology is concerned. Because Carbon-13 weighs a fraction more, it cannot move as fast as Carbon-12. On a planetary surface, Carbon-12 tends to move away over time and escape into space, meaning that the Martian atmosphere has more Carbon-13 remaining compared to Carbon-12, which will have tended to float off. Carbon also has an unstable isotope, Carbon-14, whose radioactive decay is often used by archaeologists here on Earth—it helped determine the true age of the Shroud of Turin, for example. Because its radioactive clock has been ticking, the presence of Carbon-14 helps calibrate when physical events have taken place.

262   X-ray spectrometer: The instrument, formally known as PIXL (Planetary Instrument for X-ray Lithochemistry), can observe the total number of elements that are contained in the samples also observed by SHERLOC.

268   meteorite specialists: These include researchers who work in the same building at the NASA Johnson Space Center where the features inside the Allan Hills meteorite were originally analyzed.

271   rolling with the punches: The irony is that, originally, ExoMars should have already arrived with American involvement. Cost overruns, cutbacks, and innumerable changes have added an extra decade before it has even reached the launch pad. Instead of duplicating efforts, sharing joint goals, instruments, and budgets is often seen as a better way forward. However, such cooperation, usually seen as a utopian joining together of like minds by participants, has its own complications, not least when one or more partners effectively renege on the deal. "The opportunity to choose our instruments came in 2003, a year before NASA's did to select those they wanted aboard Curiosity," Jorge Vago says. "This gives you an idea of how complicated the life of ExoMars has been. In

the end, NASA picked their payload later, but they flew before us."
When NASA pulled out, the European Space Agency agreed with
the Russian state space research organization, Roscosmos, that it
would provide the launch vehicle.

273 **The rest failed completely:** Russia's last major attempt to land
on the Red Planet was in the mid-1990s. In November 1996, it
planned a mission far more complex and ambitious than the two
American ones that were dispatched in the same launch window,
Pathfinder and the Mars Global Surveyor. The whole Mars '96
mission weighed over six tons. By comparison, the Global Surveyor weighed one ton and *Pathfinder* a measly 1,967 pounds (894
kilograms). The Mars '96 mission was hoping to find important
clues about the possibilities for biological evolution, with instruments including shell-like penetrators that would drill into the surface on landing. In any event, it crashed into the Pacific when its
upper stages malfunctioned. Intriguingly, Moscow has also had a
greater interest in the moons of Mars. The arrival of *glasnost* saw
twin missions aimed for the larger Martian moon, Phobos, that
were a last hurrah for the Soviet Union in the summer of 1988.
Both failed, but not before *Phobos 2* entered orbit around the Red
Planet in January 1989. An even more ambitious mission to return
rock and dust samples from Phobos, with the odd-sounding name
Phobos-Grunt (*grunt* is Cyrillic for "ground"), was launched successfully in November 2011. Once again, its engines failed to fire
it toward the Red Planet.

273 **Trace Gas Orbiter:** In the interim, the orbiter has been moved
into its current "operational" orbit. For its first year at the Red
Planet, the orbiter was still in a highly elliptical "parking" orbit,
which has been reduced—by aerobraking—to a near-circular one.
It is now orbiting roughly 250 miles (400 kilometers) above the
surface, which takes two hours per orbit to complete.

274 **Rosalind Franklin who was overlooked:** Partly it was because of
her gender, but also because she died tragically young and, unlike
Crick and Watson, that meant she was not eligible for a Nobel
Prize.

276 **dominated by certain clays:** Technically, they are known as
iron-magnesium philo-silicate clays, which means they have
formed in the crust of the planet at a specific point in Martian
history.

278   **3.8 billion years ago:** Geologists refer to this as the "Noachian era," which has a vague connection to biblical floods. In the southern hemisphere of Mars, there is an area called Noachis Terra—the land of Noah—which exhibits telltale signs of flooding in the ancient past on Mars.

Chapter 11: Return to Sender

283   **tried and tested technology:** Though the lander will be heavy, JPL is looking toward powered descent with a larger version of the InSight mission rather than a sky crane, for they need to put more mass down than even a sky crane can drop on Mars.

287   **solar electric propulsion:** This propulsion system works as follows. Two identical, nearly fifty-foot-long (fifteen-meter-long) solar arrays collect energy from the Sun and transform it into electricity. Then the propulsion system gets to work. Inactive xenon gas moves into the thrusters, where its outermost electrons are removed. This creates an electric charge on the xenon atoms, which are also called ions. A high-voltage grid system then uses electric charges to move these xenon ions out of the thrusters at 165,000 feet (50,000 meters) per second.

288   **limit exposure to the Martian atmosphere:** Some will act as "witnesses," samples taken at various stages of the rover's journey to Mars and on the surface. This will allow as accurate as possible a background signal on conditions that were present. The rover will also monitor the evolution of any contamination that may have taken place.

289   **cleanest drill:** The drill bit has been sterilized "even more so than the rest of the rover," says one engineer. To date, all landers on Mars have only scratched the surface. The Franklin rover will be capable of reaching six feet (two meters) below the surface "so we can drill into solid bedrock," says Jorge Vago. Technically speaking, the rover has a suite of instruments that will investigate what is called "the shallow subsurface environment."

289   **mole on InSight:** Originally, the seismometer and the mole on InSight were to have flown on the first iteration of the ExoMars mission in the early 2000s, which would have seen NASA and ESA collaborate closely on the whole program. Prototypes were developed, but when there were subsequent changes to the program, JPL decided to build its own mission. As a result, the Franco-British

seismometer package and the German mole were added to the
InSight mission at no cost to the American taxpayer. That is why
there are more European investigators on an American-managed
mission.

290 **bite off more than they can chew:** If the tip of the drill gets
stuck, the operators would be able to jettison the drill so the rover
wouldn't end up tethered to a rock. That would be done only as a
last resort, because if it happened, there would be no possibility of
taking any more samples.

290 **MicrOmega:** In this case, "*les glaces*" refers to ice, not ice cream,
though in Jorge Vago's telling phrase, when the mission gets to
work, "the first lick of the ice cream is a good lick at all three fla-
vors," meaning the three instruments in the laboratory drawer
(Raman spectrometer, a laser desportion mass spectrometer, and
a gas chromatograph) will return useful information on their very
first sample.

293 **phosphates:** Phosphorus may seem an unusual ingredient in the
elixir of life. In the form of phosphates, it is a prime constituent
of many biologically important molecules, most notably nucleic
acids. Most available phosphate at Earth's surface is locked up as
calcium phosphate (most familiar as bone), which is insoluble in
water. How the rare phosphate molecule came to be so important
is unknown, but it has been suggested that mild acidity would have
liberated the necessary phosphates from rocks to create life.

294 **improved version of the SAM oven:** One part of the SAM team
at NASA Goddard is involved with the European experiment. As
a result, this son of SAM has benefited from changes in technology. It is also being housed in a smaller space than the Curiosity
instrument.

295 **desorption:** Technically, this occurs where something is released
from, or through, a surface layer. With respect to a laser in mass
spectrometry, the term means pulling apart some of the ions from
larger molecules to minimize the fragmentation of the original.
The large organic molecules that are important in biology tend to
be fragile and fragment when ionized by more conventional methods. Laser desorption represents the best of both worlds: it reveals
what a complex molecule is made of but does not turn all the evidence into irretrievable fragments.

299 **Shown in the laboratory:** The year 1953 was an annus mirabilis

for studies of the origins and the mysterious inner workings of life. Stanley Miller, then a graduate student of Harold Urey at the University of California in San Diego, carried out an experiment that was, for its time, astounding. In a five-hundred-milliliter glass flask, Miller created a putative early atmosphere of methane, ammonia, hydrogen, and water that flowed throughout in the form of vapor after it was heated. By use of an electric discharge, which acted as lightning, within a few days as much as ten percent of the carbon had become converted to organic compounds. At least one in fifty were in the form of amino acids, the chemical basis for proteins. When Urey and Miller published their findings in *Science* as "A Production of Amino Acids under Possible Primitive Earth Conditions" on May 15, 1953, it followed hot on the heels of another extraordinary paper in its British equivalent *Nature*. "A Structure for Deoxyribose Nucleic Acid" by Crick and Watson had been published three weeks earlier.

301 **"informational" molecule:** Another way of looking at DNA is that it encodes all the genetic information from the blueprint from which all life is created. That information, which precedes the creation of life, is stored in syntax, a code, in the molecule's physical shape. In the long term, DNA acts as a kind of biological thumb drive that passes those blueprints on to the next generation. In that sense, RNA acts as the reader that decodes the information on the thumb drive. The reading process takes many stages. Some say that the real problem in the origin of life is not chemistry, but information. But, at best, information theory can only look at small pieces of the puzzle.

303 **chirality may well be the clincher:** It should be noted that chirality, too, could take place nonbiologically. If, however, it is found in material that is some sort of biosignature, it will hint at a very different evolution.

303 **a handful of other spacecraft:** Just before this, a Japanese mission called Hayabusa retrieved samples from an asteroid. As of this writing, *Hayabusa 2* is at asteroid Ryagu, and NASA's OSIRIS-REx is at asteroid Bennu. Both missions will return samples to Earth in the next few years. Another NASA mission called *Stardust* returned samples of interplanetary material in 2006. China plans to return samples from the Moon in the 2020s. In the 1970s, the Soviet

Union did the same with the robotic *Luna 16, 20,* and *24* craft. Six Apollo missions returned samples from 1969 to the end of 1972.

307 *Viking* **data:** As noted earlier, there were signs of perchlorates in the data from the 1970s. *Viking 1* had measured chloromethane at Chryse Planitia just 560 miles (1,000 kilometers) west of where ExoMars will land. In Utopia, *Viking 2* discovered dichloromethane.

307 **ultraviolet radiation:** Ultraviolet radiation from the Sun provided sufficient energy to fire the very esoteric chemical reactions needed for life to form. Ultraviolet bathed the surfaces of both Earth and Mars, although there is a curious paradox. Without ultraviolet radiation, life could not have evolved and yet, with prolonged exposure, life can quite easily be destroyed. On Earth, a direct consequence of the oxygen given off by life-forms was the formation of the ozone layer that protects us from the Sun. Although trifling amounts of ozone have been discovered above Mars, most recently by the MAVEN mission, the best shield for any life is the surface itself.

309 **expertise in microbiology:** Although the NASA Ames Research Center has the requisite experience in biology and the Johnson Space Center has expertise in geochemistry and lunar sample curation, broader scientific knowledge would be required to do as good a job as possible. As of this writing, it is unclear which NASA center will be lead for any samples that are returned.

SOURCES

Chapter 1: Frozen in Time

The primary information for this chapter came from interviews with Chris McKay, Jack Holt, Stefano Nerozzi, Rich Zurek, and Frances Butcher. They have all provided further information by additional interviews and email clarifications.

Overview

There are very few popular-level summaries of the state of research into the Martian poles. A very good technical summary may be found in "The Polar Deposits of Mars" by Shane Byrne of the University of Arizona, *Review of Earth and Planetary Science* 2009, 535–560. We have found a number of papers useful. The first laser altimetry results were reported by Maria Zuber et al., "Observations of North Polar Region of Mars from MOLA," *Science* 282 (1998), 5396; see also "NASA Findings Suggest Jets Bursting From Martian Ice Caps," JPL news release, May 30, 2006; C. J. Hansen et al., "HiRISE Observations of Gas Sublimation-driven Activity in Mars's Southern Polar Regions," *Icarus*, 2009; A. M. Bramson et al., "A Study of Martian Mid-Latitude Ice Using Observations and Modeling of Terraced Craters,"*46th Lunar & Planetary Science*

Conference, LPI Contribution No. 1832 (2015), 1,565; C. Stuurman et al., "SHARAD Detection and Characterization of Subsurface Water Ice Deposits in Utopia Planitia, Mars," *Geophysical Research Letters* 43 (September 26, 2016), 9484–9491.

The diaries from Ross's various expeditions are published online, but the original book was *A Voyage of Discovery and Research in the Southern and Antarctic Regions during the Years 1839–1843*, published by John Murray in 1847; further information on these remarkable pioneers may be found in *Polar Pioneers: John Ross and James Clark Ross* by Maurice Ross (Montreal: McGill-Queen's University Press, 1994); and *The Polar Rosses, John and James Clark Ross and Their Explorations* by Ernest S. Dodge (London: Faber & Faber London, 1973).

The most recent work by Jack Holt and Stefano Nerozzi may be found in a May 22, 2019, press release from the University Texas at Austin entitled "Massive Martian Ice Discovery Opens a Window into Red Planet's History," based on their paper "Buried Ice and Sand Caps at the North Pole of Mars" in *Geophysical Research Letters*, May 22, 2019. Elizabeth Howell's interview with Elena Petanelli may be found in various stories on Space.com, especially "Could Volcanic Activity Allow Water Beneath the Martian Polar Ice?" February 14, 2019.

Chapter 2: Inside Out

The primary information for this chapter came from interviews with scientists directly involved with the InSight mission as well as the magnetometer researchers on earlier spacecraft (notably Daniel Winterhalter in October 1997). In particular, Bruce Banerdt and Anna Horleston read through multiple drafts and shared their thoughts and information by telephone and email as the manuscript took shape. We would also like to thank Professor Bruce Jakosky with regard to the results from the MAVEN mission (also discussed in chapter 7).

Elizabeth Howell covered the launch of InSight in May 2018 for Space.com and has covered updates since. She had a lengthy interview with Bruce Banerdt at that time and then for this book in September 2019. Two further long interviews were carried out over the telephone with Sue Smrekar in October 2019 as well as with Troy Lee Hudson just after Thanksgiving 2019. Nicholas Booth spoke with Anna Horleston at the University of Bristol on several occasions in the summer and fall of 2019.

Overview

There were two major review meetings from the InSight mission in 2019 while this chapter was being written: in Paris in July and in Pasadena in October. A number of participants spoke to us about the "feeling" of the meetings ahead of the first scientific papers, which did not appear until late February 2020. A timely overview is the "perspectives" paper by W. B. Banerdt et al., "Initial Results from the InSight Mission on Mars," in *Nature Geoscience* 13 (February 24, 2020), 183–189. See the piece on Space.com by Elizabeth Howell, "Mars Lander Reveals New Details about the Red Planet's Strange Magnetic Field," February 26, 2020; "Magnetic Field at Martian Surface Ten Times Stronger Than Expected," University of British Columbia news release, February 24, 2020; and the piece by Maya Wei-Haas, "Mars Is Humming. Scientists Aren't Sure Why," *National Geographic*, February 24, 2020.

A good review of the mission's investigation may be found in Panning, M. P. et al., "Planned Products of the Mars Structure Service for the InSight Mission to Mars," *Space Science Review*, vol. 211, nos. 1–4 (2017), 611–650, and Clinton, J. D. et al., "The Marsquake Service: Securing Daily Analysis of SEIS Data and Building the Martian Seismicity Catalogue for InSight," *Space Science Review*, 214:133 (2018). A number of early abstracts were presented at the Ninth International Conference on Mars held in August 2019 in Pasadena.

The JPL press kit for the Mars InSight launch contains a great deal of useful information. The AP backgrounder on the mission was published with different headlines in both the *New York Times* and the *Wall Street Journal* on April 30 and May 5, 2018, respectively. See also Witze, Alexandra, "Mars Quakes Set to Reveal Tantalizing Clues to Planet's Early Years," *Nature*, April 26, 2018; "Gravity Assist Podcast, Insight with Bruce Banerdt," *NASA and Astrobiology*, June 19, 2018; and "NASA Launches InSight Spacecraft to Explore the Insides of Mars," *Guardian*, May 5, 2018. The "tough mission" quote is from the preflight press conference.

The announcement of the first-ever marsquake came in a JPL press release, April 23, 2019: "NASA's InSight Detects First Likely 'Quake' on Mars." See also: Bartels, Meghan, "Marsquake! NASA's InSight Lander Feels Its 1st Red Planet Tremor," *Science & Astronomy*, April 23, 2019; Barbuzano, Javier, "What the First Marsquake Means for NASA's InSight," *Sky and Telescope*, April 26, 2019. For background on Homestead Hollow, see Dvorsky, George, "New Images from Mars Show NASA's InSight Landed on an Absolutely Glorious Spot," *Gizmodo*, December 7, 2018.

The early progress of the mission after landing can be followed in the various JPL press releases: "InSight Is Catching Rays on Mars," November 26, 2018; "NASA InSight Lander Hears Martian Winds," December 7, 2018; "NASA's InSight Places First Instrument on Mars," December 19, 2018; "InSight Lander Completes Seismometer Deployment on Mars," February 4, 2019. See also the Agence France Presse story "NASA's Martian Quake Sensor InSight Lands at Slight Angle," December 1, 2018.

The saga of the mole has played out on the blog published by the German Aerospace Center (DLR), which built the instrument. Throughout 2019, the story was covered as follows: Meghan Bartels for Space.com, "Robot 'Mole' on Mars Begins Digging into Red Planet This Week," February 26, 2019; and "Tests for the InSight 'Mole' by DLR," April 12, 2019; Jeff Foust, "Troubleshooting of Mars InSight Instrument Continues," May 15, 2019; JPL press release "InSight's Team Tries New Strategy to Help the 'Mole,'" June 5, 2019; "Common Questions about Insight's Mole," InSight Mission News, posted July 1, 2019 on JPL Mars Insight website. See also the article by Marina Koren, *Atlantic*, "A Robot Has Been Stuck on Mars for Months," June 21, 2019.

The MAVEN mission is managed by the Laboratory for Atmospheric and Space Physics (LASP) in Boulder. Its website at the University of Colorado, Boulder, has a chronological series of press releases that we have used: "Maven Sets Its Sights beyond Mars," April 29, 2019; "MAVEN Shrinking Its Orbit to Prepare for Mars 2020 Mission," February 11, 2019; "Mars Terraforming Not Possible Using Present-Day Technology," July 30, 2018, based on *Nature Astronomy* paper by B. Jakosky & J. Edwards, "Inventory of CO_2 Available for Terraforming Mars," same date; "MAVEN Finds That 'Stolen' Electrons Enable Unusual Auroras on Mars," July 23, 2018.

Chapter 3: Curiosity

The primary information for this chapter came from lengthy interviews that Elizabeth Howell carried out over the summer of 2018 by telephone, and then in person at the Jet Propulsion Laboratory in October 2019. Primarily, these were with Adam Steltzner, Ashwin Vasavada, Abigail Fraeman, and Matt Golombek. Nicholas Booth spoke with Professor John Bridges at Leicester University in November 2019.

Overview

Given the many hundreds of scientists involved with the Mars Science Laboratory, as the Curiosity mission is more formally known, it is very difficult to summarize all the results since 2012. We have, however, used the following for the overview of the subject of this chapter: Grotzinger, J. P. et al., "Deposition, Exhumation, and Paleoclimate of an Ancient Lake Deposit, Gale Crater, Mars," *Science*, vol. 350, no. 6257 (October 2015); Grotzinger, J. P. et al., "A Habitable Fluvio-lacustrine Environment at Yellowknife Bay, Gale Crater, Mars," *Science*, vol. 343, no. 6169 (January 24, 2014); Webster, C. et al., "Background Levels of Methane in Mars' Atmosphere Show Strong Seasonal Variations," *Science*, vol. 360 (June 8, 2018), 1093–1096; Eigenbrode, J. et al., "Organic Matter Preserved in 3 Billion-Year-Old Mudstone at Gale Crater, Mars," *Science*, vol. 360 (June 8, 2018), 1096–1099. A scientific paper presented at the Ninth International Mars Conference in the summer of 2019, "The Paradox of Martian Methane" by Kevin Zahnle of NASA Ames, is a more technical consideration.

A very fine review of the current understanding of water on Mars is by Ashwin Vasavada: "Our Changing View of Mars," *Physics Today*, March 2017. For findings in Yellowknife Bay, see Kerr, R. A., "New Results Send Mars Rover on a Quest for Ancient Life," *Science*, December 2013, and Ken Kremer, "Curiosity Celebrates 1st Martian Christmas at Yellowknife Bay," *Universe Today*, 2014. The background on Spirit and Opportunity's findings may be found in Koren, Marina, "NASA Mars Rovers Curiosity and Opportunity Are Having a Rough Year," *Atlantic*, September 18, 2018, and Wall, Mike, "How NASA's Opportunity and Spirit Rovers Changed Mars Exploration Forever," Space.com, February 13, 2019. Professor Stephen Squyres was project scientist; see his interview with Miles O'Brien, "Miles to Go," October 23, 2018. See also Matt Steecker, "Cornell Professor Remembers Working on NASA's Now Defunct Opportunity," *Ithaca Journal*, February 14, 2019, and Evan Gough, "The Global Dust Storm That Ended Opportunity Helped Teach Us How Mars Lost Its Water," *Universe Today*, May 3, 2019.

All information on the instruments that Curiosity carries may be found on the NASA MSL website: https://mars.nasa.gov/msl/spacecraft/instruments/summary. For further background information on the SAM instrument, see Schirber, Michael, "SAM I Am," *Astrobiology Magazine*, December 5, 2011, and "Sols 2536–2537: SAM Wet Chemistry Experiment," Mission Update, Mars Exploration Program, September 24, 2019. The cautionary quote regarding the organics is taken from Ian Sample, "Nasa Mars Rover Finds Organic Matter in Ancient Lake Bed," *Guardian*, June 7, 2018. Background on what happened in 2014 may be found in Wall, Mike, "Curiosity Rover on Mars Stalled by 'Hidden Valley' Sandtrap," Space.com, August 19, 2014. Further background information may be found in Wall, Mike, "After 5 Years on Mars, NASA's Curiosity Is Still Making Big Discoveries," Space.com, August 5, 2017.

The saga of methane on Mars may be found in Chang, Kenneth, "NASA Rover on Mars Detects Puff of Gas That Hints at Possibility of Life," *New York Times*, June 22, 2019. A good earlier review of the background to methane discoveries is Peplow, Mark, "The Search for Life on Mars," *Nature*, July 27, 2004. For more recent stories, see Koren, Marina, "A Startling Spike on Mars," Atlantic, July 3, 2019, and Amos, J., "So Where Did the Mars Methane Go?" *BBC Online*, April 10, 2019. A good background piece is Grady, Monica, "Methane on Mars: A New Discovery or Just a Lot of Hot Air?" *The Conversation*, April 1, 2019; and for the mystery as seen by the Trace Gas Orbiter, "ESA's Mars Orbiter Did Not See Latest Curiosity Methane Burst," ESA news release, November 13, 2019. For the more recent oxygen finding, see "With Mars Methane, A Mystery Unsolved: Curiosity Serves Scientists a New One: Oxygen," NASA Goddard news release, November 12, 2019, referring to paper in November 12, 2019, *JGR Planets*.

For the most recent tranche of papers referring to salt lakes and Old Soaker, see also "Mars Once Had Salt Lakes Similar to Earth," Texas A&M release, October 14, 2019; "NASA's Curiosity Rover Finds an

Ancient Oasis on Mars," JPL press release, October 2019, referring to paper in *Nature Geosciences* by W. Rapin et al.; and John Wenz, "We Now Have Good Idea When Martian Ocean Started Dying," *Popular Mechanics*, October 7, 2019.

Chapter 4: The Road to Utopia

The primary information for this chapter came from conversations Nicholas Booth had with people involved with the first wave of JPL Mars missions, including Gerry Soffen, Bud Schurmeier, Al Hibbs, Norm Horowitz, Carl Sagan, John Guest, Frank Colella, and Conway Snyder (all of whom have subsequently passed away) as well as Gil Levin, Peter Staudhammer, Ben Clark, James Lovelock, Jim Cutts, Geoff Briggs, Michael Carr, and James Tillman at various times from 1982 to 1999. James Tillman's daughter, Rachel Tillman, who runs the Viking Mars Missions Education and Preservation Project website, shared a number of helpful insights.

In 1996, Nicholas Booth interviewed Gerry Soffen over the telephone at his office in Goddard and later while he was at the ISU in Strasbourg; a year later, he talked to Norm Horowitz in Pasadena. As noted in the text, Horowitz's book, *To Utopia and Back: The Search for Life in the Solar System* (1986), is a very fine scholarly account. It is interesting to look at the original scientific papers and more recent interpretations, which may be found in Klein, H. P. et al., "The Viking Mission Search for Life on Mars," *Nature* 262 (1976), 24–27, and a more recent analysis from 1998, "The Search for Life on Mars: What Was Learned from Viking," *Journal of Geophysical Research* 103, 28463–66. Another useful summary is the paper by Conway Snyder, "The Planet Mars as Seen at the End of the Viking Mission," *Journal of Geophysical Research*, 84, B14 (December 30, 1979), 8487–8519.

History

The NASA book *On Mars* by Edward Ezell and Linda Ezell (Special Publication 4212) is a solid history of the *Viking* project; *Mars at Last!* by Mark Washburn (New York: Putnam, 1977) is a vivid contemporaneous chronicle of the mission. JPL's various Viking Mission Bulletins, edited by Duke Reiber, are another fantastic resource. Two further NASA Special Publications, *The* Viking *Orbiter Views of Mars* and *The Martian*

Landscape, contain a wealth of information and some often-overlooked photographs.

Because of the level of interest in the project at the time, the saga of *Viking's* investigations on Mars was told in a pre-landing profile of Carl Sagan in *The New Yorker*; there was also coverage in two other issues of *The New Yorker* from 1979. That material was collected and published in book form by Henry S. F. Cooper, *The Search for Life on Mars: Evolution of an Idea* (New York: Henry Holt, 1980). Dismayed by American television coverage of *Viking*, Sagan conceived the idea of his thirteen-part television series *Cosmos* with Gentry Lee. His chapter "Blues for a Red Planet" in the book of that series is a useful guide.

A number of key *Viking* personnel have also recorded oral histories for NASA and the American Institute of Physics in College Park, Maryland.

Chapter 5: The Measure of Mars

The prime source material in this chapter is from two lengthy interviews Nicholas Booth conducted with Dr. Michael Malin in 1998 and 1999. Further information has come from other interviews with Dr. Phil Christensen, Ashwin Vasavada, and others quoted in this chapter. Nicholas Booth covered the Mars Observer mission at the time for British newspapers; see his articles "Watch the Birdie, Mars," *Observer*, August 22, 1993, and "Now the Forecast for Mars," *Times* (London), September 22, 1992.

A very good popular introduction to the topic is Oliver Morton's *Mapping Mars: Science, Imagination, and the Birth of a World* (New York: Picador, 2002). See also "The Brains behind the Eyes on Mars," Lisa Grossman, *Slate.com*, September 30, 2012. Information on all the instruments in this chapter may be found on relevant university and NASA websites. For HiRise, for example, see uahrise.org, and for THEMIS, Themis.asu.edu.

Other information

A very fine technical review paper is "An Overview of the 1985–2006 Mars Orbiter Camera Science Investigation," *International Journal of Mars Science & Exploration*, January 2010, Issue 5. More recent papers include Malin, M. C. et al., "Camera Investigation on board the Mars Reconnaissance Orbiter," *Journal of Geophysical Research* 112 (2007), article E05S04, 10.1029; see also McEwen, A. S. et al., "Mars Reconnaissance Orbiter's High-resolution Imaging Science Experiment (HiRISE)," *Journal of Geophysical Research* 112 (2007), article E05S02, https://doi.org/10.1029/2005je002605.

Further information has come from "Prolific MRO Completes 50,000 orbits," JPL press release, March 30, 2017, and another press release called "What on Mars Is a High Thermal-Inertia Surface," High Resolution Imaging Experiment (HirISe) at LPL website, April 8, 2015. A more technical explanation is available at the Planetary Science Institute Website, sc.psi.edu, with a mass of information by Nathaniel Putzig. For further information on CRISM, see "Martian Diaries," JPL website, 2014, interview with Scott Perl; "CRISM Joins Mars' Water Detectives," *Astrobiology Magazine*, August 19, 2005; and "Instrument Development," crism.apu.edu.

Chapter 6: The Pathfinder

The prime source material in this chapter comes from the many interviews Nicholas Booth carried out with Donna Shirley from 1996 to 1998 on repeated visits to JPL, including one in the von Kármán TV studio for a JPL history where Al Hibbs interviewed Donna Shirley in October 1997.

Other interviewees were Chuck Weisbin, Matt Golombek, and the late Jake Matijevic. See the Booth stories "Mars Marathon for Daughter of Donna," *Times*, October 29, 1997, and for a sense of how "silly" some people thought rovers were, a cover story for the "Interface" section, "Small Computers Take Giant Steps to Explore Mars," October 23, 1996.

As noted in the text, Donna Shirley's autobiography, *Managing Martians* (New York: Broadway Books), was published in 1998. More recently, her oral histories—Dr. Donna L. Shirley, NASA JSC Oral History Project, July 17, 2001, and her Oral History, Special Collections, Oklahoma State University, March 14, 2001—have described what happened next. The authors asked for further information from Donna, who provided it for this chapter.

See also Nathalia Holt, "How Sexism Held Back Space Exploration," *Atlantic*, June 11, 2016, and Simon Worrall, "The Secret History of the Women Who Got US beyond the Moon," *National Geographic*, May 8, 2016.

Chapter 7: Waterworld

Two of the more prominent experts on water on Mars are Michael Carr and Victor Baker. In 1979, Carr took a sabbatical from the US Geological Survey to write the first version of his encyclopedic *The Surface of Mars* (Yale University Press), a very fine review of where understanding of Martian geology was in the immediate aftermath of the *Viking* missions. An updated version appeared in 2006. In 1996, Carr produced *Water on Mars* (Oxford University Press) as a more recent discussion of themes covered in this chapter. Victor Baker's *The Channels of Mars* (University of Texas Press) first appeared in 1982. Their writings were the wellsprings of the debate that followed. Nicholas Booth's original interviews from 1996 and 1997 are included here.

Elizabeth Howell has interviewed Bruce Jakosky on several occasions, as did Booth in the 1990s. Here we benefited from the insights from his most recent work as project scientist for the ongoing MAVEN mission, which is bringing together a whole range of issues concerning water and volatile losses on Mars. His very fine and scholarly book *The Search for Life on Other Planets* (Cambridge University Press) appeared in 1998. Professor Jakosky has written more extensively on the issue of water and volatiles.

For this chapter, we spoke with a number of researchers, including Zach Dicheson, Peter Grindrod, Frances Butcher, Ashwin Vasavada, and Matt Golombek. With regard to hydrothermal activity, Nicholas Booth interviewed Jack Farmer in 1997 and David Des Marais in 1999.

Scientific Papers

Given its fairly complicated role in the evolution of Martian climate and possible biology, most papers about water on Mars are increasingly labyrinthine. A very timely review that weaves all the results from the latest missions, and already cited for chapter 3, is "Our Changing View of Mars" by Ashwin Vasavada, *Physics Today*, vol. 70, no. 3 (March 2017),

34–41. A good starting point for the original notion of an ocean on Mars is Victor Baker et al., "Ancient Oceans, Ice Sheets, and the Hydrological Cycle on Mars," *Nature* 352 (1991), 589, with a further discussion by Baker in *Bulletin of American Astronomical Society* (1999), 1133. See also Edgett, K. S. and T. J. Parker, "Water on Early Mars: Possible Subaqueous Sedimentary Deposits Covering Ancient Cratered Terrain in Western Arabia and Sinus Meridiani," *Geophysical Research Letters*, vol. 24, no. 22 (1997), 2897–2900; Beaty, D. W. et al., "Key Science Questions from the Second Conference on Early Mars: Geologic, Hydrologic, and Climatic Evolution and the Implications for Life," *Astrobiology*, vol. 5, no. 6 (2005), 663–689; and Clifford, S. M. and T. J. Parker, "The Evolution of the Martian Hydrosphere: Implications for the Fate of a Primordial Ocean and the Current State of the Northern Plains," *Icarus* 154 (2001), 40–79. A more recent discussion of the valley networks may be found in Fassett and Head, "Valley Network-fed, Open-basin Lakes on Mars: Distribution and Implications for Noachian Surface and Subsurface Hydrology," *Icarus* 198 (2008), 37. Much of the underlying work on the "smoothness" of the northern hemisphere of Mars was based on the readings from the laser altimeter known as Mars Orbiter Laser Altimeter (MOLA) and was originally published in the 1999–2000 time frame. See Zuber, M. T. et al., "Internal Structure and Early Thermal Evolution of Mars from Mars Global Surveyor Topography and Gravity," *Science* 287 (2000), 1788–1793; Smith, D. E. et al., "The Global Topography of Mars and Implications for Surface Evolution," *Science* 284 (1999), 1495–1503; Zuber, M. T., "Snapshots of an Ancient Cover-up," *Nature* 397 (1999), 560–561; Head, J. W. et al., "Oceans in the Past History of Mars: Tests for Their Presence Using Mars Orbiter Laser Altimeter (MOLA) Data," *Geophysical Research Letters* 25 (1998), 4401–4404; and Smith, D. E. et al., "The Global Topography of Mars and Implications for Surface Evolution," *Science*, 284 (1999), 1,495–1,503. See also Smith, D. E. et al. "Mars Orbiter Laser Altimeter (MOLA): Experiment Summary after the First Year of Global Mapping of Mars," *Journal of Geophysical Research* 106 (2001), 23689–23722.

The faint young Sun paradox was one of the more important contributions to the debate by Carl Sagan, originally discussed in Sagan, C. and G. Mullen, "Earth and Mars: Evolution of Atmospheres and Surface

Temperatures," *Science* 177 (1972), 4043. A more recent discussion may be found in Kasting, J. F., "Faint Young Sun Redux," *Nature* 464 (2010), 687–689. For discussions of Mars having been "wild, wet and windy" see Pollack, J. B. et al., "The Case for a Warm, Wet Climate on Early Mars," *Icarus* 71 (1987), 203–224, and Kasting, Toon, and Pollack, "How Climate Evolved on the Terrestrial Planets," *Scientific American* 256 (1988), 90–97. More recent work in this area may be found in McKay, C. P., O. B. Toon, and J. F. Kasting, "Making Mars Habitable," *Nature* 352 (1991), 489–496; Squyres, S. W. and J. F. Kasting, "Early Mars: How Warm and How Wet?," *Science* 265 (1994), 744– 749; Mischna, M. and J. F. Kasting, "CO_2 Clouds and the Climate of Early Mars: Effect of Cloud Height and Optical Depth, " *Icarus* 145 (2000), 546–554; Ramirez, R. M., et al., "Warming Early Mars with CO_2 and H_2," *Nature Geoscience* 7 (2014), 59–63 (2014); and Ramirez, R. M. and J. F. Kasting, "Could Cirrus Clouds Have Warmed Early Mars?" *Icarus* 281 (2017), 248–261.

The Hypanis Valles work came from a paper by Joel Davis and Peter Fawdon, "The Hypanis Valles Delta: The Last Highstand of a Sea on Early Mars?" *Earth Planetary Science Letters*, vol. 500, October 15, 2018, 225–241. Joel Davis is quoted in Williams, Matt, "Was This Huge River Delta on Mars the Place Where Its Oceans Finally Disappeared?" *Universe Today*, September 17, 2018.

Work by Bruce Jakosky may be found in Jakosky, B. M. and M. T. Mellon, "Water on Mars," *Physics Today*, 57 (2004), 71–76, and Jakosky, B. M. and E. L. Shock, "The Biological Potential of Mars, the Early Earth, and Europa," *Journal of Geophysical Research*, 103 (1998), 19359–19364.

Reporting on recurring slope lineae may be found in Chang, Kenneth, "Mars Shows Signs of Having Flowing Water, Possible Niches for Life, NASA Says," *New York Times*, September 28, 2015, and Kerr, Richard A., "Is Mars Weeping Salty Tears?" *Science*, August 4, 2011. Source material for the hydrothermal vents section may be found in "Volcanic Rock in Mars's Gusev Crater Hints at Past Water," JPL press release, 2004; Choi, Chris, "Habitable Hotspots on Mars? Volcanic Vents May Be Signs," Space.com, October 31, 2010; Payton, M., "Evidence of Past Life on Mars Could Be Found in Northern Chile, Says Scientist," *Independent*, November 24, 2016; and "Gusev Crater Once Held a Lake

After All, Says ASU Mars Scientist," Arizona State University release, April 9, 2014.

Chapter 8: Claims

Much of the material at the start of this chapter comes from Percival Lowell's own writings. These include Lowell's books on the Red Planet, *Mars and Its Canals, Mars,* and *Mars as the Abode for Life,* as well as his books about the East Asia, *The Soul of the Far East* and *Noto: An Unexplored Corner of Japan.* William Graves Hoyt's *Lowell and Mars* is probably the standard reference to his work in Flagstaff along with William Sheehan's *The Planet Mars: A History of Observation and Discovery,* which puts everything into perspective. Here they are buttressed by the biography by Lowell's brother as well as some contemporary newspaper reports (see below). Further information comes from chapter 3 in Mark Washburn's excellent *Mars at Last!: Canals and Controversy* (New York: Putnam, 1977), along with Alfred Russel Wallace's *Is Mars Habitable?* (Whitefish, MT: Kessinger, 2010). Our thanks to Samantha Thompson, former curator at the Lowell Observatory and now at the Smithsonian Air and Space Museum, for her reading of the section about Lowell and his work.

Contemporaneous newspaper coverage reveals interesting nuances to the story of Lowell and Mars. From the *Times* (London): "The Canals in Mars," September 29, 1905; "An Astronomical Pilgrimage," September 30, 1910; "The Approaching Opposition of Mars," July 19, 1909; "The Atmosphere of Mars," September 18, 1909; "The Investigation of Mars," March 31, 1910; "Professor Lowell in London," April 9, 1910; "Professor Lowell on Mars," January 3, 1910; "The Problem of Mars," September 11, 1909; *The Observer:* "Are the Planets Inhabited?," April 6, 1913; "Mars and Its Canals," January 2, 1910; "Science in the Coming Year," January 3, 1909; "Is There Life on Mars?," December 15, 1907; and "Watching Mars," July 7, 1907. From the *Manchester Guardian:* "Signor Schiaparelli," Obituary, July 5, 1910; "The Atmosphere of Mars," September 10, 1909; "Life on Mars," September 13, 1909; "The Canals of Mars," April 7, 1904; "Latest News from Mars," September 29, 1909; "The Weather on Mars," November 28, 1911.

The Loss of Mars Observer

Nicholas Booth interviewed Glenn Cunningham, David Evans, Arden Albee, and several others at JPL in the early 1990s a number of times before the launch and after the loss of Mars Observer. See "Now the Forecast for Mars," *Times* (London), September 22, 1992, and "Watch the Birdie, Mars," *Observer* (London), August 22, 1992. By late 1993, the sense of shock at the loss was still palpable; the quotes from Mr. Cunningham were recorded then. The Michael Malin quotes are from 1999. A quick web search will reveal some of the more ridiculous theories about the "Face on Mars."

Gil Levin

Nicholas Booth interviewed Gil Levin at length in 1997, a shorter version of which appeared in the *Times* (London) as "The Secret of Mars." Background information on the Labeled Release experiment may be found in contemporary reports (op. cit. above in chapter 4) and a special Science report, "Controversy over Life on Mars Revived" in the August 7, 1986 issue of the *Times* (London); more recently, in an interview with Patricia Ann Straat by Clara Moskowitz in the online version of *Scientific American*, "Looking for Life on Mars: Viking Experiment Team Member Reflects on Divisive Findings," April 2, 2019. In October 2019, Gil Levin penned an opinion piece, "I'm Convinced We Found Evidence of Life on Mars in the 1970s," *Scientific American*, Opinion, October 10, 2019. A more scientific survey of their work may be found in Levin and Straat, "The Case for Extant Life on Mars and Its Possible Detection by the Viking Labeled Release Experiment," *Astrobiology* 16 (2016).

Perchlorates

In 2008, *Aviation Week* carried the first stories about the perchlorate discovery. Elizabeth Howell interviewed Melissa Guzman in the summer of 2018, with follow-up emails for this book and another interview in December 2019 by Nicholas Booth. See also "Martian Life or Not? NASA's Phoenix Team Analyses Results," JPL, August 6, 2008; "White House Briefed on Potential for Mars Life," *Aviation Week*, August 1, 2008; O'Neill, Ian, "The White House Is Briefed: Phoenix about to Announce 'Potential for Life on Mars'"; Wadsworth and Cockell, "Perchlorates on Mars: Enhance the Bacteroidal Effects of UV Light," *Nature*, July 6, 2017; Archer et al., "Perchlorate Enhancing the Rate of Bacterial Death on Mars," Edinburgh University website; Glavin, Daniel, "Oxychlorine Species on Mars: Implications from Gale Crater Samples," 47th LPI Conference, abstract, 2947 (2013); and "Evidence for Perchlorates and the Origin of the Chlorinated Hydrocarbons Detected by SAM at the Rocknest Aeolian Deposit in Gale Crater, Mars," *Journal of Geophysical Research Planets* 118, 1,955–1,973.

For the original GCMS analysis, see Biemann et al., "GCMS Results," *Journal of Geophysical Research* 82 (1977), 4641–4658. Details about Professor Biemann are contained in obituaries in *American Society for Mass Spectrometry*, vol. 27, 10 (October 2016), 1583–1589, and "Klaus Biemann, Professor Emeritus of Chemistry Dies at 89," *MIT News*, June 9, 2016.

Chapter 9: Reactions

Nicholas Booth followed the ALH84001 saga as a technology editor on the *Times* in London. He was in a unique position to interview Monica Grady, Everett Gibson, and Colin Pillinger a number of times in the 1996–1999 period. Fiona Gammie and Annabel Gilling both originally provided transcripts of the *BBC Horizon* program on which they worked in 1996, "Aliens from Mars," which was broadcast that November. Others interviewed at the time were David Des Marais, Christopher McKay, and Jack Farmer. In particular, we thank Professor Charles Cockell and the JPL media staff, especially Veronica Macgregor, for their time.

Two books provided very useful background to the story: Donald Goldsmith's *The Hunt for Life on Mars* (Collingdale, PA: Diane Publishing, 1997) and Kathy Sawyer's *The Rock from Mars* (New York: Random House, 2006) show how the story changed over the first ten years. Here, we are grateful for further insights from Kathie Thomas-Keprta and Professor David Barber, who read through drafts of this story. Our thanks to Professor Barber's daughter, Dr. Ros Barber, for facilitating that contact.

The "organics" story in Martian meteorites was told in newspapers at the time. On ALH84001: Hawkes, Nigel, "Space Rock Sheds Light on Martian Surface," *Times*, December 15, 1994, and Nuttall, Nick, "Martian Meteorite Shows Glimmer of Life on Red Planet," *Times*, March 21,

1995. On the EETA79001 meteorite: Gee, Henry, "Meteorites Made Mars a Dry Planet," Nature Times News Service, April 10, 1989, and Gee, Henry, "Martian Life Debate Renewed," Nature Times News Service, July 20, 1989.

The original leaked story was by Len David, "Meteorite Find Incites Speculation on Mars Life," *Space News*, August 5, 1996; the story of the escort was detailed in "Dick Morris Resigns in Wake of Scandal," *Wall Street Journal*, August 23, 1996.

After the initial press conference, which was front-page news in August 1996, see also Hawkes, Nigel, "Second Meteorite Gives Hint of Martian life," *Times*, October 14, 1996; Hawkes, Nigel, "Martian Life in Doubt," Science Briefing, *Times*, August 17, 1998; Henderson, Mark, "New Evidence of Life Found in Mars Rock," *Times*, December 16, 2000; Crenson, Matt, "After Ten Years Few Believe Life on Mars," Associated Press, August 6, 2006; "Mars Life? 20 Years Later, Debate over Meteorite Continues"; Ward, Peter, "Hunting Life in Martian Rocks," *New Scientist*, August 23, 2006.

Scientific Papers

The original paper for the meteorite discovery was by McKay, D. S. et al., "Search for Past Life on Mars: Possible Relic Biogenic Activity in Meteorite ALH8400," *Science*, 54 (1996), 924–930. A more recent summary may be found in Gibson, E. et al., "Life on Mars: Evolution of the Evidence within Martian Meteorites ALH 84001, Nakhla and Shergotty," *Precambrian Research*, 106 (2001), 15–34. A useful guide to the developments after that may be found in a blog maintained by G. Jeffrey Taylor at psrd.hawaii.edu, which is a fine repository of scientific papers.

The magnetite aspect has, as indicated in the chapter, been played out in specialist journals. A key paper is by Edward Scott and David Barber, "Resolution of a Big Argument about Tiny Magnetitic Minerals in Martian Meteorite," May 13, 2002, on the psrd.hawaii.edu site. See also Buseck, P. R. et al., "Magnetite Morphology and Life on Mars," *Proceedings of the National Academy of Sciences*, vol. 98, no. 24 (November 20, 2001); Bell, M., "Experimental Shock Decomposition of Siderite and the Origin of Magnetite in Martian Meteorite ALH84001," *Meteoritics and Planetary Science* 42, no. 6 (2007), 935–949; Thomas-Keprta, K. et al., "Magnetofossils from Ancient Mars: A Robust Biosignature in the Martian Meteorite ALH84001," *Applied and Environmental Microbiology*, 68 (8) August 2002; and Barber, D. J., and E. R. D. Scott, "Origin of Supposedly Biogenic Magnetite in the Martian Meteorite Allan Hills 84001," *Proceedings of the National Academy of Sciences*, 99 (2002), 6556–6561.

Chapter 10: Signatures

Elizabeth Howell interviewed many of the Mars 2020/Perseverance personnel for various stories in Space.com in the 2016–2018 time frame. As is clear from the text, she made a dedicated trip to JPL in October 2019 for this book and was able to speak at length with Adam Steltzner, Matt Golombek, Rich Rieber, Ashwin Vasavada, and Abigail Fraeman. She also spoke extensively with both Luther Beegle and Professor Ken Farley at Caltech via telephone. Nicholas Booth spoke with many of the European ExoMars scientists and engineers in the summer of 2019. They include Jorge Vago at ESTEC, Paul Meacham at Airbus, Professor Charles Cockell at Edinburgh, Dr. Claire Cousins at St. Andrews, Dr. Peter Grindrod at the Natural History Museum in London, and Professor John Bridges at Leicester University in November. He rounded off his interviews with a Skype interview with Professor Bethany Horgan in early December 2019.

General

Adam Steltzner's book *The Right Kind of Crazy: A True Story of Teamwork, Leadership, and High-Stakes Innovation*, cowritten with William Patrick, has more on his philosophy of teamwork. For general background, see Wall, Mike, "Four Mars Missions Are One Year away from Launching to the Red Planet in July 2020," Space.com, July 25, 2019, and Haynes, Korey, "Scientists Gear Up to Look for Fossils on Mars," *Discover*, May 21, 2019. NASA and JPL's Mars Program websites have a wealth of detail on Mars 2020/Perseverance.

For information on SHERLOC, see Beegle, Luther, "Life as We Don't Know It," *Analytical Scientist*, September 2, 2018; Beegle et al., "SHERLOC, An Investigation for 2020," 11th International Geo Raman Conference (2014); Nowakowski, Tomasz, "SHERLOC Could Solve the

Mystery of Life on Mars," *Space Flight Insider,* March 17, 2017; Patel, Neel, "NASA's 2020 Rover Will Search Mars Signs of Life," *popsci.com,* November 21, 2018; "SHERLOC to Micro-map Mars Minerals and Carbon Rings," JPL news release, 2014–254; and "SHERLOC Technical Specification Paper," 5th International Geo-Raman Conference.

A good review of the ExoMars program is Vago, Jorge et al., "ESA's Next Steps in Mars Exploration," *ESA Bulletin* 155 (August 2013). A good review of the state of play with regard to life on Mars is Lovett, Richard A., "Is There Life on Mars? Let's Assess the Evidence," *Cosmos Magazine,* March 8, 2019. A technical overview is Vago et al., "Habitability on Early Mars and the Search for Biosignatures with the ExoMars Rover," *Astrobiology,* vol. 17, nos. 6 and 7 (2017). Claire Cousins has also written a less technical overview, "Our Rover Could Discover Life on Mars: Here's What It Would Take to Prove It," *The Conversation,* January 4, 2018. See also Kaufman, Marc, "Where Should We Look for Ancient Biosignatures on Mars in 2020?" *Manyworlds.space* website (also reproduced on the NASA Astrobiology website).

Perseverance/Mars 2020

See "NASA's Mars 2020 Will Hunt for Microscopic Fossils," JPL press release, referencing Horgan et al., "The Mineral Diversity of Jezero Crater: Evidence for Possible Lacustrine Carbonates on Mars," *Icarus* 2019.113526 (November 1, 2019). Jezero Crater landing site: "NASA Announces Landing Site for Mars 2020 Rover," JPL news release, November 19, 2018. A good discussion of the other candidate sites is David, L., "Scientists Double Down on Landing Sites for Sample-Collecting Mars Rover," *Scientific American,* October 22, 2018, and "Overview of Improvements on the Mars 2020, the Entry, Descent & Landing Technologies," JPL Mars website. Information on the Sayh al Uhaymir meteorite: "A Piece of Mars Is Going Home," JPL press release, February 13, 2018, and "From the Museum to Mars: A Meteorite Returns Home," NHM release, February 14, 2018.

ExoMars

Information has come from Jorge Vago, ExoMars project scientist, interviewed on May 29 and July 30, 2019, and Paul Meacham, interviewed on June 5, 2019. See Wall, Mike, "Meet 'Kazachok': Landing Platform for ExoMars Rover Gets a Name," Space.com, March 21, 2019; "ESA's Mars Rover Has a Name, Rosalind Franklin," ESA press release, February 7, 2019; "ExoMars Landing Platform Arrives in Europe with Name," ESA release, March 21, 2019; Amos, J., "Mars: The Box Seeking to Answer the Biggest Questions," *BBC Online*, May 17, 2019; Archer, Joseph, "Martian Robot Will Explore the Red Planet with Mind of Its Own," *Daily Telegraph*, January 2, 2019; Hignett, Katherine, "Is There Life on Mars? This Rover Wants to Find Out," *Newsweek,* May 19, 2018; Wilks, Jeremy, "Will ExoMars Be the Mission to Find Life on Mars?" *Euronews*, November 23, 2018; Wilks, J., "Mars Rovers: The Red Planet Welcomes Careful Drivers," *Engineering and Technology*, May 25, 2017; and "Mars Doesn't Need Our Microbes: How to Keep the Red Planet Pristine," *Discover*, June 2019. MOMA is described in detail in https:// exploration.esa.int/web/mars; Todd, Iain, "Science Instruments Installed on UK-built Mars Rover Rosalind Franklin," BBC *Sky at Night* magazine, August 22, 2019; "Moving on Mars," ESA Feature, May 30, 2019; and Amos, J., "Rosalind Franklin Mars Rover Assembly Continued," *BBC News Online*, August 27, 2019.

The saga of the parachutes was covered in "ExoMars Parachute Testing Continues," ESA website, "Our Activities," August 12, 2019, and Rincon, Paul, "ExoMars: Parachute Test Failure Threat to Launch Date," *BBC News Online*, August 13, 2019.

Chapter 11: Return to Sender

Primary sourcing for this chapter came from interviews with Charles Cockell, Chris McKay, Luther Beegle, Claire Cousins, Ken Farley, Melissa Guzman, Briony Horgan, Brian Muirhead, Paul Meacham, Rich Reiber, Adam Steltzner, Jorge Vago, and Sanjay Vijendran. Anita Heward provided useful background to much of the discussion.

A good popular-level overview for sample return may be found in the article by Alexandra Witze in the January 19, 2017, issue of *Scientific American* entitled "The $2.4-Billion Plan to Steal a Rock from Mars." See also Gough, Evan, "How Will NASA and ESA Handle Mars Samples When They Get Them Back to Earth," *Universe Today*, June 4, 2019; Taylor, Nola, "Mars Doesn't Need Our Microbes: How to Keep the Red

Planet Pristine," *Discovery* (June 2019); and Foust, Jeff, "NASA Unlikely to Return Mars Samples in the 2020s," *Space News*, February 26, 2019.

Background information may be found in Kaufman, Marc, "Where Should We look for Ancient Biosignatures on Mars in 2020?" *Astrobiology*, April 28, 2017; Retherford, Bill, "European Rover to Probe the 'Third Dimension' of Mars," *Forbes*, June 13, 2019; David, Leonard, "The Search for Life on Mars Is about to Get Weird," *Scientific American*, May 9, 2017; Amos, J., "Fetch Rover! Robot to Retrieve Mars Rocks," *BBC News Online*, July 6, 2018; Choi, Charles, "Habitable Hotspots on Mars? Volcano Vents May Be Signs," Space.com, October 31, 2010; and Retherford, Bill, "Looking for Life on Mars? Go Underground, Say Scientists," *Forbes Science*, May 14, 2019.

Papers

The architecture—the exact details of how samples will be returned from Mars—is still being defined, and the final missions may well be different from those that are being scoped and defined now. At the International Astronautics Congress in Washington, DC, held in October 2019, various ideas for how it will all be achieved were discussed. See Brian Muirhead et al., "Mars Sample Return Mission Concept Status," International Astronautical Congress-19-A3.3A. An ESA presentation entitled "MSR Sample Retrieval Lander Major Element Concepts" and labeled "predecisional information, for planning and discussion purposes only" was very useful.

On the general question of life on Mars, see Cockell et al., "Habitability: A Review," *Astrobiology* 16 (2016), 89–117; Cousins, C. et al., "Selecting the Geology Filter Wavelengths for the ExoMars Panoramic Camera Instrument," *Planetary and Space Science* 7, 80–100; Faire, A. G. et al., "Astrobiology through the Ages of Mars: The Study of Terrestrial Analogues to Understand the Habitability of Mars," *Astrobiology* 10 (2010), 821–843; and Georgiou, C. D. and D. W. Deamer, "Lipids as Universal Biomarkers of Extraterrestrial Life," *Astrobiology* 14 (2014), 541–549.

Professor Ken Farley's work is discussed in "First Rock Dating Experiment Performed on Mars," Caltech news release, December 10, 2013. The paper itself is Farley et al., "In Situ Radiometric and Exposure Age

Dating of the Martian Surface," *Science* 343 (2014). Further details on MOMA may be found in "Scientists Shrink Chemistry Lab to Seek Evidence for Life on Mars," NASA Goddard press release, June 3, 2018. See also Goesmann, F. et al., "The MOMA Instrument: Characterization of Organic Material in Martian Sediments," *Astrobiology* 17 (2017), 655–685, and Goetz, W. et al., "MOMA: The Challenge for Organics and Biosignatures on Mars," *International Journal of Astrobiology* 15 (2016), 239–250.

ILLUSTRATION CREDITS

Unless otherwise stated, the photographs in the insert are public domain images from NASA/JPL.

Plate 1, top: The original frames from *Mariner 4* have been assembled and processed by Ted Stryk (@tsplanets), whose work may be found at planetimages.blogspot.com.

Plate 2 top: Chris McKay, NASA photo from *Astrobiology* magazine.

Plate 2, bottom: McMurdo Sound and Dry Valleys from the US Long Term Environmental Research Station in Antarctica, supported by the National Science Foundation, Office of Polar Programs.

Plate 3, top: The HiRise dunes image was taken in January 2020 and is credited to NASA/JPL/University of Arizona.

Plate 3, bottom: The midlatitude glacier image was taken HiRise and generated by Seán Doran, @_theseaning from imagery produced by NASA/JPL/University of Arizona.

Plate 4, bottom: The mission control photo of Bruce Banerdt at the InSight landing was taken by Bill Ingalls for NASA.

Plate 5, top left: Cerberus Fossae images is from HiRise NASA/JPL/University of Arizona.

Plate 5, top right: The recurring slope lineae image was taken by HiRise and is credited to NASA/JPL/University of Arizona.

Plate 5, bottom: The image of Tory Hudson and the Mole is a NASA JPL photo.

Plate 6, bottom: Adam Steltzner at the Curiosity landing was taken by Bill Ingalls for NASA.

Plate 8, top: the surface and wheels images are credited to NASA/JPL/Malin Space Science Systems.

Plate 8, bottom: Abby Fraeman, courtesy of Dr. Fraeman.

Plate 9, top: The *Mariner 6/7* spectrometer is credited to the Department of Physics, University of California, Berkeley.

Plate 9, bottom: The Gerry Soffen image is from NASA Langley archives, collated at: https://crgis.ndc.nasa.gov/historic/Viking_Archives_Collection#Personnel.

Plate 10, bottom: The Jim Martin image is from NASA Langley archives, collated at: https://crgis.ndc.nasa.gov/historic/Viking_Archives_Collection#Personnel.

Plate 11, bottom: The image of Pathfinder on the surface is credited to NASA/JPL/University of Arizona.

Plates 12, top: The so-called "Face on Mars" is shown in detail from a HiRise image (NASA/JPL/University of Arizona); the inset is from the original Viking image taken on July 25, 1976.

Plate 12, bottom: Dr. Malin and camera from NASA/JPL/Malin Space Science Systems.

Plate 13, top and bottom: Percival Lowell in South Korea and the Canals of Mars are from the Lowell Observatory Archives, @signalfrommars.

Plate 14, top: the "Pearl Harbor moment" from NASA Press Conference in August 1996, taken by Bill Ingalls for NASA.

Plate 14, bottom: "The worm" is from the original NASA press kit.

Plate 15, top and bottom: The Perseverance rover and test bed images are from NASA/JPL.

Plate 16, top: The Rosalind Franklin rover is from ESA/Airbus.

Plate 16, bottom: The MOMA image courtesy of the MOMA team at NASA Goddard.

INDEX

Isaak, Chris, 63, 323n
Isidis Planitia, 254, 289
isotope, 73, 82, 111, 127, 183, 188, 230, 231, 232, 260, 281, 330n, 333n, 336n
isotopic fractionation, 261, 285

Jakosky, Bruce, 47, 48, 51, 52, 53, 183, 189, 191, 322n, 330n, 346n, 354n
Jagger, Mick, 43
James Webb Space Telescope (JWST), 94
Japan, 199, 200, 355n
jarosite, 68
Jet Propulsion Laboratory (JPL), 12, 29, 35, 40, 43, 44, 46, 57, 58, 59, 62, 63, 64, 68, 69, 71, 84, 85, 94, 95, 96, 105, 107, 108, 117, 118, 130, 134, 137, 138, 146, 147–154, 156, 158, 161, 164–167, 173, 178, 196, 207, 208–213, 246, 247–248, 249–251, 254, 256, 261, 263, 272, 282, 283, 292, 296, 300, 327n, 331n
  Caltech origins, 86, 87
  energy business, 150
  history, 86–90, 91–93, 95–96, 99
  mavericks, 87, 150
  machinists, 251
  Mars Yard, 66–67
  Miss Guided Missile, 148
  missiles, 87
  mission design section, 150
  misogyny and sexual misconduct, 145, 167, 352n
  relations with NASA, 70, 252
  Section 23 "human computers," 147, 148
  Visitor Center, 86
  women at JPL, 144, 147, 148, 167
Jezero Crater, 84, 254–258, 261, 263, 265, 266, 276, 279, 282, 287, 288, 289, 292, 360n
Johnson, Catherine, 50
Johnson, Lyndon Baines, 117
JPL. See Jet Propulsion Laboratory

Johnson Space Center (JSC), 150, 224, 233, 240, 336n, 341n
Jung, Carl, 204
Jupiter, 14, 115, 226, 326n, 327n

Kasparov, Garry 155
Kazachok lander, 272, 273, 275, 277, 361n
Kazakhstan, 155
Kelvin, Lord, 304
Kennedy, John F., ("JFK") (President), 88, 151, 224
kerogen, 268, 269
Koch, Christina, 169, 329n

Labeled Release (LR) experiment, 102, 108, 110, 111, 112. See also Gil Levin
  controversy over Viking results, 112, 213–18, 322n, 356n
  sensitivity of, 216
lacustrine deposits, 83, 347n, 350n
landslides, 30, 37
Langley, Virginia. See also NASA Langley
laser altimetry, 17, 179, 180, 322n, 343n, 353n
laser absorption mass spectrometer, 238
laser ranging, 17
LATMOS (Paris), 22'
launch window, 88, 96, 208, 284, 287, 296
Lawson, Janez, 148
layered terrain, 13, 20, 184, 195
Leeds, 175, 227
Leighton, Robert, 90, 93
Levin, Gilbert, 110, 112, 213–19
  background, 213-14, 356n
  claim to have discovered life, 213, 216, 217
  disagreement over consensus, 216–20, 332n
  Gulliver instrument, 94, 102, 215
  Labeled Release (LR) experiment, 102, 108, 110, 111, 112
  work on Viking mission, 110–12